Optical and Electronic Fibers

Optical and Electronic Fibers

Emerging Applications and Technological Innovations

Edited by Lei Wei

WILEY-VCH

Editor

Prof. Lei Wei
Nanyang Technological University
50 Nanyang Avenue
Singapore
SN, 639798

Cover Design: Wiley
Cover Image: © Connect world/Shutterstock

Library of Congress Card No.: applied for

British Library Cataloguing-in-Publication Data: A catalogue record for this book is available from the British Library.

Bibliographic information published by the Deutsche Nationalbibliothek
The Deutsche Nationalbibliothek lists this publication in the Deutsche Nationalbibliografie; detailed bibliographic data are available on the Internet at <http://dnb.d-nb.de>.

Print ISBN 978-3-527-35091-9
ePDF ISBN 978-3-527-83930-8
ePub ISBN 978-3-527-83931-5
oBook ISBN 978-3-527-83932-2

Typesetting Straive, Chennai, India

Printed and bound by CPI Group (UK) Ltd, Croydon, CR0 4Y

C9783527350919_260924

Contents

Preface

Advanced functional fibers represent a convergence of multiple disciplines, spanning optical waveguides, optoelectronics, electronics, micro-/nanofabrication, biointerfaces, and wearable devices. This book highlights the emerging applications and technological innovations of diverse optical and electronic fibers, distinguished by their functional structures and materials. The book is organized into ten chapters. A brief description of each of the chapters follows:

- Chapter 1 provides a comprehensive review of the recent advances in silica fiber-2D-materials integration technology that has allowed high-performance photonic and optoelectronic applications. The basic properties of several mainstay 2D materials and their heterostructures, the design principles of fiber-2D-materials integrations, the current and potential applications, and the challenges and opportunities in this field are discussed.
- Chapter 2 introduces the common methods used to produce semiconductor optical fibers that are amenable to optoelectronic and nonlinear optical applications. The versatility of laser processing offers additional functionality, such as bandgap engineering and microstructuring, while tapering can tailor the core diameter for realizing nonlinear applications.
- Chapter 3 presents the precision sensing technologies relying on processed optical fiber-based interrogation methods to achieve wearable sensors. The viability of such wearable optical sensors has motivated extensive research toward developing 2D flexible optical components and exploring other sensing utilities for building a healthy environment and revolutionizing healthcare services.
- Chapter 4 reviews the in situ and continuous monitoring of the activities of clean energy devices over the whole power transfer chain. In-fiber grating-assisted plasmonic sensors for real-time electrochemical and photochemical sensing are summarized, leading to the understanding and evaluation of the operation quality of energy storage devices in active service.
- Chapter 5 discusses fiber optofluidic laser as an exciting area of lab-on-fiber technology that integrates optical fiber, laser, and microfluidics to achieve high-performance biochemical sensing. It not only inherits the advantages of lasers, including high sensitivity, high signal-to-noise ratio, and narrow linewidth, but also the unique characteristics of optical fiber, such as easy integration, high reproducibility, and low cost.

- Chapter 6 presents 2 μm ultrafast fiber lasers using Tm^{3+} or Ho^{3+} as the gain medium with the output wavelength at around 2 μm and the pulse width within ps or fs. Such laser has broad application prospects in the fields of space optical communication, mid-infrared pumping, lidar, emerging non-metallic material processing, and medical and life sciences.
- Chapter 7 covers the development of optogenetic technology with the basic principles of optogenetic modulation of neural activities and the design considerations of fiber-based probes. The advances of several main types of fiber-based probes for optogenetics, such as glass fiber probes, nonstretchable polymer fiber probes, and stretchable polymer fiber probes, are summarized.
- Chapter 8 reports a comprehensive review of the state-of-the-art fiber-based neural probes and their applications in neuroscience. Thermally drawn multimaterial fibers, including the design principle, fabrication methods, and their versatile applications, are overviewed.
- Chapter 9 discusses the recent development of a scalable manufacturing approach to the realization of fiber-embedded integrated photonics and optoelectronic circuitry. Drawing inspiration from very-large-scale integration (VLSI), VLSI-for-Fibers harnesses 3D printing, and melt-processing of multimaterial fiber preforms to materialize arbitrarily complex architectures typical of integrated circuitry in fibers.
- Chapter 10 presents the recent development of inorganic thermoelectric fibers made by different fabrication methods such as electrospinning, two-step anodization process, solution-phase deposition method, focused ion beam, and self-heated 3ω method. Some techniques, such as thermal drawing, are also discussed to achieve high-performance fiber-based thermoelectric devices for wearable devices and smart electronics.

I thank all the chapter authors for their outstanding contributions. Moreover, I thank Nanyang Technological University for its constant support. Finally, express my sincere gratitude to my family and friends for their support and encouragement.

July 2023 Lei Wei

1

Optical Fiber with Two-dimensional Materials Integration for Photonic and Optoelectronic Applications

Jin-hui Chen[1, 2] *and Fei Xu*[2]

[1] *Xiamen University, Institute of Electromagnetics and Acoustics, Siming South Road, Xiamen, 361005, China*
[2] *Nanjing University, College of Engineering and Applied Sciences, Xianlin Road, Nanjing, 210023, China*

1.1 Introduction

The booming development of long-haul optical communications has been built on the ultralow-loss optical fiber network. Inspired by the seminal study by Kao and Hockham [1] that the optical loss of silica fibers can be significantly mitigated by reducing the impurity concentrations, the invention of low-loss silica fibers (~20 dB/km) marked the debut of modern optical communications [2]. The standard optical fibers are mainly made of silica materials with concentric core and cladding structures. Since the fiber core has a higher refractive index than the encircling cladding, the optical fiber usually confines light field in the core region through the total internal reflection at the interface. With the advances in materials science and manufacturing, state-of-the-art optical fibers show a transmission loss of 0.154 dB/km, which means that the light signal intensity only suffers from half loss after propagating 20 km along the fiber [2]. While the optical fibers simply transmit the light signals in the optical communications system, they have also found extended applications in various fields, such as optical imaging, distributed sensing, fiber lasers, and nonlinear optics.

In the past 20 years, it has witnessed the great progress of fiber optics with the development of fundamental science and advanced technology. From microstructures, the one-dimensional (1D) photonic crystals are implemented to realize the fiber Bragg gratings, and later many other kinds of grating structures, such as long-period gratings and tilted fiber Bragg gratings, are proposed and demonstrated for the optical sensing and signal processing. The photonic crystal fibers (PCFs) are emerged with two-dimensional (2D) structures running along the fiber length. These unique photonic structures overcome the limitations of total internal reflection principle and permit light guidance in a hollow core by a photonic bandgap, which provides a paradigm shift in fiber optics, and has stimulated

Optical and Electronic Fibers: Emerging Applications and Technological Innovations, First Edition.
Edited by Lei Wei.

important technological and scientific applications [3]. For example, PCFs have been implemented for in-line controlling the wavelength, modes, polarization, and orbital angular momentum of the light fields. There is continuous progress in designing the low-loss microstructured fibers for optical communications, and recently developed hollow-core conjoined-tube negative-curvature fiber shows the potential of low loss beyond the standard optical fibers [4].

The photonic integrations with small footprints are always in pursuit of goals for improved performance, power, volume, and cost. By tapering from the bulk fiber/materials, the miniaturized microfibers are demonstrated with many intriguing properties, such as the strong light field confinement, large surface evanescent fields, tunable dispersion, and mechanical configurability. These micro-/nano-scale waveguides bridge the fiber optics and nanotechnology for enhanced light–matter interactions. The mechanical flexibility and strength of microfibers also enable various photonic structures for the proposed sensing and optoelectronic applications [5–7]. On the other hand, the merging of materials and microstructures into a single fiber is proposed and demonstrated to realize optical, electronic, and mechanical functions that can sense and communicate. By sophisticatedly stacking multimaterials and co-drawing the preforms, they yield kilometers of functional fiber devices. While the multifunctional multimaterial fibers show the great potential of lab-on-fiber technology, there are very limited materials for joint integration, considering the interface-energy-driven capillary break-up effect [8]. Moreover, they are still challenging to be connected with the standard optical fiber systems with low loss for the intrinsic mode-field mismatch and splicing difficulty. As a result, there is a rising interest in hybrid integration of functional materials to the structured silica fiber platform by the well-developed post-processing techniques. In the beginning of the twenty-first century, the rise of Dirac fermionic graphene materials has attracted great research interest among many fields from the seminal work of Novoselov et al. [9]. Graphene is demonstrated with many fascinating properties, including the ultrahigh carrier mobility, thermal conductivity, broadband light absorption, and mechanical strength. The fast development of graphene-based 2D materials brings new opportunities in the fiber integration, as shown in Figure 1.1, since they have excellent optical and optoelectronic properties. The strong in-plane mechanical strength of the typical 2D materials with naturally passivated bonds enables the tight attachment to the fiber structures either in the flat or curved surface without the need for lattice match. Besides, the atomic thickness of 2D materials keeps the optical modes almost intact along with the enhanced light–matter interactions. Consequently, 2D materials can be fully complementary to the conventional passive silica fibers and achieve the functions of light emitting, modulating, sensing, and detecting.

The optical fiber integration with 2D materials for optics and optoelectronics is a highly interdisciplinary research field that includes condensed matter physics, optics, and mechanics. This field is still in its early stages of development, and there have been substantial advances recently. This chapter will first review the development of optical fibers. Second, the basic properties of mainstay 2D materials, i.e. graphene, transition metal dichalcogenides (TMDC), and black phosphorous (BP), and their heterostructures are introduced. Third, the 2D-materials-fiber design strategy and fabrications are described. Fourth, recent progress of fiber integrations for

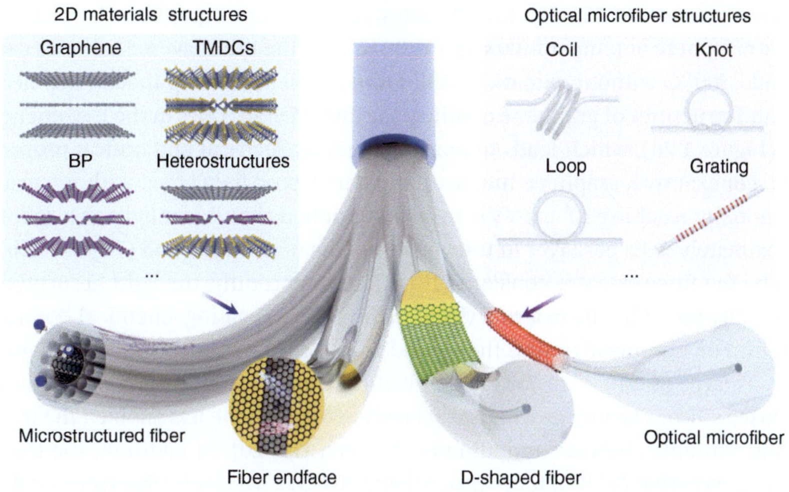

Figure 1.1 Schematic of 2D materials integration to optical fiber platform. TMDCs: transition metal dichalcogenides, BP: black phosphorous. Source: Chen et al. [10] / Springer Nature / CC BY 4.0.

optics and optoelectronics is discussed, mainly in the scope of polarimetric device, light sources, modulators, detectors, nonlinear optics, and fiber-optic sensors. Finally, the challenges and prospects of this field are briefly overlooked.

1.2 Fiber-integrated 2D Materials for Photonics and Optoelectronics

Basically, 2D materials are defined as crystalline materials composing single- or few-layer atoms, most of which are formed by the strong intralayer bonds and the weak interlayer van der Waals force. Benefited from the graphene research, the library of 2D materials is rapidly being increased from the mono-elements to compounds. Different from their three-dimensional (3D) counterparts, 2D materials can strongly interact with light, and their optical responses can be modulated with high flexibilities, such as by electrical gating, optical excitations, chemical doping, and strain gauge [10, 11]. Note that the fiber integration with 2D materials is still limited to a few mature 2D materials, and there is plenty of room to explore in this emerging field. As following, the basic optical and optoelectronic properties of the mainstay 2D materials are discussed.

1.2.1 Basic Properties of 2D Materials

1.2.1.1 Graphene

Monolayer graphene consists of carbon atoms arranged in a 2D honeycomb lattice, and each carbon atom is connected to the three nearest neighbors by a strong σ bond. Consequently, graphene shows an ultrahigh Young's modulus of ~1.0 TPa and an intrinsic strength of 130 GPa [12]. Besides, the monolayer graphene is stable in oxygen atmosphere at temperatures up to 300 °C, and the multilayered samples can withstand ~500 °C without oxidation [13]. From the tight-binding model, the electronic band structures of graphene manifest the linear Dirac cones in the low-energy regime (Figure 1.2a), which leads to many intriguing electrical and optical properties [14]. For example, graphene has an ambipolar electric field effect with intrinsic carrier mobility reaching 10^6 $cm^2/V s$. In optics, graphene has linear light absorption of approximately 2.3% per layer in the visible and near-infrared band (Figure 1.2b), defined by the fine structure constant [18]. More importantly, the light absorption can be readily tuned by the external fields such as electric gating, chemical doping, and strain. For nonlinear optics, the typical third-order nonlinearity ($\chi^{(3)}$) of graphene is as large as ~10^{-19} m^2/V^2, which is three orders of magnitude larger than that of fused silica. The ultrafast carrier relaxation of graphene has enabled the low-threshold saturable absorber and all-optical switch [19, 20]. In addition, the third harmonic generation (THG), four-wave mixing (FWM), saturable absorption (SA), and Kerr effect of graphene have been observed based on the third-order nonlinearity. It is revealed that the THG and FWM processes are highly dependent on the graphene Fermi level and the input light frequency [21]. The high harmonic generation (HHG) in graphene is observed and enhanced by elliptically polarized light excitation, which opens up the possibility of exploring strong-field and ultrafast nonlinear dynamics in massless Dirac fermions. Note that the intrinsic second-order nonlinearity is electric–dipole forbidden in perfect graphene films due to their centrosymmetry. Nevertheless, various methods have been proposed, such as applying strain, electric current, and chemical doping, to induce the second harmonic generation (SHG).

1.2.1.2 Transition Metal Dichalcogenides

The TMDCs, in the form of MX_2, where M is a transition metal atom (such as Mo and W) and X is a chalcogen atom (such as S, Se, or Te), have a long research history before the graphene discovery. Different from the gapless graphene, most of the TMDCs are semiconductors. Here, the MoS_2 or WS_2 are mainly introduced, and similar properties can be found in other TMDCs. The evolution of the band structures of MoS_2 is first studied, which reveals the transition from indirect bandgap to direct bandgap with the layer numbers (Figure 1.2c), because of the strong interlayer coupling [15, 22]. It is found that the photoluminescence quantum yield of a suspended monolayer MoS_2 is increased by four orders of magnitude compared with the bulk crystals [22]. The reduced dielectric screening and enhanced Coulomb interaction in TMDCs lead to the tightly bound excitons, charge excitons (trions), and biexcitons with large binding energies. For instance, the typical binding energies of A excitons in TMDCs are measured in the range of 0.2–0.8 eV. Thus, the optical properties of layered TMDCs are dominated by the excitonic effect and exhibit considerably strong light–matter interactions of light absorption up to ~20% in the visible/near-infrared spectra [23].

Another intriguing property of monolayer MoS_2 is the valleytronics. The monolayer MoS_2 is in trigonal prismatic (2H) phase, and the valence band maximum

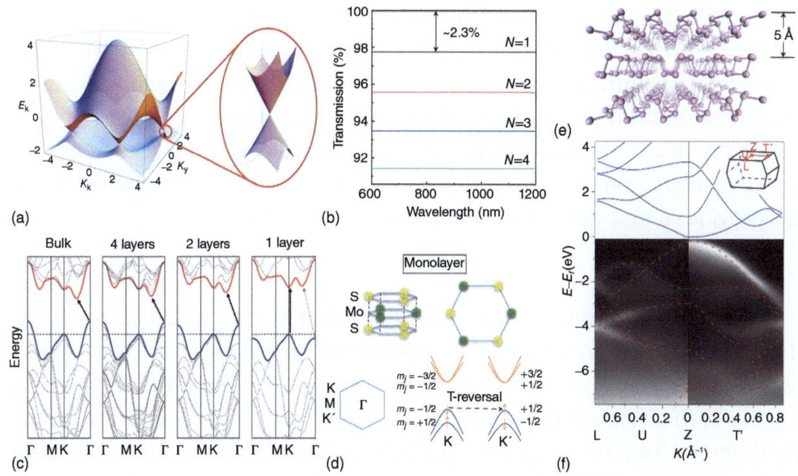

Figure 1.2 (a) The electronic band structures of monolayer graphene. Inset shows the Dirac cone structure at the edge of Brillouin zone [14]. (b) Calculated transmittance spectra of free-standing graphene (in air) of different layer numbers ($N = 1-4$). (c) Calculated band structures of 2H-MoS_2 for different layer numbers [15]. (d) Upper panel: the honeycomb lattice structures of monolayer MoS_2. Lower panel: the lowest-energy conduction bands and the highest-energy valence bands labeled by the z-component of their total angular momentum. The spin degeneracy at the valence-band edges is lifted by the spin–orbit interactions [16]. (e) Atomic structure of black phosphorus (BP) [17]. (f) Band structures of bulk BP mapped out by angle-resolved photoemission spectroscopy (ARPES) measurements [17]. Source: (a) Reproduced with permission from Neto et al. [14], American Physical Society. (c) Reproduced with permission from Splendiani et al. [15], American Chemical Society. (d) Reproduced with permission from Mak et al. [16], Springer Nature. (e, f) Reproduced with permission from Li et al. [17], Springer Nature.

and the conduction band minimum are located at the two inequivalent points K and K′. The inherent inversion symmetry breaking along with spin–orbit interactions contribute to coupled spin and valley physics in monolayer MoS_2 (Figure 1.2d) [16, 24], which are also observed in other group-VI dichalcogenides. The optical pumping of circular polarization is implemented to achieve complete dynamic valley polarization in monolayer MoS_2 [16], which has stimulated many valley-based optoelectronic applications in plasmonic, photonic crystal, and waveguide platforms. For nonlinear optics, the TMDCs have intrinsic second-order nonlinearity ($\chi^{(2)}$) due to the broken inversion symmetry. The typical $\chi^{(2)}$ of MoS_2 or WS_2 is reported to be ~1 nm/V, which is much higher than the commercial nonlinear crystals, such as $LiNbO_3$. Although the absolute conversion efficiency in most of the TMDCs devices is far less than the bulk crystals due to their atomically thin light–matter interactions, the unique tunability or modulation of TMDCs nonlinear optics by the excitonic effect, field effect, edge state, and stacking order can promise novel optoelectronic devices [25]. To boost the efficiency of SHG, the integration of TMDCs and photonic structures, such as plasmonic lattices, photonic crystals, Mie resonators, whispering gallery cavities, and optical waveguides, is demonstrated. The third-order nonlinear susceptibilities of MoS_2 is comparable to those of graphene, and THG, SA, FWM , and two-photo absorption (TPA) are also extensively studied. Note that the TPA coefficient of monolayer MoS_2 is as high as 7.6×10^{-8} m/W, which is three orders of magnitude higher than that of the conventional semiconductors (GaAs, CdS, and ZnO) [26]. The nonperturbative HHG from a monolayer MoS_2 is demonstrated with even and odd harmonics up to the 13th order. The combination of strong many-body Coulomb interactions and tunable electronic properties in 2D materials provides a new testbed for attosecond science [27].

1.2.1.3 Black Phosphorus

The BP is a single-elemental layered orthorhombic material, in which the intralayer phosphorus atoms are arranged in a puckered honeycomb lattice, as shown in Figure 1.2e. The low-symmetry crystal structure of BP determines their strong anisotropy in both electrical and optical properties (Figure 1.2f), which are fundamentally different from the graphene and TMDCs (such as MoS_2 and WS_2). Since the first demonstration of few-layer BP transistors, BP has regenerated interest for its high carrier mobility (~1000 cm^2/V s) and on/off ratio (10^5) for devices of ~10 nm thickness at room temperature [17, 28]. In addition, BP fills the energy gap between graphene and TMDCs, and the strong interlayer coupling leads to the wide bandgap tunability with layer numbers. For example, the direct optical bandgap of bulk BP is ~0.35 eV, and it is increased up to ~1.73 eV for monolayer [29, 30]. Consequently, the thin BP film can cover wide optical spectra in visible, near-infrared, and mid-infrared, which indicates important applications in thermal imaging, optical communications, and medicine. Due to the puckered lattice structure, BP has demonstrated unique strain-dependent properties. For example, the bandgap of thin-film BP exhibits a modulation rate of 1.70 μm/% by strain covering 2.3–5.5 μm, which can be readily used to detect multiple gases such as CO_2, CH_4, and H_2O [31]. The widely gate-variable mid-infrared light emission in thin-film BP is also realized [32].

On the other hand, BP shows strong anisotropy in the optical properties. For example, the infrared and Raman spectroscopy reveal the linear-polarization-dependent response of BP, which can be implemented to characterize the crystal orientations and fabricate polarization-control devices. Intriguingly, the excitonic photoluminescence is highly polarized along the light effective-mass direction regardless of the input laser polarizations [33]. Recently, broadband electro-optic polarization conversion with tri-layer BP integrated in a Fabry–Pérot cavity is realized based on the strong birefringence effect [34]. For nonlinear optics, BP has comparable third-order nonlinearity (10^{-19} m^2/V^2) with graphene, and various nonlinear effects such as THG, FWM, and SA are reported. Note that although bulk BP is stable under ambient conditions, few-layer BP is found to degrade quickly in environments with oxygen, moisture, or at high temperatures (>60 °C). The physical coating materials such as graphene, aluminum oxide, polymethylmethacrylate (PMMA), and the covalent functionalizations are successfully utilized to suppress the oxidations at different time scales [35]. It is in great demand to improve the stability of BP for the ultimately practical applications.

1.2.1.4 Other 2D Materials and the Heterostructures

In recent years, the 2D materials research is rising quickly, and many different kinds of 2D materials are being discovered. It is predicted that more than 600 stable 2D materials exist, and there are still a plethora of opportunities to explore in 2D materials along with their heterostructures. On the other hand, researchers are tending to explore the specific materials with designed properties. For example, hexagonal boron nitride (hBN) is composed of boron and nitrogen atoms arranged in the honeycomb structures similar to graphene. The hBN has strong mechanical strength with chemical inertness, and it does not exhibit any optical absorption in the visible spectra due to the large bandgap in the ultraviolet (UV) range. Thus, hBN is very suitable as an encapsulating film or substrate for 2D material devices [36]. The group-IV monochalcogenides such as SnS, GeS, SnSe, and GeSe are isoelectronic with BP, and they have the orthorhombic structures. Due to the broken inversion symmetry, they feature spin-orbit splitting and piezoelectricity. They also show potential for valleytronics' applications for their twofold degenerate valence and conduction valleys, and the valley pairs can be selected using linear polarized light [36]. The natural layered oxides, i.e. α-MoO_3, are revealed to demonstrate phonon polaritons with elliptic and hyperbolic in-plane dispersion, which promises an unprecedented potential for the directional control of light in the mid-infrared spectra [37].

The hetero-integrations of semiconductors, dielectrics, or metals are essential for modern electronics and optoelectronics. In particular, integrating different materials with pristine interfaces is important for creating functional devices, which has been long pursuit of materials science and technology [36, 38]. The 2D materials offer a unique platform for assembling different materials in one vertical stack held by the weak van der Waals force, which is promising for the atomic-scale material design and application. The van der Waals integration is radically different from the conventional chemical epitaxial growth, such as molecular beam epitaxy or physical vapor deposition, which require the interfacial lattice match. When stacking different 2D layered materials together, many exciting demonstrations of proof-of-concept devices are realized, such the Moire lattices, surface reconstruction, atomic-layer light-emitting diodes, and tunneling devices [36]. Very recently, the magic-twist graphene layers are reported for superconductor and correlated insulator [39, 40], which open a new horizon in twisted van der Waals materials. Note that most of the 2D heterostructures are achieved in the planar substrate for condensed matter physics and optoelectronic applications; the optical fiber integration for the cutting-edge applications still remains to be explored.

1.2.2 Optical Fiber-2D-material Integrations

The unique structures of silica optical fibers determine that there is no universal strategy in fiber-2D-material integration. Overall, the conventional fiber structures mainly include fiber endface (tip), microfiber, D-shaped fiber (DSF), and microstructured fiber, and each has its own advantages and disadvantages for integration. The fiber endface is an emerging microscopic platform for light–matter interactions, which shows miniature footprint and low insertion loss. Nevertheless, it suffers from the short light–matter interaction distance and small area. In the past years, many methods based on top-down or bottom-up techniques have been developed by the functional materials and photonics structures to enhance and manipulate the light fields on the fiber endface [41]. The fiber endface with 2D materials is straightforward, and the hybrid devices can be achieved by (polymer) film laminating, drop casting, and micro-transfer depending on the states of 2D materials. The hybrid graphene-fiber endface structure is first investigated for saturable absorption and used in ultrafast fiber lasers [19, 42]. Very recently, the van der Waals hetero-integration on fiber endface has attracted great attention and demonstrated applications in high-performance photodetection [43–45], chemical sensors [46], and polarimeters [47]. With the synergetic combinations of advanced photonic structures and 2D materials, it is promising to achieve novel all-fiber optoelectronics. On the other hand, the intriguing optomechanics of 2D materials can be interrogated on a fiber capillary tip, which can be used for all-fiber microelectromechanical system (MEMS) [48, 49].

Microfiber is a miniaturized optical fiber with diameters in the scale of the guided light wavelength, which is usually fabricated by flame tapering or chemical etching from a standard fiber (~125 μm). In the past years, optical microfibers have attracted interest for a range of photonic applications since they demonstrate many intriguing properties, such as strong field confinement, large evanescent fields, and structural reconfigurability [5]. For the hybrid integration, the 2D materials are conventionally deposited around the lateral wall of a microfiber by drop casting the nanoflakes solutions, which have important applications in optical sensors and saturable absorption for pulsed fiber lasers [50]. While more intriguing photonic applications are based on the intrinsic response of crystalline 2D materials integrated by either mechanical transfer or the chemical vapor deposition (CVD) method. For example, the ultrafast all-optical switch [20], tunable light absorption [51], reconfigurable exciton emission, and harmonic generations [52] have been demonstrated recently. On the other hand, the ingenious lab-on-a-rod technique realizes the 3D integration of coil microfiber and the functional materials or photonic structures onto a rod, as shown in Figure 1.3, which shows the great flexibility in tailoring light–2D materials interactions [54, 55] and can be applied for optical sensors, tunable fiber lasers, and polarimetric devices [53, 55–57].

The DSF structure is usually fabricated by mechanically polishing the side wall and exposing the fiber core, which can significantly enhance the surface evanescent interactions. The unique flat surface along the propagation in DSF not only enables the tight attachment of 2D materials but also increases the light–matter interaction length. The fabrication of hybrid DSF 2D materials is quite similar to that of the fiber endface, since both of them have the flat surfaces. The graphene–DSF is first reported for broadband fiber polarizer, and the high-performance all-optical modulator (AOM) [58], chemical sensors [59], and tunable fiber lasers [60] have been investigated recently. Microstructured optical fiber (MOF) with the air holes in the cladding or core region such as PCF is another emerging platform for accommodating 2D materials with potential of mass production. While MOF has been demonstrated to

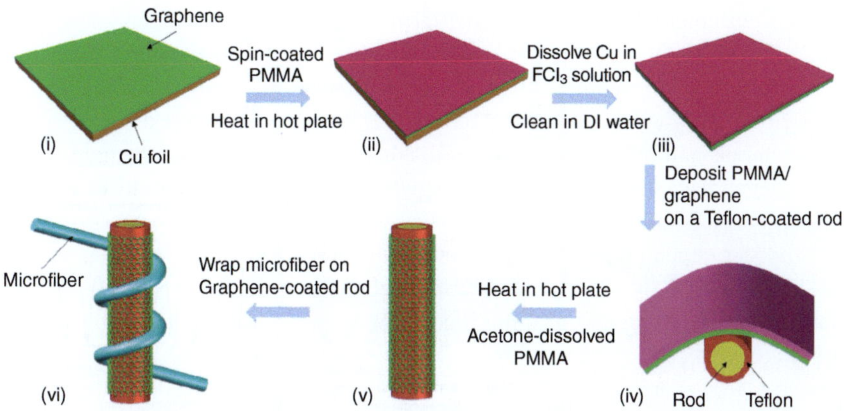

Figure 1.3 Fabrication process of the hybrid graphene-microfiber device. Panels (i)–(vi) demonstrate the detailed steps for transferring chemical-vapor-deposition growth graphene onto a surface modified rod (coated with low-index Teflon polymer), and panel 6 shows a schematic structure of the microfiber-on-graphene-rod device. Source: Reproduced from Chen et al. [53], Nature Publishing Group.

infiltrate gases, liquids, semiconductors, or metals for optoelectronics, it is a nontrivial task to fabricate high-quality 2D materials in their tiny microfluidic channels. Recently, a direct CVD-growth method was reported for the meter scale of graphene-PCF [61] and MoS_2-PCF production [62], which are applied for optical modulators and nonlinear optics. Note that different from the conventional multimaterial fibers, the atomic thickness of 2D materials keeps the optical modes of silica PCF intact and brings extraordinary optical functions. These works indicate the great potential for the ultimate commercial applications of all-fiber-2D-material devices.

1.2.3 Photonic and Optoelectronic Applications

In the past years, the fast development of hybrid fiber-2D-material integration has been witnessed, which has generated various novel fiber devices. In this section, recent development of fiber devices, i.e. polarimetric devices, light sources, modulators, photodetectors, nonlinear optics, and fiber-optic sensors, based on the 2D materials is discussed.

1.2.3.1 Polarimetric Devices

The optical polarizer is an optical filter that allows light waves of a specific polarization to pass through while blocking light waves of other polarizations. In fiber optics, an in-line fiber polarizer, based on polarization-selective coupling between evanescent fields and birefringent materials or plasmonic metals, has important applications in the coherent communication systems, optical gyroscopes, and interferometric sensors. Bao et al. [63] demonstrated the operation of a broadband polarizer in a graphene–DSF structure with an extinction of 27 dB, as shown in Figure 1.4a–c. The fiber polarizer showed much smaller loss in transverse-electric (TE) mode than in

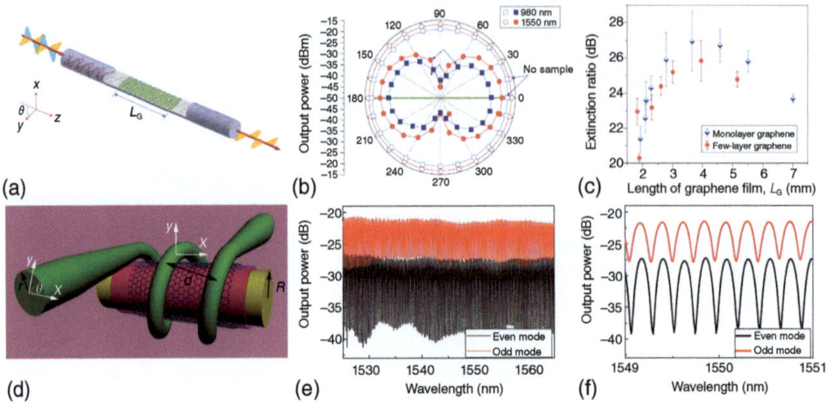

Figure 1.4 Polarimetric graphene-fiber devices. (a) Schematic model of graphene-D-shaped-fiber for polarizer. L_G, propagation distance (length of covered graphene film). Polarization angle θ is defined as the angle between the polarization direction of the analyzer (in x–y plane) and the graphene plane (y–z plane) [63]. (b) Polar image measured at 980 nm ($L_G = 3$ mm) and 1550 nm ($L_G = 2.1$ mm) [63]. (c) Polarization extinction ratio as a function of L_G measured at 1550 nm [63]. (d) Schematic structure of a stereo graphene-coil-microfiber in-line polarizer [55]. (e) Transmission spectra of the graphene-based resonator for two orthogonal modes [55]. (f) Enlarged view of the spectra at around 1550 nm from (e) [55]. Source: (a–c) Reproduced with permission from Bao et al. [63], Springer Nature. (d–f) Kou et al. [55] / Optical Society of America / CC BY 4.0.

transverse-magnetic (TM) mode since the graphene layer supported weakly damped TE mode. Different from the conventional fiber polarizer, the extinction ratios or even the polarization effect (TE-pass or TM-pass) in graphene polarizer is controllable by gate voltage (Fermi level) [60] or photon doping [53]. There are continuous efforts in optimizing the polarimetric performance through the polymer coating and structural modifications [64–66]. For example, Zhang et al. [65], reported a high extinction ratio of ~37.5 dB in the polymer–graphene–DSF structures with a low device loss ~1 dB. Kou et al. [55] presented a stereo graphene–coil–microfiber structure for polarization manipulation using a lab-on-a-rod technique (Figure 1.4d). Intriguingly, by tailoring the near-field coupling effect, they further demonstrated a high-Q single-polarization microresonator, which can be used for interferometric sensors with suppressed polarization cross-talk, as shown in Figure 1.4e–f. This design strategy can be readily used for other microfiber resonating structures. Recently, the 2D materials with strong optical dichroism effect such as ReS_2 and BP are investigated, and they show potential in polarimetric fiber devices [67].

1.2.3.2 Light Sources

The light-emitting devices are conventionally based on the direct radiation transition of carriers through the electrical or optical pumping. Consequently, the silica fiber (dielectric materials)-emitting devices require integrating external functional materials, such as semiconducting nanowires, quantum dots, rare earth ions, and fluorescence molecules. The emerging 2D materials with direct electronic bandgap offer unique alternatives. There have been considerable advances of planar light-emitting devices based on TMDCs; on the contrast, the electrical pump light sources in fiber devices are lacking. Chen et al. [52] reported mechanical transfer of crystalline WS_2 film onto the surface of a submicrometer microfiber, and they observed enhanced excitonic absorption and emission in the evanescent-coupling configuration (Figure 1.5a), which is free from the limitations of the atomic thickness of WS_2. For example, the light absorption of the waveguide-coupled A-exciton was measured of ~97%, which was much larger than the free-space coupling of ~13%. From the in-fiber strain engineering, the peak wavelength of the photoluminescence and absorption spectra was dynamically tunable, as shown in Figure 1.5b,c; and this reconfigurable light source is difficult to achieve in conventional bulk materials. Liao et al. [69] investigated enhanced monolayer photoluminescence on silica micro-/nanofibers for low-threshold lasing. They found that the

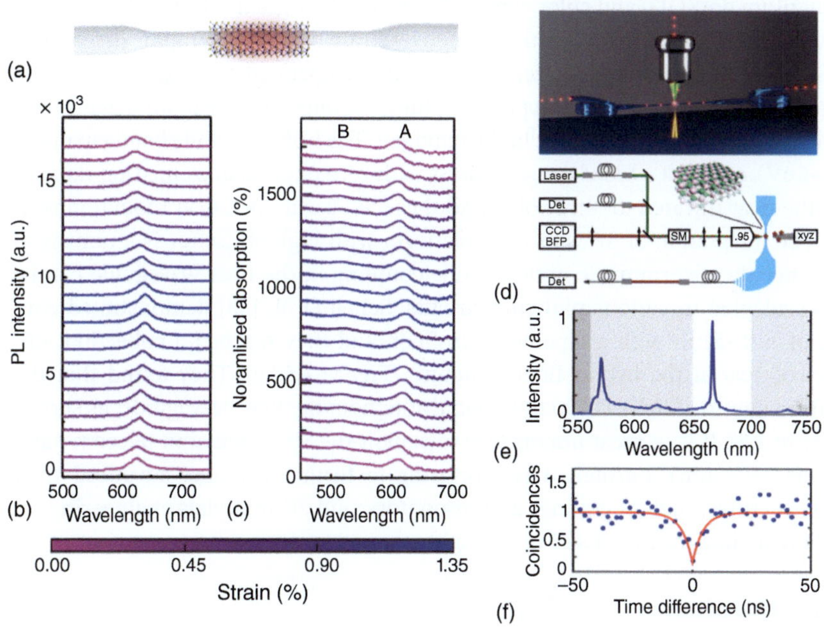

Figure 1.5 2D-materials-microfiber light emission sources. (a) Schematic hybrid WS_2-microfiber structure for enhanced light matter interactions [52]. In-line strain tuning of (b) the photoluminescence spectra and (c) absorption spectra of hybrid WS_2-microfiber. A and B are excitonic type of WS_2. (d) Upper panel: illustration of coupling quantum emitters in hBN to microfiber; lower panel: experimental setup. The experimental system consists of a homebuilt confocal microscope equipped with a piezo-driven three-axis manipulator (*xyz*) for attaching hBN flakes to a microfiber (colored blue). The microscope uses an objective lens with a numerical aperture of 0.95, and raster scanning is achieved using a scanning mirror assembly (SM) [68]. (e) Optical spectrum collected through the fiber, and (f) anti-bunching measurements of the light collected through the fiber. Value at zero time difference as extracted from the fits (red lines) is $g^{(2)}(0) = 0.15 \pm 0.20$ through the fiber [68]. Source: (a–c) Chen et al. [52] / Springer Nature / CC BY 4.0. (d–f) Reproduced with permission from Schell et al. [68], American Chemical Society.

high-density oxygen-dangling bonds released by photoactivation could be used to passivate the defects in MoS_2 or $MoSe_2$, and consequently improved the photoluminescence quantum yields. Furthermore, they demonstrated ultralow laser threshold of 5 W/cm^2 in MoS_2/microbottle device.

In addition to the classical light emission, a compelling aspect of the 2D materials is their applications for single-photon emitters (SPEs). The SPEs are isolated quantum systems that emit one photon at a time per excitation cycle, which can be triggered optically or electrically [70]. For the quantum science applications, the stable SPEs with high brightness, single-photon purity, and indistinguishability are in demand. The brightness of the SPEs characterizes the maximum single-photon emission rate. Single-photon purity represents the multiphoton emission probability, which can be quantified by the second-order autocorrelation function. The photon indistinguishability is conventionally characterized in the Hong–Ou–Mandel two-photon interference [71]. There are basically two kinds of SPEs in 2D materials, i.e. quantum dot (QD) and color center. The QD SPE originates from a bound exciton, which is confined in a potential formed by strain or crystallographic defect. The QD SPEs are first revealed in layered WSe_2 and are operated at cryogenic temperatures [72]. For the color-center type SPEs, they are impurities or point defects with electronic states that exist deeply in the bandgap. The hBN, as a wide bandgap material (~6 eV), can host a broad range of SPEs with zero phonon line emission ranging from the near-infrared to ultraviolet, and these SPEs can operate at room temperature. The integration of SPEs in fiber platform with high efficiency is a nontrivial task, since the deterministic optical mode coupling in the fiber waveguide can be further adapted in various photonic circuits. Schell et al. [68] demonstrated coupling of hBN-SPEs with a tapered microfiber, and they reached a collection efficiency of 10% in the hybrid fiber structure (Figure 1.5d,e). They found that the quantum nature of light was maintained throughout the fiber waveguide, as shown in Figure 1.5f. It is of great importance to exploit the SPEs that emit telecom band photons (~1.5 μm). Further, it is promising to further combine advanced fiber microstructures, such as gratings and resonators, and 2D materials engineering for the ultimate integrated SPEs of high brightness and purity.

1.2.3.3 Modulators

An optical modulator is a device used for controlling the amplitude, phase, or polarization state of a light beam. Based on the supreme optoelectronic functions of 2D material, the optical modulators have achieved substantial advances, which operate in the electromagnetic spectra spanning from ultraviolet to microwave. In this section, the fiber-integrated modulators for visible/infrared applications are discussed. In the past years, the all-optical modulator (AOM) and electric-optic modulator (EOM) have been extensively explored for their importance in optical communications network. The AOM uses a switching light beam to control the signal light beam, which can realize ultrafast modulation speed free from the electrical bottleneck. There have been various AOM devices designed based on Pauli-blocking effect, Kerr effect, and opto-thermal effect in 2D materials [10, 73], and their modulation bandwidth are different. Liu et al. [74] reported a graphene-covered-microfiber AOM device, with modulation depth and speed to be 13 dB and 1 MHz, respectively. The moderate performance is probably limited by the quality of CVD-growth graphene and the large footprint of the device. Li et al. [20] showed an ultrafast graphene-clad microfiber with a modulation depth of 38% and a response time of ~2.2 ps (Figure 1.6a,b), approaching the intrinsic carrier relaxation time of graphene. This work unambiguously demonstrates the superiority of graphene-based modulators for optical signal processing. Later, Yu et al. [75] reported a graphene-clad microfiber with a Mach–Zehnder interferometer to covert Kerr phase shift to intensity modulation. This method can realize high modulation depth and optical transmission, avoiding the dilemma of a saturable absorption device, while the spectra bandwidth is compromised. Note that it is still challenging to handle and encapsulate such sub-wavelength microfiber (~1 μm diameter) for practical applications. Later, Chen et al. [53] proposed a robust stereocoiled graphene-microfiber structure for polarization-dependent light modulation with a maximized modulation of ~7.5 dB and a modulation efficiency of ~0.2 dB/mW. Zhang et al. [58] developed a polymer–graphene–DSF structure for enhanced modulation with an operation bandwidth of 0.5 THz. On the other hand, Gan et al. [76] demonstrated an all-fiber phase shifter and switch assisted by graphene's photothermal effect. They obtained a phase shift exceeding 21π with a nearly linear slope of 0.192 π/mW when pumped by 1540 nm

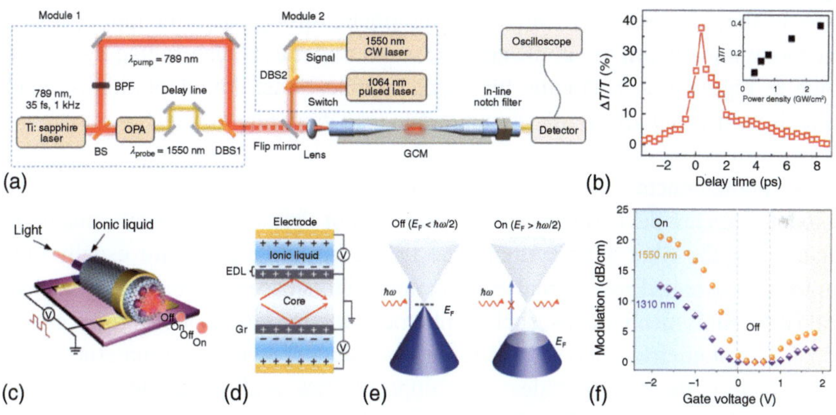

Figure 1.6 Optical fiber modulators with graphene integration. (a) Schematic illustration of experimental setups. Module 1: light source for pump–probe measurement. Module 2: light source for nanosecond-pulse modulation of a CW light [20]. (b) Differential transmittance of the probe light through a graphene-microfiber waveguide as a function of the pump–probe time delay. The inset shows the dependence of the modulation depth on pump intensity [20]. (c) Schematic of a graphene–PCF-based electro-optic modulator. The gate voltage between the ionic liquid and graphene controls light transmission through the graphene-PCF. (d, e) The working principle of the graphene–PCF electro-optic modulator, where an electrical double layer (EDL) forms at the interface between the graphene (dark gray rectangle) and ionic liquid (blue area). The ionic liquid-gating tunes the graphene's Fermi level and switches on and off the optical absorption in the graphene [61]. (f) Modulation curves at 1310 and 1550 nm show a transition between the "on" and "off" state with large modulation depth [61]. Source: (a, b) Reproduced with permission from Chen et al. [20], American Chemical Society. (c–f) Reproduced with permission from Chen et al. [61], Springer Nature.

light. Recently, there have been continuous advances in AOM devices beyond graphene, such as BP, TMDCs, and MXenes (transition metal carbides, carbonitrides, and nitrides). As for the ultimate commercial applications, the performance metrics of fiber AOM such as the control power consumption, switch time, and insertion loss are needed to be optimized and balanced [10]. In addition, the effective encapsulation of fiber devices is in demand.

The EOM is an optical device that implements direct-current (DC) or low-frequency electric field to control the light beams. There have been many advances in on-chip EOM based on graphene and other 2D materials [73]; on the contrast, the high performance of fiber-integrated EOM is less explored. Xu et al. [77] proposed a high-speed traveling-wave EOM based on a graphene/microfiber structure. The all-in-fiber electro-optic modulation was achieved by changing Fermi level of graphene, and $V\pi$ was designed of 4.9 V with 82 GHz 3 dB bandwidth. While, until now, the high-speed graphene-based fiber-EOM is not realized experimentally. Lee et al. [60] reported electrically tunable in-line graphene field effect transistors on a DSF. This device featured gate-variable electrical transport and optical transition at monolayer and multilayer graphene, and exhibited impressive optical transmission change >90%. Since the ionic liquid acts as efficient gating medium, this modulation speed is very slow. Chen et al. [61] demonstrated graphene-photonic-crystal-fiber with a length up to half a meter, which was produced using a modified CVD method. Similarly, they showed electrically gating graphene through an ionic liquid, with a large modulation depth of ~20 dB/cm at 1550 nm under a low gate voltage (~2 V), as shown in Figure 1.6c–f. For the high-speed applications, the adaptation of solid gating-dielectric and systematic circuit design is in demand [78, 79].

1.2.3.4 Photodetectors

Photodetectors are devices used for the detection of light or other electromagnetic radiation. In the fiber-optic systems, the transport light signals are usually out-coupled and detected by external bulk photodetectors based on various photoelectrical conversion mechanisms, such as photoconductive effect, photovoltaic effect, and photo-bolometric effect. The fast development of planar 2D-material-based photodetecting devices provides new opportunities and guidelines for fiber-integrated photodetectors. For the evanescent-field-coupling photodetectors, Sun et al. [80] reported broadband photodetection in a microfiber-graphene device, with a maximum photocurrent responsivity of ~2.81 mA/W in 1500–1600 nm. Jin et al. [81] investigated clean-transfer 2D InSe photodetectors onto a stripped fiber with a fast response time (~67 μs) and a responsivity of 769 mA/W (@~10 mW/cm^2). They also demonstrated a proof-of-concept in-line detection of a binary image signals. Recently, Zhuo et al. [82] demonstrated a hybrid carbon nanotubes (CNT)/graphene film in a DSF for enhanced responsivity of ~1.48×10^5 A/W (@0.23 μW/cm^2). On the other hand, there is an increasing interest in fiber endface photodetectors (FPD) for the simplified techniques and structural flexibility, as shown in Figure 1.7. Chen et al. [83] developed a mechanical scratch method to fabricate electrodes in fiber endface, and the few-layer MoS_2 was bonded to the silica substrate by van der Waals force. They measured a considerably high responsivity of

~600 mA/W (@5.5 mW/cm^2) at a bias of 4V. Intriguingly, they also observed a responsivity of 10 mA/W under the bias-free conditions, which was attributed to the spatial asymmetry electrodes for the imbalanced photocarriers' transport. Later, they presented multi-materials integration by combining perovskite nanocrystals ($CsPbBr_3$) and multilayered graphene, and achieved a high responsivity of 2×10^4A/W (@75 μW/cm^2) with a power detection limit of 10^{-11} W [43]. Xiong et al. [44] reported a graphene–MoS_2–WS_2 heterostructure FPD with a synergetic photogating mechanism, as shown in Figure 1.7a. They found that the photodetectors exhibited an ultrahigh responsivity of ~6.6×10^7 A/W (@6.35 nW/cm^2) and a fast response time of ~7 ms at visible band. Moreover, they revealed that due to the type-II band alignments in MoS_2–WS_2, the photodetectors displayed a responsivity of ~17.1 A/W (@25.4 mW/cm^2) at 1550 nm. Recently, they demonstrated an ultracompact seven-core fiber De-multiplexer by integrating a patterned graphene photodetector array on a single-fiber endface [84], which shows great promise in multicore fiber optics.

In the past years, the figures of merit for fiber photodetectors are significantly improved, benefiting from the heterointegration strategy, as shown in Table 1.1. Nevertheless, the high-speed photodetecting components based on 2D-materials-fiber structure for GHz bandwidth are still lacking until now, which is highly in

Table 1.1 Figures of merit for 2D-materials-fiber photodetectors.

2D-Materials-fiber	V_{ds} (V)	λ (nm)	P (mW/cm^2)	Responsivity (mA/W)	Noise equivalent power (W/Hz$^{1/2}$)	Rise time (ms)	References
Graphene-microfiber	1	1550	9.0×10^6	2.81	~10^{-9}	—	[80]
InSe-microfiber	0.4	543	10–100	769	—	0.045	[81]
CNT/Graphene-DSF	0.3	1550	2.29×10^{-4}	1.48×10^8	6.76×10^{-12}	75	[82]
MoS_2-fiber endface	4	400	5.5	600	—	7.1×10^3	[83]
Graphene-$CsPbBr_3$-fiber endface	0.2	400	7.5×10^{-2}	2×10^7	3.9×10^{-16}	3.1×10^3	[43]
Graphene-MoS_2-WS_2-endface	−3	400	6.35×10^{-6}	6.6×10^{10}	—	7	[44]
		1550	25.4	1.71×10^4	—	10	
Graphene-COF-fiber endface	3	473	3.3×10^{-6}	3.2×10^{10}	~10^{-16}	1.14	[45]
BP-fiber endface	0.5	1550	~79	~8.52×10^3	~1.5×10^{-4}	8×10^{-4}	[47]

demand for the optical communication networks. In another aspect, researchers are exploring the cutting-edge applications of the developed fiber photodetectors. Xiong et al. [45] developed ultrahigh-responsivity photodetectors of 2D covalent organic frameworks (COFs) and graphene with a photoresponsivity of ~3.2×10^7 A/W. Based on the high surface area and the polarity selectivity of COFs, they demonstrated the gas sensing potentials with the fiber phototransitors, as shown in Figure 1.7b. Recently, Xiong et al. [47] for the first time demonstrated twisted BP-based van der Waals stacks for fiber-integrated polarimeters (Figure 1.7c). The real-time in-line detection of the light polarization is crucial in communication, imaging, and sensing fields. The conventional polarimeters use a series of opto-electro-mechanical elements, including lenses, polarizers, waveplates, and photodetectors, which are discretely arranged in an optical path and are highly cumbersome. Recent work ingeniously stacked three photodetection units based on six-layer van der Waals materials, including one isotropic Bi_2Se_3 for power calibration, two twisted BP layers for polarization detection, and three hBN for encapsulation. Intriguingly, the proposed device enabled polarimetric imaging to acquire spatial polarization information by Hadamard single-pixel imaging technique, as shown in Figure 1.7d [47]. The ultracompact device structure with the deterministic and accurate van der Waals materials integration, free from external optical and mechanical modules, may inspire the development of miniaturized optical and optoelectronic systems.

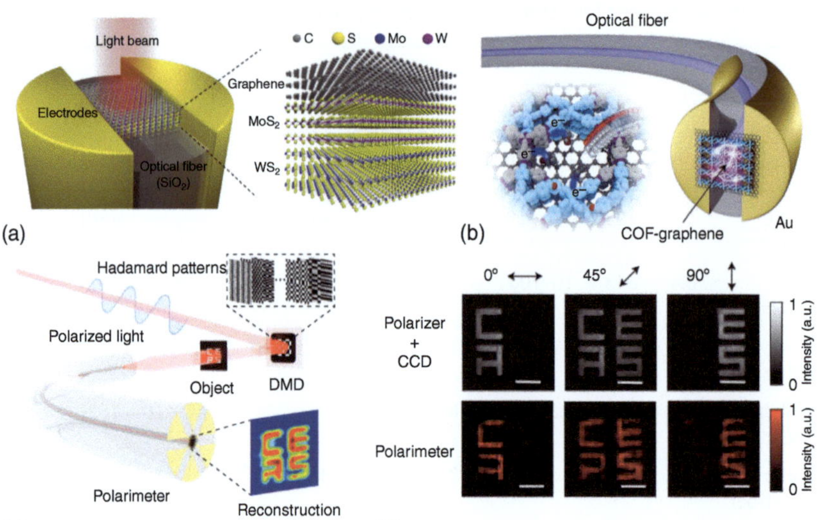

Figure 1.7 Optical fiber photodetectors with 2D-materials integrations. (a) Schematic view of the fiber endface photodetector (FPD) of the graphene-MoS_2-WS_2 heterostructure [44]. (b) Optoelectronic properties of FPD with of the covalent-organic-frameworks (COF) modified graphene film, regulated by gas molecules [45]. (c) Schematic of single-pixel polarimetric imaging by the FPD-based polarimeter. (d) Comparison of the images recorded by the CCD with a polarizer and the images (normalized intensity) reconstructed by the fiber polarimeter device under different linearly polarized illuminations [47]. Scale bars, 1 mm. Source: (a, b) Reproduced with permission from Xiong et al. [44, 45], JOHN WILEY & SONS, INC. (c, d) Reproduced with permission from Xiong et al. [47], American Association for the Advancement of Science - AAAS.

1.2.3.5 Nonlinear Optics

Nonlinear optics is the study of the intense light field in nonlinear media, where the optical response of materials responds nonlinearly to the electric field of light. The optical polarization $\boldsymbol{P}(t)$ of a material to the applied fields can be represented by a power series of the electric field strength $\boldsymbol{E}(t)$, as following:

$$\boldsymbol{P}(t) = \varepsilon_0 \left(\chi^{(1)} \boldsymbol{E}(t) + \chi^{(2)} \boldsymbol{E}(t)^2 + \chi^{(3)} \boldsymbol{E}(t)^3 + \cdots \right) \tag{1.1}$$

where ε_0 is the vacuum permittivity, $\chi^{(n)}$ is the nth order nonlinear susceptibility ($n \geq 2$), and $\chi^{(1)}$ is the linear susceptibility. From Eq. (1.1), it readily indicates that the nonlinear polarization terms can become significant when the light intensity is high. The typical nonlinear optical effects include second-order and third-order nonlinearity, which correspond to the second and third terms in the right side of Eq. (1.1), respectively. From the process of light–matter interactions, the nonlinear effects can also be classified as parametric and nonparametric processes. The parametric process indicates no occurrence of net transfer of energy or momentum between light field and the physical system [85], and it usually includes harmonic generations, sum/difference frequency generation, parametric oscillation, and optical Kerr effect, etc. On the contrast, the nonparametric process indicates that the quantum state of the medium is changed by the process, which includes Raman scattering, Brillouin scattering, two-photon absorption, saturable absorption (SA), and optical limiting, etc. In the past years, there is emerging research interest of 2D-materials nonlinear optics, since they have so many intriguing nonlinear optical properties, such as tunable nonlinearity from stacking order, strain and doping, phase-matching free by the atomic thickness and excitonic resonance [86]. Besides, the nonlinear spectroscopy has been shown to be an effective tool to characterize the fundamental properties of 2D materials. Considering the moderate nonlinear susceptibility of silica materials, for example, the small third-order nonlinearity and zero second-order nonlinearity from centrosymmetry, 2D materials integrations can go beyond the silica fiber nonlinear optics.

The SA is a third-order nonlinear effect of materials, where the light absorption is decreased with increasing light intensity (Figure 1.8a). The SA effect of graphene-based 2D materials has been extensively explored for pulsed fiber lasers spanning the spectra from visible to mid-infrared, along with the Q-switched or mode-locked state. In particular, the interplay of broadband gain media, SA effect, and the balance of linear and nonlinear effects in the fiber laser circuit can enable the ultrashort pulse duration. Bao et al. [19] showed that the modulation depth of light absorption could be tuned from 66.5% to 6.2% by varying the graphene layers, as shown in Figure 1.8b. They further reported the use of atomic-layer graphene in a mode-locked fiber laser (Figure 1.8c) for the generation of ultrashort soliton pulses (756 fs) at the telecommunication band. Sun et al. demonstrated a graphene–polymer composite for saturable absorber, and they showed a passively mode-locked fiber laser at 1559 nm, with 5.24 nm spectra bandwidth and ~460 fs pulse duration. Compared with commercial semiconductor saturable absorber mirrors (SESAMs), the graphene is shown to have an ultrawide spectra operation and tunable

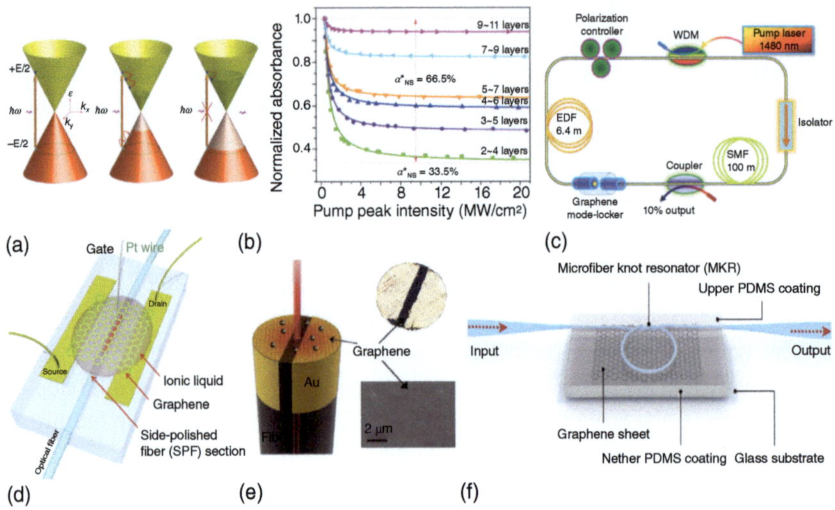

Figure 1.8 Optical fiber integrated with graphene saturable absorber. (a) Schematic of optical interband transition in graphene (left panel), the photogenerated carriers relaxation process (middle panel), and Pauli blocking effect at high excitation intensity (right panel) [19]. (b) Nonlinear absorption of graphene films with different number of layers [19]. (c) Laser configuration constituting a ring cavity. This schematic shows the standard fiber-optic components such as wavelength division multiplexer (WDM), polarization controller, coupler, optical isolator, erbium-doped fiber (EDF), and single-mode fiber (SMF) [19]. (d) Schematic diagram of gate-variable all-fiber graphene device [60]. (e) Schematic of graphene-fiber endface integration device, manipulating by electrical current induced Ohmic heat. Upper right panel shows the microscopic image of the sample. Lower right panel is the scanning electron microscope image of the graphene [87]. (f) Schematic of a high-Q microfiber ring resonator (with fiber pigtails) attached to multilayer graphene film and coated by PDMS [88]. Source: (a–c) Reproduced with permission from Bao et al. [19], JOHN WILEY & SONS, INC. (d) Lee et al. [60] / Springer Nature / CC BY 4.0. (e) Reproduced with permission from Li et al. [87], JOHN WILEY & SONS, INC. (f) Reproduced with permission from Ding et al. [89], Elsevier Ltd.

modulation depth, thus the output state of fiber lasers can be dynamically manipulated. Lee et al. [60] reported active control of transmission for graphene–DSF devices with electrical gating (Figure 1.8d), which enabled electrically tunable fiber laser at various operational regimes. Li et al. [87] proposed and demonstrated the state-variable fiber laser by thermally engineering the Fermi–Dirac distributions of graphene films on a fiber endface based on the simple Ohmic heating method, as shown in Figure 1.8e. Similarly, Qin et al. [88] demonstrated tunable laser microcomb dynamics by leveraging the electrical tunability of the graphene device incorporated in a fiber cavity. Recently, Ding et al. [89] realized a multifunctional all-fiber mode-locked laser based on a graphene-integrated polarization-dependent microfiber resonator (Figure 1.8f). They considered the SA mechanism, nonlinear polarization rotation effect, and high-Q knot resonator, and observed the Q-switched pulse, single soliton mode-locked pulse, harmonic mode-locked pulse, and high-repetition-rate dissipative-FWM pulse depending on the experimental conditions.

Inspired by the graphene research, many other 2D materials, such as BP, TMDCs, and MXenes, are revealed to have SA effect and thus can be used for pulsed fiber lasers. Note that most of the previous work focuses on the spectral and temporal properties (repetition rate and pulse duration) of fiber lasers depending on the various 2D materials and the fiber cavities, while the noise characteristics (intensity noise, timing jitter, and comb-line frequency noise) of the output fiber lasers are less investigated, which is important for optical science and photonic technology [90].

The waveguide-integrated 2D-materials for enhanced parametric processes are attracting attentions, since these nonlinear effects have broad applications in novel light sources, optical sensors, optical metrology, and signal processing. FWM is an intermodulation phenomenon from third-order nonlinearity, and it describes interactions between two and three wavelengths producing two or one new wavelengths. Wu et al. [91] demonstrated the cascaded FWM phenomenon with a large spectral range over 15 nm in a graphene-coated microfiber waveguide. Recently, An et al. [59] reported tunable FWM in graphene–DSF configuration for individual gas molecule detection (Figure 1.9a,b). They showed that the FWM conversion efficiency could change steeply when the graphene Fermi level was tuned approximately at 0.4 eV for ~1.5 μm pump and seed pulse. At this condition, the gas adsorption on graphene with modulated Fermi level would be sensitively detected from the Stokes light signal. SHG is a nonlinear optical process in which two photons with the same frequency are combined to generate a new photon with twice the energy of the initial photons. Chen et al. [52] systematically studied a hybrid WS_2-microfiber structure for broadband and tunable enhancement of light–matter interactions. Based on the high second-order nonlinearity of deposited WS_2, they showed that the SHG conversion in a modified microfiber was ~20 times larger than that of a bare microfiber (surface nonlinearity from symmetry-breaking). They also revealed the strong polarization-dependent SHG of hybrid waveguide due to the interactions between evanescent fields and WS_2, as shown in Figure 1.9c,d. Intriguingly, the dynamic control of SHG by strain gauge was also demonstrated based on the tuning of phase matching (Figure 1.9e), which can be applied for tunable visible light sources. Jiang et al. [92] reported high-efficiency SHG and sum-frequency generation in a microfiber assisted with few-layer GaSe with only submilliwatt continuous-wave pump lasers, and the SHG intensity was enhanced by more than four orders of magnitude. Recently, Li et al. [93] demonstrated deterministic all-optical generation and electrical control of multiple plasmon polaritons in a hybrid graphene-DSF via difference frequency generation. The electrically tunable Fermi level of graphene enables the generation of broadband plasmon polaritons, which provides a new way to high-speed logic operations.

Generally, 2D-material-enhanced optical nonlinearity depends on the trade-off between optical loss and interaction length. The unintentionally generated defects absorption/scattering during 2D materials transfer processes can significantly influence the ultimate nonlinear conversion efficiency in the post-processed fiber devices [10]. Zuo et al. developed a direct growth of MoS_2 onto the internal walls of a silica PCF [62]. By using the as-fabricated 25 cm long fiber, both the SHG and THG

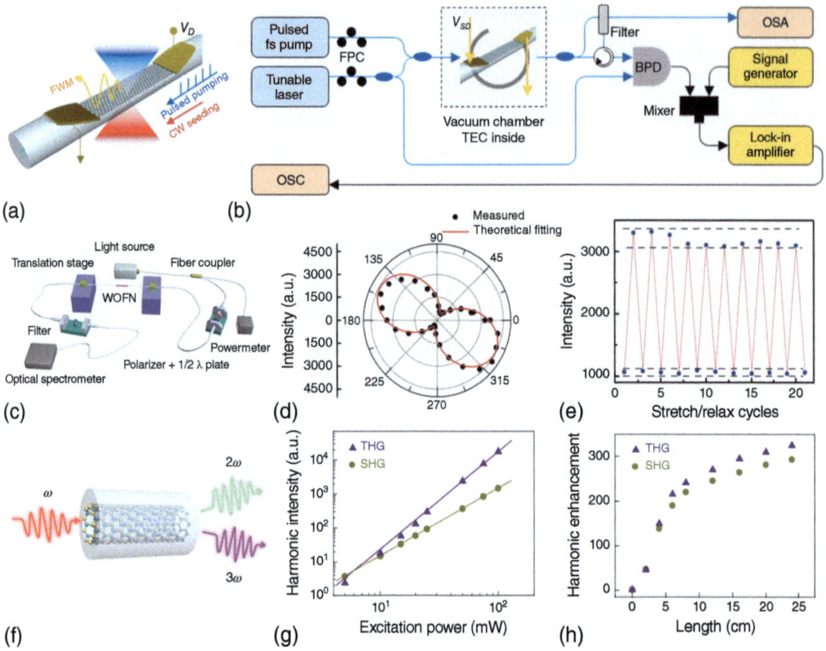

Figure 1.9 Optical fiber integrated with 2D materials for nonlinear optics. (a) Sketch of the electrically tunable four-wave-mixing (FWM) process in graphene–DSF [59]. (b) Experimental setup for gas sensing with based on gate-tunable FWM in graphene. BPD, balanced photodetector; OSA, optical spectrum analyzer; OSC, oscilloscope [59]. (c) Experimental setup for strain control of second harmonic generation (SHG) in the hybrid WS_2-microfiber [52]. WOFN: WS_2 optical fiber nanowire. (d) Polar image of the measured SHG intensity, as a function of the linear polarization of the pump light with a fixed input power [52]. (e) Cycling tests of strain modulation of the SHG of the WS_2-microfiber. Each stretch/relax step corresponds to a strain value of 0.45% [52]. (f) Schematics of SHG (2ω) and third harmonic generation (THG, 3ω) in MoS_2-embedded hollow-core fiber [62]. (g) Excitation-power dependence of SHG and THG intensities for MoS_2-HCF, showing the as-expected quadratic and cubic laws [62]. (h) Fiber-length-dependent SHG and THG enhancements in MoS_2-HCF with respect to MoS_2/silica, showing a general monotonic increase and saturation behavior with the fiber length [62]. Source: (a, b) An et al. [59] / American Chemical Society / CC BY 4.0. (c–e) Chen et al. [52] / Springer Nature / CC BY 4.0. (f–h) Reproduced with permission from Zuo et al. [62], Springer Nature.

were enhanced by ~300 times compared with monolayer MoS_2/silica, as shown in Figure 1.9f–h. This work shows the great potential of mass production of 2D materials fiber devices and can stimulate versatile nonlinear applications in the near future.

1.2.3.6 Fiber-optic Sensors

The fiber-optic sensor uses optical fibers as the sensing element, which exhibits many advantages over the electrical sensors, such as light weight, immune to electromagnetic interference, and remote sensing. In the past years, the 2D-material-integrated fiber-optic sensors attracted research interest, since the photonic/

optoelectronic properties of 2D materials can be strongly modulated by the external stimuli, which can be used to strengthen the performance of the silica fiber-optic sensors. Without loss of generality, here the graphene-based fiber sensors are focused in the scope of chemical and physical sensing, as shown in Table 1.2.

For chemical sensors, the graphene-fiber scheme has appealing features compared with the pure silica fiber structures, since graphene has ultrahigh surface-to-volume ratio, large adsorption capacity, and ultrafast carrier mobility [10, 102]. Generally, the adsorption of molecules changes the dielectric properties of graphene, which indirectly modulated the transported light fields by the amplitude, phase, polarization, or wavelengths. To achieve photonic resonance-enhanced sensing, researchers often integrate the graphene in the evanescent-coupling scheme of DSF, microfibers, gratings, interferometers, and cavities, as shown in Figure 1.10. Wu et al. [94] demonstrated an ultrasensitive gas sensor with a graphene-coated microfiber Bragg grating. They obtained sensitivities of −4 and 2 pm/ppm for NH_3 and xylene gas, respectively, which were 10 times larger than that of a bare microfiber grating. Similarly, Yao et al. [95] reported graphene-enhanced evanescent field in microfiber multimode interferometer (MMF) for gas sensing, and they achieved sensitivities of ~8 pm/ppm (~4 pm/ppm) for NH_3 (H_2O), which were 80 times higher than those of a bare MMF. Xiao et al. [98] studied reduced graphene

Table 1.2 Typical graphene-fiber sensors and their figures of merits.

2D-Materials-fiber	Sensing parameters	Maximal sensitivity	Detection limit	Response time	References
Graphene-microfiber-grating	NH_3	−4 pm/ppm	0.2 ppm	~5 min	[94]
	Xylene	2 pm/ppm	0.5 ppm	~10 min	
Graphene-multimode-microfiber	NH_3	8 pm/ppm	0.1 ppm	~1 s	[95]
	H_2O (vapor)	4 pm/ppm	0.2 ppm		
rGO-microfiber-grating	NO_2	~5 pm/ppm	0.5 ppm	~10 min	[96]
GO-long-period-fiber-grating	NO	~63.6 pm/ppm	0.36 ppm	23.6 min	[97]
Graphene-DSF	Humidity	0.31 dB/%RH	—	~15 s	[98]
Graphene-HCF	Refractive index	365.9 dB/RIU	2.73×10^{-6} RIU	—	[99]
Graphene-fiber-ferrule	Pressure	1.1 nm/Pa	~60 μPa/$Hz^{1/2}$[a]	—	[100]
MoS_2-fiber ferrule	Pressure	89.3 nm/Pa	—	—	[101]
Graphene-etched-fiber-tip	Electrical current	2.2×10^{5} nm/A^2	0.3 mA	0.25 s	[48]
Graphene-coiled-microfiber	Electrical current	~6.7×10^{4}/A^2	0.5 mA	—	[56]

a) Noise equivalent pressure.

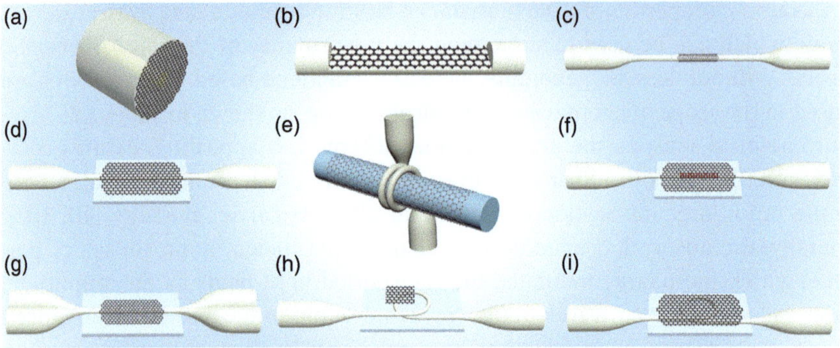

Figure 1.10 Typical optical microstructured fibers integrated with graphene-based 2D materials for sensors. (a–c) Graphene deposited on fiber endface, D-shaped fiber flat surface and tapered microfiber lateral wall. (d–g) Tapered microfiber on graphene substrate, coiled microfiber on graphene-coating rod, microfiber gratings on graphene and microfiber coupler on graphene substrate. (h, i) Microfiber ring partially and fully coated with graphene, respectively.

oxide (rGO)-coated DSF for humidity sensing with a sensitivity of 0.31 dB/% RH (relative humidity) in 75–95% RH (relative humidity). Huang et al. [99] reported plasma-enhanced CVD-growth graphene on hollow-core fibers (HCF), and they obtained a refractive sensitivity of 365.9 dB/RIU and a detection limit of 2.73×10^{-6} RIU. Xu et al. [97] reported ultrasensitive and selective detection of NO gas based on graphene oxide (GO)-coated long-period grating with a sensitivity of 63.6 pm/ppm. They showed that adsorption of NO molecules by GO was the combined effect of physical adsorption and chemical adsorption. In addition to the gases and vapors related sensors, the 2D-materials-fiber-optic sensors have been also implemented for detection of volatile organic compounds and heavy metal ions [103]. Note that there are continuous efforts to improve the chemical sensing sensitivity by optimizing the fiber structures and 2D materials (such as TMDCs, Mxenes, and BP); nevertheless, the sensing performance is fundamentally limited by the simply passive spectral sensing methods.

The active or nonlinear spectral sensing is promising for the ultimate single-molecule detection contributed by the high optical gain and electronic gain technique. Cao et al. [104] reported fluorescent resonance energy transfer in fiber-microfluidic resonator (Figure 1.11a). Besides, by measuring the intermode interference through noise-canceled beat notes and locked-in heterodyne detection with Hz-level precision, they achieved individual molecule sensitivity for dopamine, nicotine, and single-strand DNA detection. Yao et al. [105] demonstrated graphene-enhanced Brillouin optomechanical microresonator for ultrasensitive gas detection (Figure 1.11b,c). Theoretically, the molecular adsorption-induced surface elastic modulation in graphene enabled the Brillouin optomechanical modes extremely sensitive in NH_3 detection. They achieved a detection limit down to 1 ppb and an unprecedented dynamic range over five orders-of-magnitude. Note that the hybrid optical-electronic sensing

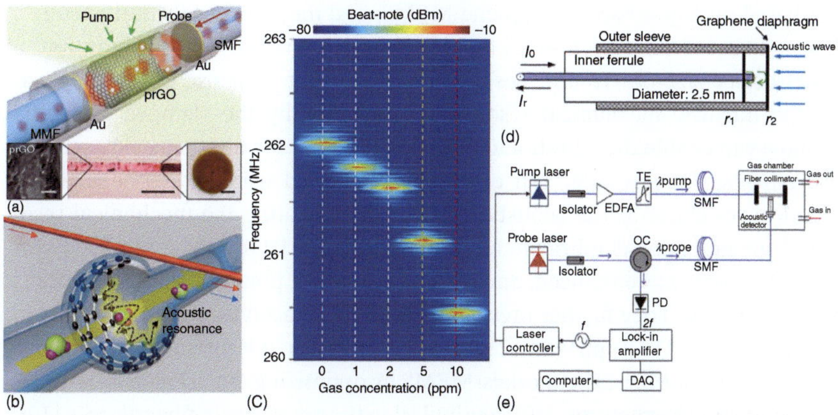

Figure 1.11 Active/nonlinear optical fiber-graphene device for ultrahigh sensitivity sensors. (a) Conceptual design and functionalization of the partially reduced graphene oxide (prGO) inner-deposited fiber sensor [104]. It composes two gold reflectors, prGO deposited inside-fiber, and Rhodamine 6G (Rh6G) as the optical-gain media. Inset from left-to-right shows the scanned electron micrograph of the prGO, the microscopic pictures of the resonator, and the fiber end with Au coverage. Inset scale bars from left-to-right are 2500 and 50 μm, respectively. (b) Schematic of graphene-enhanced Brillouin optomechanics of a microbubble cavity [104]. In this microcavity, the pump mode (red arrow) generates the Stokes mode (blue arrow) and the acoustic mechanical resonance mode (black arrow) [105]. (c) Colored map of the beat note spectra from the cavity optomechanical resonance, varies with the NH_3 gas concentration [104]. (d) Schematic of the fiber Fabry–Pérot interferometer with graphene diaphragm [106]. (e) Setup of photoacoustic spectroscopy system for acetylene detection [105]. Blue line, optical fiber; black line, electrical cable; TF, tunable filter; OC, optical circulator; PD, photo-detector; DAQ, data acquisition; DFB, distributed feedback laser (the pump); ECDL, external-cavity diode laser (the probe); EDFA, erbium-doped fiber amplifier. Source: (a) Cao et al. [104] / Springer Nature / CC BY 4.0.
(b, c) Reproduced with permission from Yao et al. [105], American Chemical Society.
(d, e) Reproduced with permission from Tan et al. [106], IEEE Publishing.

platform can obtain ultrahigh sensitivity, while the system complexity is also significantly increased compared with the conventional passive sensing methods.

For physical fields sensing, the 2D material-based MEMS are attracting great attention for their supreme optomechanical properties [12]. Ma et al. [49] reported a fiber-tip pressure sensor using a few-layer graphene as a diaphragm (diameter of 25 μm), which formed a low-finesse Fabry–Pérot (FP) interferometer. They realized a high pressure sensitivity over 39.4 nm/kPa. Further, they implemented multilayered graphene as a diaphragm of diameter 125 μm and obtained an improved pressure sensitivity of 1100 nm/kPa. They showed that this kind of graphene-FP cavity could be used for acoustic sensors, exhibiting a flat frequency response in the range of 0.2–22 kHz and a noise-equivalent acoustic single of ~60 $\mu Pa/Hz^{1/2}$ [100]. Later work showed that higher pressure sensitivity was achieved in few-layer MoS_2 diaphragm sensors, probably due to their relatively low Young's modulus of ~0.3 TPa and higher transferred film quality [101]. By integrating the metal electrodes on a fiber endface, Zheng et al. [48] demonstrated a miniature optical fiber current sensor based on a quasi-static graphene MEMS, and they achieved simultaneously high sensitivity of 2.2×10^5 nm/A^2, a short response time of ~0.25 seconds, and a compact device size of ~15 μm. Besides the quasi-static-response MEMS applications, the intrinsic mechanical resonating properties by the clamped graphene membrane can enable the vibrational fiber-optic sensors for high-precision sensing measurements. For example, Tan et al. [106] reported using a graphene nanomechanical resonator as the acoustic detector, by operating the mechanical resonance of the graphene diaphragm, as shown in Figure 1.11d,e. The sensitivity for acoustic detection was enhanced, and a noise equivalent pressure of 2.11 μPa/Hz$^{1/2}$ was demonstrated. They further presented detection of acetylene gas and achieved a detection limit of 119.8 ppb with 123.9 mW pump power. It is promising to improve the performance of fiber-2D materials-MEMS by optimizing the 2D materials' quality and geometrical structures, and manipulating the pre-stress in fiber devices [10].

1.3 Conclusion

In this chapter, we have reviewed the basic optical properties of several mainstream 2D materials and their integrations to silica optical fibers for advanced optics and optoelectronics applications. We lay emphasis on the optical fiber devices for light field generation, modulation, and detection. Currently, most of the applications remain as proof-of-concept demonstrations, and the issues to realize scalable production with reliable packaging are waiting to be solved before any commercial implement. Recent development of CVD-growth method to produce polycrystalline 2D materials [61, 62], for example graphene and MoS_2 in hollow-core PCF, has provided a unique way to the massive production of hybrid fiber-2D-material devices. As for the future development of fiber-2D-materials integrations, more cutting-edge applications are in demand to explore by merging the extraordinarily properties/functions of 2D materials and the advanced photonic structures. In addition to the classical optics applications, it is very promising to explore the quantum optics in 2D materials for fiber-integrated quantum sources and even the quantum information processing.

References

1 Kao, K.C. and Hockham, G.A. (1966). *Proc. Inst. Electr. Eng.* 113: 1151–1158.
2 Peng, G.-D. (ed.) (2019). *Handbook of Optical Fibers*. Springer.
3 Russell, P. (2003). *Science* 299: 358.
4 Gao, S.-F., Wang, Y.-Y., Ding, W. et al. (2018). *Nat. Commun.* 9: 2828.
5 Brambilla, G., Xu, F., Horak, P. et al. (2009). *Adv. Opt. Photon.* 1: 107.
6 Chen, J.-H., Li, D.-R., and Xu, F. (2019). *J. Lightwave Technol.* 37: 2577.
7 Zhang, L., Tang, Y., and Tong, L. (2020). *iScience* 23: 100810.
8 Abouraddy, A.F., Bayindir, M., Benoit, G. et al. (2007). *Nat. Mater.* 6: 336.
9 Novoselov, K.S., Geim, A.K., Morozov, S.V. et al. (2004). *Science* 306: 666.

10 Chen, J.-H., Xiong, Y.-F., Xu, F., and Lu, Y.-Q. (2021). *Light Sci. Appl.* 10: 78.
11 Dai, Z., Hu, G., Ou, Q. et al. (2020). *Chem. Rev.* 120: 6197.
12 Lee, C., Wei, X., Kysar, J.W., and Hone, J. (2008). *Science* 321: 385.
13 Liu, L., Ryu, S., Tomasik, M.R. et al. (2008). *Nano Lett.* 8: 1965.
14 Neto, A.C., Guinea, F., Peres, N.M. et al. (2009). *Rev. Mod. Phys.* 81: 109.
15 Splendiani, A., Sun, L., Zhang, Y. et al. (2010). *Nano Lett.* 10: 1271.
16 Mak, K.F., He, K., Shan, J., and Heinz, T.F. (2012). *Nat. Nanotechnol.* 7: 494.
17 Li, L., Yu, Y., Ye, G.J. et al. (2014). *Nat. Nanotechnol.* 9: 372.
18 Nair, R.R., Blake, P., Grigorenko, A.N. et al. (2008). *Science* 320: 1308.
19 Bao, Q., Zhang, H., Wang, Y. et al. (2009). *Adv. Funct. Mater.* 19: 3077.
20 Li, W., Chen, B., Meng, C. et al. (2014). *Nano Lett.* 14: 955.
21 Jiang, T., Huang, D., Cheng, J. et al. (2018). *Nat. Photonics* 12: 430.
22 Mak, K.F., Lee, C., Hone, J. et al. (2010). *Phys. Rev. Lett.* 105: 136805.
23 Li, Y., Chernikov, A., Zhang, X. et al. (2014). *Phys. Rev. B* 90: 205422.
24 Xiao, D., Liu, G.-B., Feng, W. et al. (2012). *Phys. Rev. Lett.* 108: 196802.
25 Wen, X., Gong, Z., and Li, D. (2019). *InfoMat* 1: 317.
26 Li, Y., Dong, N., Zhang, S. et al. (2015). *Laser Photonics Rev.* 9: 427.
27 Liu, H., Li, Y., You, Y.S. et al. (2017). *Nat. Phys.* 13: 262.
28 Ling, X., Wang, H., Huang, S. et al. (2015). *Proc. Natl. Acad. Sci.* 112: 4523.
29 Yang, J., Xu, R., Pei, J. et al. (2015). *Light Sci. Appl.* 4: e312.
30 Li, L., Kim, J., Jin, C. et al. (2017). *Nat. Nanotechnol.* 12: 21.
31 Kim, H., Uddin, S.Z., Lien, D.-H. et al. (2021). *Nature* 596: 232.
32 Chen, C., Lu, X., Deng, B. et al. (2020). *Sci. Adv.* 6: eaay6134.
33 Wang, X., Jones, A.M., Seyler, K.L. et al. (2015). *Nat. Nanotechnol.* 10: 517.
34 Biswas, S., Grajower, M.Y., Watanabe, K. et al. (2021). *Science* 374: 448.
35 Wang, X. and Lan, S. (2016). *Adv. Opt. Photon.* 8: 618.
36 Novoselov, K.S., Mishchenko, A., Carvalho, A., and Castro Neto, A.H. (2016). *Science* 353: aac9439.
37 Ma, W., Shabbir, B., Ou, Q. et al. (2020). *InfoMat* 2: 777.
38 Liu, Y., Huang, Y., and Duan, X. (2019). *Nature* 567: 323.
39 Cao, Y., Fatemi, V., Fang, S. et al. (2018). *Nature* 556: 43.
40 Cao, Y., Fatemi, V., Fang, S. et al. (2018). *Nature* 556: 80.
41 Xiong, Y. and Xu, F. (2020). *Adv. Photonics* 2: 064001.
42 Sun, Z., Hasan, T., Torrisi, F. et al. (2010). *ACS Nano* 4: 803.
43 Chen, J.-H., Jing, Q., Xu, F., and Lu, Y.-Q. (2017). *Optica* 4: 835.
44 Xiong, Y.-F., Chen, J.-H., Lu, Y.-Q., and Xu, F. (2019). *Adv. Electron. Mater.* 5: 1800562.
45 Xiong, Y., Liao, Q., Huang, Z. et al. (2020). *Adv. Mater.* 32: 1907242.
46 Wang, H., Xiong, Y., Wu, C. et al. (2022). *Opt. Express* 30: 6818.
47 Xiong, Y., Wang, Y., Zhu, R. et al. (2022). *Sci. Adv.* 8: eabo0375.
48 Zheng, B.-C., Yan, S.-C., Chen, J.-H. et al. (2015). *Laser Photonics Rev.* 9: 517.
49 Ma, J., Jin, W., Ho, H.L., and Dai, J.Y. (2012). *Opt. Lett.* 37: 2493.
50 Luo, Z.-C., Liu, M., Guo, Z.-N. et al. (2015). *Opt. Express* 23: 20030.
51 Chen, J.-H., Luo, W., Chen, Z.-X. et al. (2016). *Adv. Opt. Mater.* 4: 853.
52 Chen, J.-H., Tan, J., Wu, G.-X. et al. (2019). *Light Sci. Appl.* 8: 8.

53 Chen, J.-H., Zheng, B.-C., Shao, G.-H. et al. (2015). *Light Sci. Appl.* 4: e360.
54 Xu, F., Horak, P., and Brambilla, G. (2007). *Opt. Express* 15: 7888.
55 Kou, J.-L., Chen, J.-H., Chen, Y. et al. (2014). *Optica* 1: 307.
56 Yan, S.-C., Zheng, B.-C., Chen, J.-H. et al. (2015). *Appl. Phys. Lett.* 107: 053502.
57 Li, C., Chen, J., Yan, S. et al. (2016). *IEEE Photon. J.* 8: 1500307.
58 Zhang, H., Healy, N., Shen, L. et al. (2016). *Sci. Rep.* 6: 23512.
59 An, N., Tan, T., Peng, Z. et al. (2020). *Nano Lett.* 20: 6473.
60 Lee, E.J., Choi, S.Y., Jeong, H. et al. (2015). *Nat. Commun.* 6: 6851.
61 Chen, K., Zhou, X., Cheng, X. et al. (2019). *Nat. Photonics* 13: 754.
62 Zuo, Y., Yu, W., Liu, C. et al. (2020). *Nat. Nanotechnol.* 15: 987.
63 Bao, Q., Zhang, H., Wang, B. et al. (2011). *Nat. Photonics* 5: 411.
64 Kim, J.T. and Choi, C.-G. (2012). *Opt. Express* 20: 3556.
65 Zhang, H., Healy, N., Shen, L. et al. (2016). *J. Lightwave Technol.* 34: 3563.
66 Chu, R., Guan, C., Yang, J. et al. (2017). *Opt. Express* 25: 13278.
67 Cui, Y., Lu, F., and Liu, X. (2017). *Sci. Rep.* 7: 40080.
68 Schell, A.W., Takashima, H., Tran, T.T. et al. (2017). *ACS Photonics* 4: 761.
69 Liao, F., Yu, J., Gu, Z. et al. (2019). *Sci. Adv.* 5: eaax7398.
70 Toth, M. and Aharonovich, I. (2019). *Annu. Rev. Phys. Chem.* 70: 123.
71 Aharonovich, I., Englund, D., and Toth, M. (2016). *Nat. Photonics* 10: 631.
72 He, Y.-M., Clark, G., Schaibley, J.R. et al. (2015). *Nat. Nanotechnol.* 10: 497.
73 Yu, S., Wu, X., Wang, Y. et al. (2017). *Adv. Mater.* 29: 1606128.
74 Liu, Z.-B., Feng, M., Jiang, W.-S. et al. (2013). *Laser Phys. Lett.* 10: 065901.
75 Yu, S., Wu, X., Chen, K. et al. (2016). *Optica* 3: 541.
76 Gan, X., Zhao, C., Wang, Y. et al. (2015). *Optica* 2: 468.
77 Xu, K., Xie, Y., Xie, H. et al. (2018). *J. Lightwave Technol.* 36: 4730.
78 Dong, L., Liu, X., Zhang, Y. et al. (2020). *ACS Appl. Electron. Mater.* 2: 447.
79 Cheng, X., Zhou, X., Tao, L. et al. (2020). *Nanoscale* 12: 14472.
80 Sun, X., Qiu, C., Wu, J. et al. (2015). *Opt. Express* 23: 25209.
81 Jin, Z., Ye, F., Zhang, X. et al. (2018). *ACS Nano* 12: 12571.
82 Zhuo, L., Fan, P., Zhang, S. et al. (2020). *Nanoscale* 12: 14188.
83 Chen, J.-H., Liang, Z.-H., Yuan, L.-R. et al. (2017). *Nanoscale* 9: 3424.
84 Xiong, Y., Xu, H., Wang, Y. et al. (2022). *ACS Photonics* 9: 1808.
85 Boyd, R.W. (2020). *Nonlinear Optics*. Academic Press.
86 Autere, A., Jussila, H., Dai, Y. et al. (2018). *Adv. Mater.* 30: 1705963.
87 Li, C., Chen, J.-H., Wang, W.-S. et al. (2017). *Adv. Opt. Mater.* 5: 1700630.
88 Qin, C., Jia, K., Li, Q. et al. (2020). *Light Sci. Appl.* 9: 185.
89 Ding, Z.-X., Chen, Y., and Xu, F. (2021). *Opt. Laser Technol.* 143: 107381.
90 Kim, J. and Song, Y. (2016). *Adv. Opt. Photonics* 8: 465.
91 Wu, Y., Yao, B.C., Feng, Q.Y. et al. (2015). *Photonics Res.* 3: A64.
92 Jiang, B., Hao, Z., Ji, Y. et al. (2020). *Light Sci. Appl.* 9: 63.
93 Li, Y., An, A., Lu, Z. et al. (2022). *Nat. Commun.* 13: 3138.
94 Wu, Y., Yao, B., Zhang, A. et al. (2014). *Opt. Lett.* 39: 1235.
95 Yao, B.C., Wu, Y., Zhang, A.Q. et al. (2014). *Opt. Express* 22: 28154.
96 Sridevi, S., Vasu, K., Bhat, N. et al. (2016). *Sens. Actuators, B* 223: 481.
97 Xu, B., Huang, J., Xu, X. et al. (2019). *ACS Appl. Mater. Interfaces* 11: 40868.

98 Xiao, Y., Zhang, J., Cai, X. et al. (2014). *Opt. Express* 22: 31555.

99 Huang, M., Yang, C., Sun, B. et al. (2018). *Opt. Express* 26: 3098.

100 Ma, J., Xuan, H., Ho, H.L. et al. (2013). *IEEE Photon. Technol. Lett.* 25: 932.

101 Yu, F., Liu, Q., Gan, X. et al. (2017). *Adv. Mater.* 29: 1603266.

102 Shivananju, B.N., Yu, W., Liu, Y. et al. (2017). *Adv. Funct. Mater.* 27: 1603918.

103 Pawar, D. and Kale, S.N. (2019). *Microchim. Acta* 186: 253.

104 Cao, Z., Yao, B., Qin, C. et al. (2019). *Light Sci. Appl.* 8: 107.

105 Yao, B., Yu, C., Wu, Y. et al. (2017). *Nano Lett.* 17: 4996.

106 Tan, Y., Zhang, C., Jin, W. et al. (2017). *IEEE J. Sel. Top. Quantum Electron.* 23: 199.

2

Postprocessing of Semiconductor Optical Fibers

Hei C.L. Tsui and Noel Healy

Newcastle University, School of Mathematics, Statistics and Physics, Newcastle Upon Tyne, NE1 7RU, United Kingdom

2.1 Introduction

There has been a significant progress in the production of semiconductor optical fibers in the last two decades. However, a common problem that still exists for all the production methods is high transmission losses. This is often due to the poor control of heating or cooling associated with the high-pressure chemical vapor deposition, molten core drawing, and pressure-assisted melt-filling processes. This leads to highly polycrystalline or amorphous core materials that scatter and/or absorb light as it is transmitted. Furthermore, drawing of small-diameter core fiber is challenging to achieve, as capillary instability can cause a discontinuous core. These reasons originally motivated research into the postprocessing of semiconductor optical fibers. However, postprocessing of semiconductor optical fibers has been extended to include more advanced applications, from longitudinally varying the core's dimensions for tailoring the mode profile or dispersion to inducing/relaxing strains to alter the optoelectronic properties. This chapter will summarize the main approaches to postprocessing semiconductor optical fibers and the applications of these processing methods.

2.2 Semiconductor Optical Fibers

2.2.1 Heat Treatments of Semiconductor Optical Fibers

Thermal annealing is a standard process used in the semiconductor industry for controlling the nucleation and growth of crystals. The process involves heating the material to induce structural relaxation and increase atom mobility so that defects are reduced and crystal grain growth is promoted [1, 2]. The annealing

Optical and Electronic Fibers: Emerging Applications and Technological Innovations, First Edition. Edited by Lei Wei.

process has been adapted for improving the crystal quality in the core material of the semiconductor optical fiber. The first demonstrations of thermal annealing of semiconductor optical fibers were used to grow crystals from an amorphous silicon core to reduce the number of dangling bonds in the amorphous allotrope. Dangling bonds add an extra energy level between the valence band and conduction band, and these act as absorption sites for low-energy photons. Pacifying dangling bonds is an important step in reducing the transmission loss of semiconductor optical fiber. Subsequent heat treatments were developed to tackle a second-order loss mechanism; that of scattering from crystal grain boundaries. It is axiomatic that to produce low-loss optical fibers with semiconductor cores that large crystallites are an important factor [2–5].

Although the principle of annealing is well-known for planar semiconductor geometries, alternative considerations are important for the "fiber" geometry. For example, the silicon in a silicon optical fiber is completely surrounded by a thick silica cladding, whereas the surface of the crystal in thin films is often exposed. In addition, the differing expansion coefficient of the glass fiber cladding and the semiconductor cores can induce stress during and after annealing. Furthermore, the semiconductor cladding interface acts as a strong source of nucleation sites [2, 6, 7].

2.2.1.1 Thermal Annealing

The first semiconductor optical fibers had silicon cores that were a few microns in diameter. These were produced via the high-pressure-assisted chemical vapor deposition process that involves first depositing the material in an amorphous state and then annealing to produce a crystalline core. Amorphous silicon-core optical fibers were annealed at temperatures ranging from 1125 to 1325 °C, the upper bound limited by the softening of the silica cladding [8]. Raman spectrometry revealed that this approach produces core materials with grain sizes up to ~1 μm and with the grain boundaries consisting of an amorphous phase. A small red shift of the Raman peak indicated the presence of residual stress, which is induced by the thermal expansion difference between silicon and silica [7, 9]. Optical transmission loss measurements on the thermally annealed fibers showed that those annealed at a higher temperature and with improved crystallinity have the lower transmission loss (Figure 2.1) with a then low measurement of 8 dB cm^{-1} at a wavelength of 1.55 μm. In order to optimize the annealing process for silicon-core fiber, a two-step process was developed that comprised of a long, low-temperature anneal followed by a short, high-temperature anneal. The first annealing step was done at 530 °C, which is close to the crystallization temperature of silicon. This multiday process promoted the nucleation and formation of a limited number of crystallite seed grains. The second annealing step, at 1300 °C for tens of minutes, is used to grow the previously nucleated grains and suppress additional nucleation. The samples that underwent the first annealing step only exhibited a high optical loss of 27 dB cm^{-1} at 1.55 μm due to the presence of a large proportion of amorphous phase and small grain size. The transmission loss improved significantly with a value of 5.2 dB cm^{-1} after the second step of the process. The reduction in loss was attributed to larger

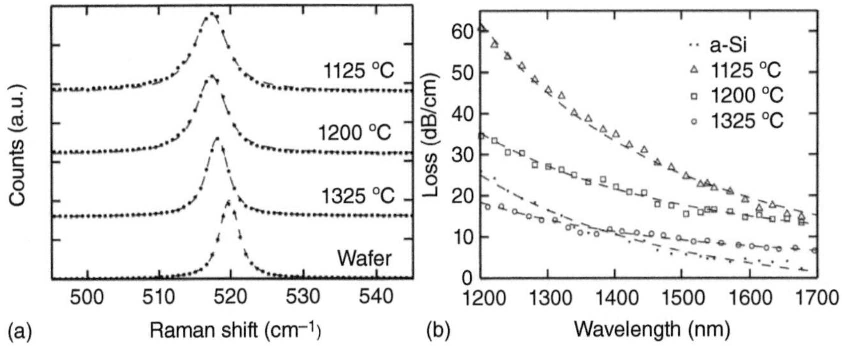

Figure 2.1 (a) Raman spectra (dashed lines are Voigt fits) and (b) transmission losses as a function of wavelengths (dashed lines are λ^{-4} fits) of annealed polycrystalline silicon-core fibers. *Source:* Reproduced from [8]. Copyright 2010 AIP Publishing LLC.

grain size and lower defect density. Transmission electron microscope images also corroborated these results (Figure 2.2).

2.2.1.2 Rapid Thermal Annealing

An alternative thermal annealing approach is rapid thermal annealing. This process uses intense flashes of ultraviolet (UV), visible, and infrared (IR) light to quickly heat and cool the semiconductor. This approach has been adapted by using a Xenon lamp (Figure 2.3) to induce a cyclic thermal process on silicon optical fibers that were produced via the molten core drawing method [10]. The IR light of the lamp produces thermal energy, which is similar to that of an annealing furnace. The UV and visible light provide additional energy for electron excitation, which helps increase the diffusion rate and also heat the silicon via phonon-assisted recombination. Crystallization is undertaken on a silicon-core fiber at 950 °C for only

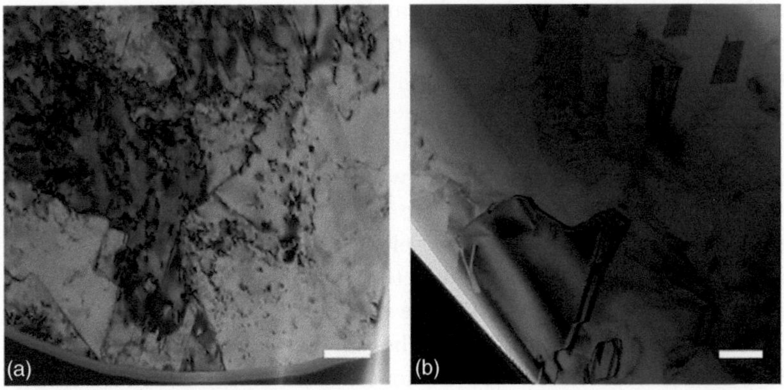

Figure 2.2 Transmission electron microscopy (TEM) images of (a) polycrystalline silicon-core fiber after annealed at 530 °C and (b) after annealed at 1300 °C. Scale bar: 250 nm. *Source:* Reprinted with permission from [6]. Copyright 2016 American Chemical Society.

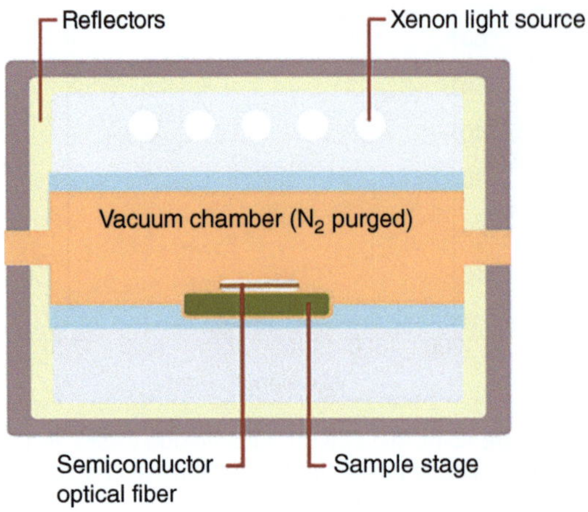

Figure 2.3 Schematic diagram for rapid photothermal annealing.

10 minutes under a nitrogen environment. The resulting length of individual crystallites within the fiber's core reached about 9 mm. Raman spectroscopy indicated that this annealing process was also a promising method of reducing the thermally induced stresses in the fiber that were introduced in the drawing process. This is likely owing to the lower annealing temperature of the approach (950 °C compared to 1300 °C in the traditional annealing process).

2.2.2 Laser Processing of Semiconductor Optical Fibers

Laser processing is an extremely versatile tool for the postprocessing of semiconductor core optical fibers. It enables precise and localized heating for recrystallization and/or compositional microstructuring of the core material. In general, the laser processing of the core of a semiconductor optical fiber can be divided into two categories: heating via electronic absorption and heating via conduction. The heating mechanism is dictated by the wavelength of the laser as is selected for a particular core material.

2.2.2.1 Electronic Absorption via the Core

Electronic absorption relies on using a laser with a wavelength, photon energy, E_p, greater than the bandgap, E_g, of the core material. Considering silicon with $E_g = 1.1\,\text{eV}$ as an example, a laser with a wavelength of 488 nm ($E_p = 2.4\,\text{eV}$) is typically used ([9, 11, 12],a). At this wavelength, the silica cladding is transparent, and all the laser's energy is absorbed by the core material. As $E_p > E_g$, the photon has enough energy to excite an electron from the valence band into the conduction band, and the electron recombines via a phonon-assisted pathway that heats the core material as silicon is an indirect bandgap semiconductor. If enough photons are absorbed, this scheme can ultimately melt the core material. This process takes just microseconds, meaning that the core can be melted while the conductive heating to the cladding is minimal (Figure 2.4) [9].

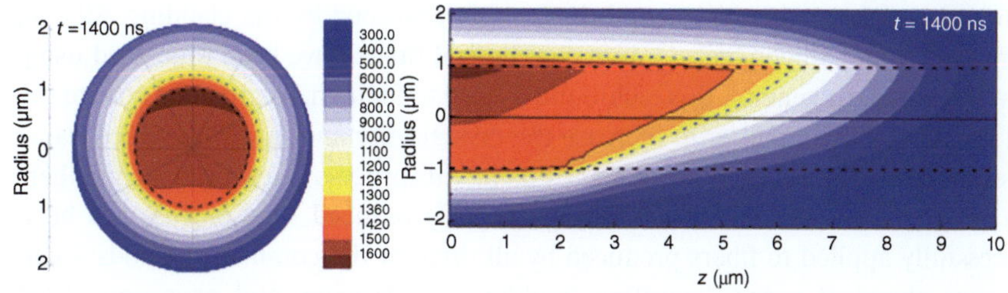

Figure 2.4 Simulated temperature profile evolution of a laser-irradiated silicon-core fiber.

For some materials such as silicon and germanium, a strong interfacial bond between the core and cladding can be formed via an oxide layer. If desired, this can be leveraged to induce large strains in the core material. For example, if using a starting material of amorphous silicon for the core, laser crystallization can, in conjunction with the strong core-cladding interface, induce strains as high as 6 GPa in the core material. This is because the cladding remains rigid throughout the process, while the core material prefers to reduce its volume [9].

Oxygen inclusion that has diffused from the cladding into the core material during the fiber production process can be treated using this recrystallization process [11]. It has been shown that the oxygen content can be reduced using laser annealing via electronic absorption, as demonstrated in Figure 2.5. Optical losses as low as $0.47\,\text{dB}\,\text{cm}^{-1}$ at $1.55\,\mu\text{m}$ for silicon-core fiber and $1.33\,\text{dB}\,\text{cm}^{-1}$ at $2\,\mu\text{m}$ for germanium core fiber have been demonstrated via this method. These low losses are comparable with the on-chip waveguides.

2.2.2.2 Conductive Core Heating via Laser Absorption by the Cladding

The alternative approach to electronic absorption for heating the core is to heat the cladding with the laser and let the core heat via conduction. For silica-cladding materials, this is typically done with a CO_2 laser emitting light at a wavelength of

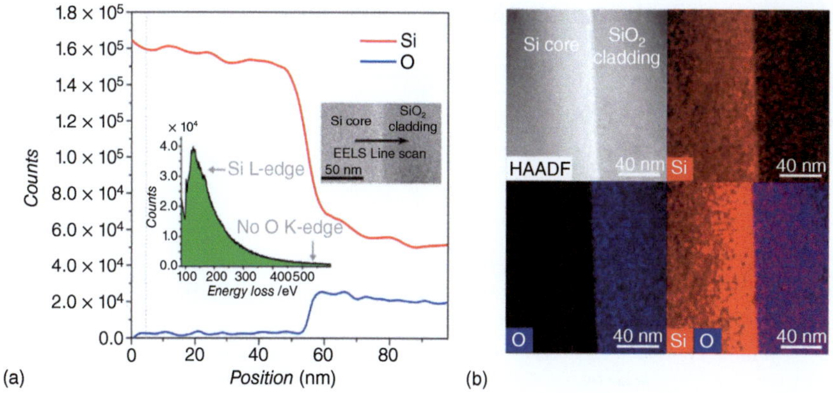

Figure 2.5 (a) Electron energy loss spectrum (EELS) and (b) Energy-dispersive X-ray (EDX) spectra map across the core-cladding boundary. *Source:* Reproduced from Ji et al. [12]. © 2017 AIP Publishing LLC.

10.6 μm, as light at this wavelength is strongly absorbed by the cladding. Silicon, germanium, and silicon–germanium core optical fibers have been processed using this technique. In general, this technique is used to melt and regrow the core material in a controlled manner to increase the crystallinity and, in turn, reduce transmission losses. It has been shown that by using this process single crystal core fibers can be produced and losses can be significantly reduced. The process has been successfully applied to fibers produced by all three of the common methods: high-pressure chemical vapor deposition, molten core drawing, and pressure-assisted molten filling [13–15]. As an example, CO_2 laser processing of silicon-core optical fibers produced using the molten core drawing method were studied [15]. This study found that the reduction in transmission losses scaled linearly with the cooling rate (see Figure 2.6). The as-drawn fibers had core sizes of 12 μm, and the grain size in the core was ~1 mm. The transmission loss of these fibers was >20 dB cm^{-1} prior to post-processing and <2 dB cm^{-1} after. The directional scanning of the laser promoted the seeding of the cooling liquid front from recently irradiated section of the core, and the crystal length was only limited by the scanning distance. Similar results were achieved for germanium core fiber, where low loss of 2.05 dB cm^{-1} at 4.7–4.9 μm was demonstrated under high-power laser annealing (Figure 2.7) [16].

This process has also been used to produce homogenous mixes of semiconductor alloys in the core. For example, the core's composition is not homogeneous post-production in the case of a Si–Ge alloy optical fiber. The germanium- and silicon-rich regions tend to segregate, owing to the uncontrolled cooling of the drawing process. A CO_2 laser processing process can remelt the core material and produce a homogeneous alloy, not achievable without postprocessing [14].

2.2.3 Applications of Laser-processed Optical Fibers

2.2.3.1 Electronic Bandgap Modulation

The strong tensile stress that can be imparted into the core of a silicon optical fiber via the electronic absorption laser crystallization process can be used to modulate the core material's electronic bandgap. The high stress is generated because the

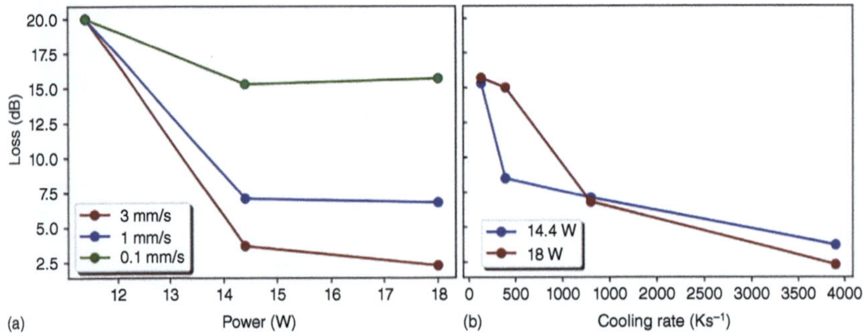

Figure 2.6 Transmission losses at 1.55 μm of CO_2 laser-treated polycrystalline silicon-core fiber as a function of (a) laser power and scan speed, and (b) cooling rate.

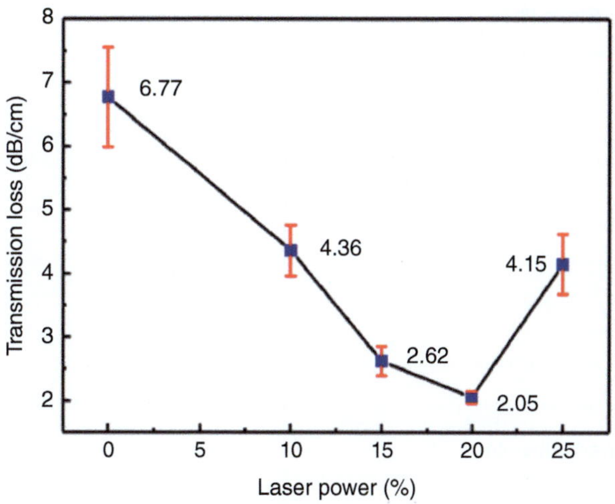

Figure 2.7 Transmission losses at 4.7–4.9 μm of Ge fibers as a function of laser power. *Source:* Zhao et al. [16] / Optical Society of America / CC BY 4.0.

density of the crystalline silicon is higher than that of its amorphous state, while the core volume is restricted as the cladding is not softened during postprocessing. Therefore, there is a large residual tensile stress built-up in the crystalline silicon core upon cooling. It has been demonstrated that the bandgap of silicon can be reduced from 1.1 to 0.59 eV by imparting an in-plane tensile strain, see Figure 2.8, which means that silicon can be used to detect near-IR light. As the stress also induces anisotropic distortion in the crystal lattice, it is suggested that this is a route to achieving a $\chi^{(2)}$ susceptibility, which is absent in centrosymmetric materials such as silicon [9].

2.2.3.2 Compositional Microstructuring

Semiconductor cores that are made of alloys, for example, SiGe, can have their composition microstuctured using CO_2 laser absorption by the cladding. For example, laser heating can produce Ge-rich regions as high as 35 at% in a $Si_{94}Ge_6$ core. By blocking the beam periodically or changing the scan speed, a periodic grating of Ge-rich regions can be achieved (Figure 2.9). Other examples include the creation of central axial enriched cores with a peak Ge concentration 15 at% higher than the exterior of the semiconductor core. Presenting the opportunity for advanced waveguiding features ([17]), the process has been adapted for the segregation and chemical reaction of metal components within a silicon core, for example, the segregation of GaSb and Si into parallel layers [18].

2.2.3.3 Capillary Instabilities

Capillary instability, also known as Plateau-Rayleigh instability, is the phenomenon where a liquid breaks into droplets because of the difference between the surface tension of a liquid and a surrounding liquid or gas upon perturbation. This is one of

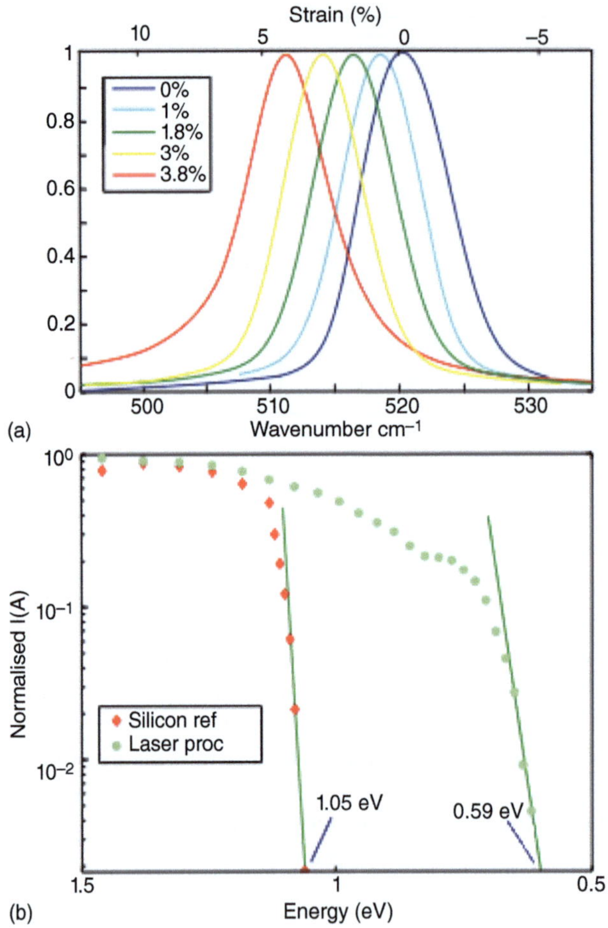

Figure 2.8 (a) Laser-induced strain in a silicon-core fiber, as measured via Raman spectroscopy, and (b), bandgap measurement of strained silicon core compared to a silicon reference.

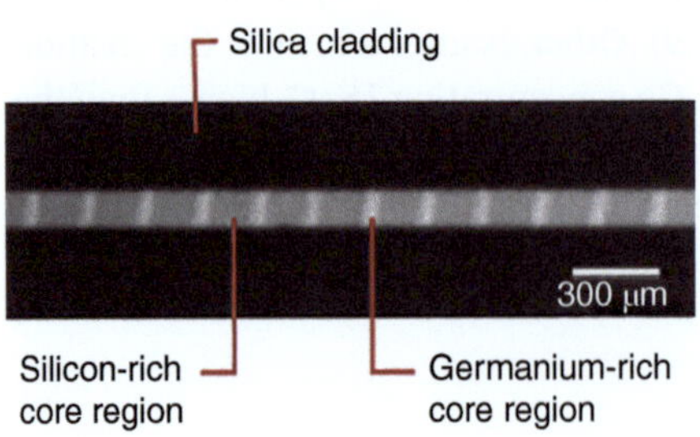

Figure 2.9 Germanium-rich grating formed in the fiber core via modulation and scanning of a CO_2 laser.

the mechanisms that can preclude the drawing of small-core semiconductor optical fibers [19]. Although usually undesirable during the drawing process, this phenomenon can be utilized to form in-fiber semiconductor devices and fabricate microspheres, microparticles, and microresonators. However, postprocessing offers a more controlled route to producing instability-induced structures. This can be achieved by furnace-based isothermal heating and oxyhydrogen flame, which generates a temperature gradient [20, 21].

Owing to its small tunable heating zone, CO_2 laser postprocessing offers the most control over capillary instabilities (Figure 2.10). Coupled with a selectable scanning speed, the various size of spheres can be fabricated precisely. Microsphere as small as 200 nm and microresonator with a resonator quality factor of 7.1×10^5 have been demonstrated using this laser-induced instability method [22]. As the silica cladding has a higher melting temperature than the semiconductor core materials, the cladding solidified while core is still in liquid state under cooling (after being heat by the laser). The solidified cladding restricts the space for the core material to break-up (due to capillary instability) and solidify, during which some semiconductors, such as silicon and germanium, expand anomalously, and therefore there is stress built in these particles [23, 24]. By adjusting the spot size of the laser, which affects the cooling rate of the process, it was demonstrated that built-in stresses occur in both the in-fiber and removed free-standing semiconductor particles with up to 1.65 and 0.3 GPa, respectively (Figure 2.11) [24].

In addition to providing conditions to sphere formation via capillary instability, the laser can also induce thermocapillary convection in which the in-fiber particles can be manipulated to specific desirable positions. This thermocapillary effect is due to the surface tension gradient induced by the CO_2 laser, as the heating liquidizes the cladding. The particles are drawn toward the laser heating spot, because the mass transfer depends on the surface tension gradient in which subjects are pulled towards regions with lower tension. This effect is particularly useful in fabricating heterojunctions within multi-material fibers. For example, the capability of relocating the particles to a designed position allowed the formation of in-fiber p–n junction (p-Ge/n-Si), see Figure 2.12 [25].

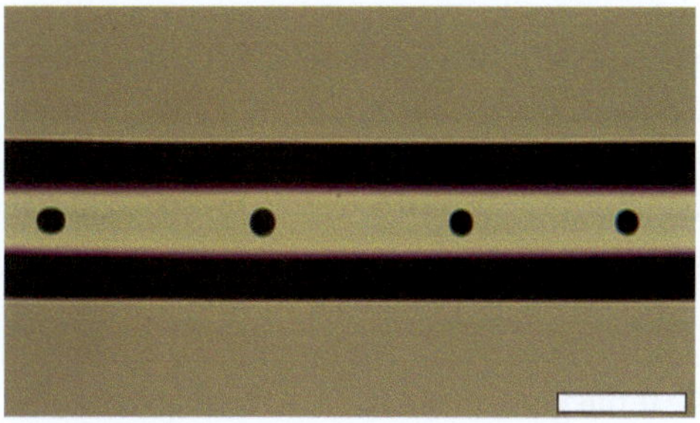

Figure 2.10 Formation of semiconductor (germanium) spheres formed by the laser-induced capillary instability method. Scale bar 100 μm.

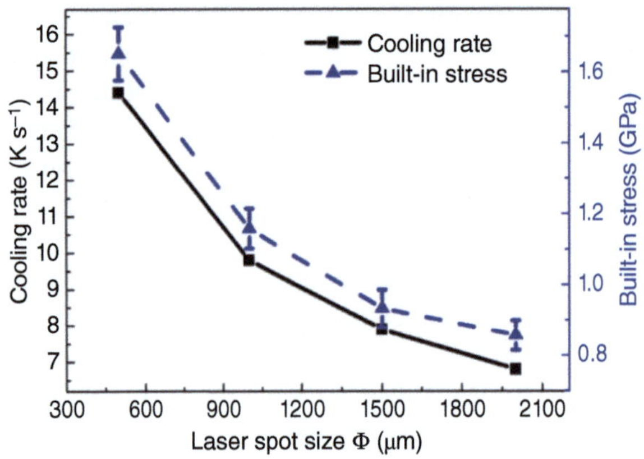

Figure 2.11 Cooling rates and the built-in stress as a result of changing laser spot size. *Source:* Reproduced from Zhang et al. [22]. © American Chemical Society.

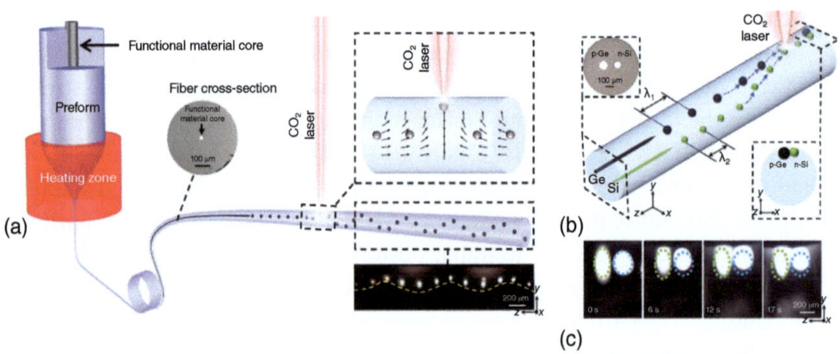

Figure 2.12 (a) In-fiber particle manipulation by laser-induced thermocapillary convection; (b) Fabrication of in-fiber heterojunction and (c) manipulation of silicon and germanium particles inside fiber at different time points. *Source:* Zhang et al. [25] / Springer Nature / CC BY 4.0.

2.2.4 Tapering of Semiconductor Optical Fibers

Tapering is a standard technique for tailoring the diameter of optical fibers for various applications, for example, mode matching, dispersion engineering, and nonlinear applications [26]. In addition to modifying the diameter of the fiber, tapering also serves as a means of improving the crystallinity of the semiconductor core material because of the high temperature used during the process.

Tapering is typically undertaken by heating the desirable part of the fiber while pulling both of its ends in opposite directions. For standard optical fiber the requirement of the heat source is that it can heat the glass to above its softening temperature, and this can be achieved using a filament, an electric arc, or a laser. An additional consideration for the semiconductor optical fiber is that the fiber is heated to above the melting temperature of the core material, and this has been achieved using all of the traditional heat sources [23, 26, 27].

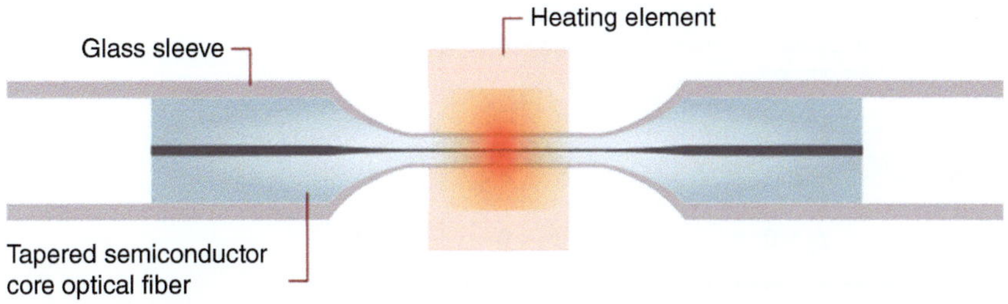

Figure 2.13 Schematic diagram of sleeved tapering process. *Source:* Franz et al. [28] / Optical Society of America / CC BY 4.0.

Owing to the handling requirements for tapering, the fiber's length needs to be of the order of 10's of cm. However, a modified approach was developed that uses a silica capillary sleeve that holds sorter lengths of semiconductor core fiber at the heat zone to improve the efficiency of the tapering of semiconductor core optical fibers (Figure 2.13). This development opened up the tapering process to fibers that are as short as 1 cm and produced tapered fibers with more robust cladding diameters.

Experimental results indicated that a higher crystallinity can be achieved with a larger taper ratio. It is believed that the number of nucleation sites is limited in a small-core fiber, which undergoes a faster cooling rate during tapering [27]. This has resulted in a means of producing low loss $<2\,\text{dB}\,\text{cm}^{-1}$ silicon-core fibers that have long single crystal regions of ($\approx$1 cm) [26–29]. Like laser processing, tapering has interesting influence on the crystallographic properties of the post-processed fiber [28, 29]. For instance, hydrostatic stress is induced during tapering, and the stress increases with decreasing core diameter. Furthermore, there is a preference for the <110> family of directions to align with the fiber's longitudinal axis after tapering. This unique change in crystal alignment was not observed in other annealing methods (furnace based and laser based).

2.2.4.1 Applications of Tapered Optical Fiber

One distinct property of semiconductor core fiber over the typical optical fiber is that the core materials possess a significantly higher nonlinear refractive index; for instance, the n_2 of silicon is about 100 times higher than silica at telecommunication wavelength [30]. Therefore, nonlinear applications, such as supercontinuum generation and four-wave mixing, have been important applications of semiconductor core fibers. Furthermore, the high refractive index of the semiconductor core enables the core sizes to be very small, while confining most of the energy to the core. The is advantageous as the energy density is an important parameter for nonlinear optics. However, this can be difficult to achieve using the molten core fiber drawing process as oxygen in-diffusion into the core becomes unacceptable [29]. Careful design of the fiber geometry can improve and enhance the nonlinear response of the waveguide, and tapering can help in realizing such designs. Supercontinuum generation (Figure 2.14) and wavelength tuning via stimulated Raman scattering (Figure 2.15) have been demonstrated via tapered silicon-core fibers [19, 27, 31]. Spectral

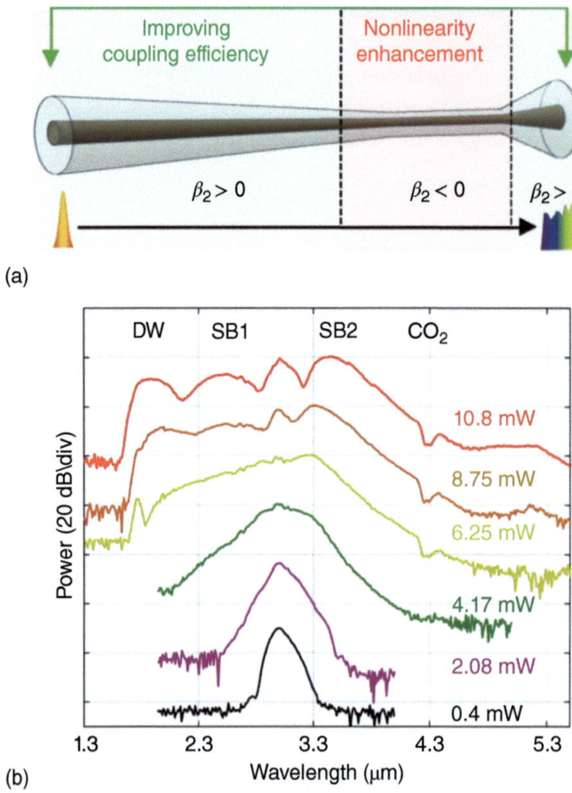

Figure 2.14 (a) Schematic diagram of asymmetric taper; (b) spectral broadening as a function of coupled average input power. *Source:* Ren et al. [19] / Springer Nature / CC BY 4.0.

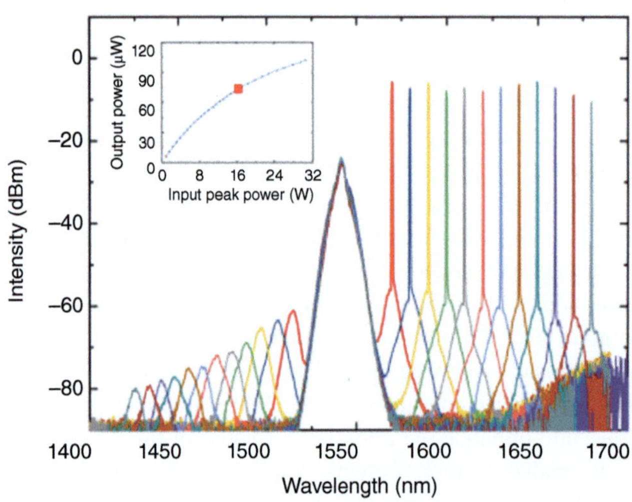

Figure 2.15 Tuning from 1570 to 1680 nm measured from a silicon-core fiber. *Source:* Reproduced from Wu et al. [31]. © 2021 IEEE.

broadening from 1.62 to 5.34 μm has been achieved by using an asymmetric taper design in which a large input core is employed to maximize the coupling into the taper waist, and minimize the losses in mode coupling and radiation, and a sharp up-taper is used to increase the output of the fiber (Figure 2.14) [19].

2.2.4.2 Dispersion Tailoring

Producing soliton in a waveguide requires the balance between the nonlinearity and group velocity dispersion. For a silicon-core fiber, tailoring the waveguide geometry is required in order to compensate the pulse broadening induced by the large values of two-photon absorption (TPA), free-carrier absorption (FCA), and free-carrier dispersion (FCD). Using realistic parameters in numerical simulations, it was shown that solitons can be achieved in silicon-core fiber by using a tapered structure; see Figure 2.16. The results showed that silicon-core fiber experiences a decrease in the magnitude of anomalous dispersion as the core diameter increases until reaching 920 nm, and this dispersion profile can help compensating the pulse broadening effect induced by the loss [33].

In addition to solitons, parabolic pulses can be created in a tapered silicon fiber. Simulations showed that the pulse shape is affected strongly by FCA in untapered fiber, while there is no distortion of the pulse observed in the nonlinear propagation when the tailored, tapered silicon-core fiber has a decreasing normal dispersion profile [32].

2.2.4.3 Mode-matched Coupling

Getting light in and out of a semiconductor core optical fiber with high efficiency is problematic, as the core has high Fresnel reflectivity owing to the semiconductor's large refractive index. Furthermore, these fibers do not tend to be splice compatible with standard single-mode fiber, as the core material does not have a "glassy" phase when heated. Tapering of the semiconductor optical fiber has been proposed to overcome this issue [34]. For this process, the semiconductor optical fiber is tapered until its core is vanishingly small, and the transition forms a "nanospike" (Figure 2.17). The fiber is cleaved at the region where the core has vanished and

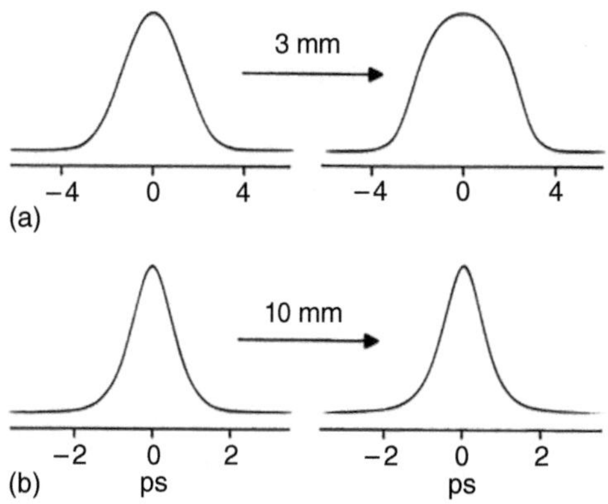

Figure 2.16 (a) Parabolic pulse generation in a 3 mm tapered fiber and (b) soliton evolution in a 10 mm taper. *Source:* Reproduced from ([32], [33]).

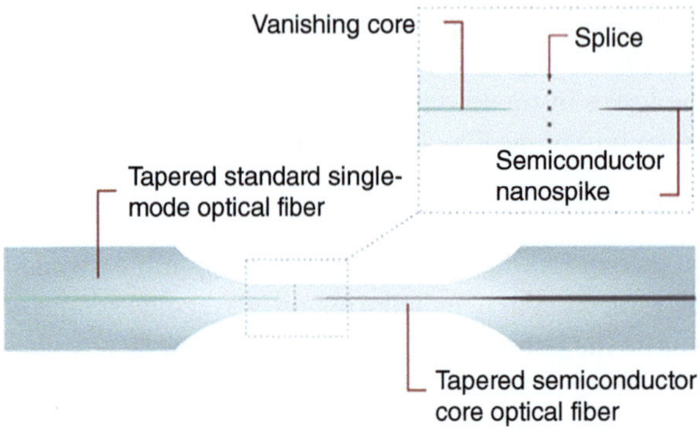

Figure 2.17 A schematic of the integration of a mode matching nanospike with a standard single-mode optical fiber.

spliced to a similarly treated standard, single-mode fiber. As the light propagates from the core of the standard single-mode fiber through the down-taper, the mode expands until it has evolved into the fundamental mode on the taper waist. The mode is matched with that at the splice point between the two fibers. As the mode is propagated through the up-tapered region, it is captured by the semiconductor core. Theoretically, coupling losses can be reduced to below 1 dB, and this technique has been used to integrate a silicon-core fiber as the active element in an all-fiber optical frequency comb [34, 35].

2.3 Conclusion

At present, postprocessing is an essential task for producing semiconductor core fibers and devices. For example, reducing the transmission losses is a key step for realizing nonlinear applications. In this chapter, many of the most important postprocessing techniques are introduced, and some applications are indicated. These range from the conventional furnace-based annealing, for improving the core material's crystallinity to using a laser for the same end result. Laser annealing/recrystallization provides new opportunities from bandgap/domain engineering to producing in-fiber microspheres from the core. Another method of engineering the fiber's properties along the direction of propagation is tapering. The high temperature used during tapering can also lead to high-quality crystal growth. While the longitudinally varying core diameter allows the dispersion to be tailored for nonlinear applications.

References

1 Roorda, S., Sinke, W.C., Poate, J.M. et al. (1991). *Phys. Rev. B* 44: 3702.
2 Spinella, C., Lombardo, S., and Priolo, F. (1998). *J. Appl. Phys.* 84: 5383.
3 Bo, X.-Z., Yao, N., Shieh, S.R. et al. (2002). *J. Appl. Phys.* 91: 2910.

4 Hatalis, M.K. and Greve, D.W. (1988). *J. Appl. Phys.* 63: 2260.
5 Thompson, C.V. (1990). *Annu. Rev. Mater. Sci.* 20: 245.
6 Chaudhuri, S., Sparks, J.R., Ji, X. et al. (2016). *ACS Photonics* 3: 378.
7 Sazio, P.J.A., Amezcua-Correa, A., Finlayson, C.E. et al. (2006). *Science* 311: 1583.
8 Lagonigro, L., Healy, N., Sparks, J.R. et al. (2010). *Appl. Phys. Lett.* 96: 041105.
9 Healy, N., Mailis, S., Bulgakova, N.M. et al. (2014). *Nat. Mater.* 13: 1122.
10 Gupta, N., McMillen, C., Singh, R. et al. (2011). *J. Appl. Phys.* 110: 093107.
11 Ji, X., Page, R.L., Chaudhuri, S. et al. (2017). *Adv. Opt. Mater.* 5: 1600592.
12 Ji, X., Lei, S., Yu, S.-Y. et al. (2017). *ACS Photonics* 4: 85.
13 Chen, H., Fan, S., Li, G. et al. (2019). *IEEE Photonics Technol. Lett.* 32: 81.
14 Coucheron, D.A., Fokine, M., Patil, N. et al. (2016). *Nat. Commun.* 7: 13265.
15 Healy, N., Fokine, M., Franz, Y. et al. (2016). *Adv. Opt. Mater.* 4: 1004.
16 Zhao, Z., Mao, Y., Ren, L. et al. (2019). *Opt. Mater. Exp.* 9: 1333.
17 Wu, W., Balci, M., Song, S. et al. (2020). *Opt. Mater. Exp.* 10: 926.
18 Song, S., Laurell, F., Meehan, B. et al. (2022). *Nat. Commun.* 13: 1.
19 Ren, H., Shen, L., Runge, A.F.J. et al. (2019). *Light: Sci. Appl.ications* 8: 105.
20 Gumennik, A., Wei, L., Lestoquoy, G. et al. (2013). *Nat. Commun.* 4.
21 Wei, L., Hou, C., Levy, E. et al. (2017). *Adv. Mater.* 29: 1603033.
22 Zhang, J., Li, K., Zhang, T. et al. (2017). *Adv. Funct. Mater.* 27: 1703245.
23 Fokine, M., Theodosiou, A., Song, S. et al. (2017). *Opt. Mater. Exp.* 7: 1589.
24 Zhang, J., Wang, Z., Wang, Z. et al. (2019a). *ACS Appl. Mater. Interfaces* 11: 45330.
25 Zhang, J., Wang, Z., Wang, Z. et al. (2019b). *Nat. Commun.* 10.
26 Healy, N., Sparks, J.R., Sazio, P.J. et al. (2010). *Opt. Express* 18: 7596.
27 Suhailin, F.H., Shen, L., Healy, N. et al. (2016). *Opt. Lett.* 41: 1360.
28 Franz, Y., Runge, A.F.J., Ren, H. et al. (2017). *Opt. Mater. Exp.* 7: 2055.
29 McMillen, C., Brambilla, G., Morris, S. et al. (2012). *Opt. Mater.* 35: 93.
30 Lin, Q., Painter, O.J., and Agrawal, G.P. (2007). *Opt. Express* 15: 16604.
31 Wu, D., Shen, L., Ren, H. et al. (2021). *IEEE J. Sel. Top. Quantum Electron.* 27: 1.
32 Peacock, A.C. and Healy, N. (2010). *Opt. Lett.* 35: 1780.
33 Peacock, A.C. (2010). *Opt. Lett.* 35: 3697.
34 Ren, H., Aktas, O., Franz, Y. et al. (2017). *Opt. Express* 25: 24157.
35 Sohanpal, R., Ren, H., Shen, L. et al. (2022). *Nat. Commun.* 13: 3992.

3

Processed Optical Fiber-based Wearable Sensors for Healthcare

Rajan Jha[1], Kalipada Chatterjee[1], and Ranjan Singh[2]

[1] *Indian Institute of Technology, School of Basic Sciences, SPARC Program, Nanophotonics and Plasmonics Laboratory, Bhubaneswar, Odisha 752050, India*
[2] *Nanyang Technological University, School of Physical and Mathematical Sciences, Division of Physics and Applied Physics, 637371, Singapore*

3.1 Introduction

Optical sensor development has encouraged extensive research in various fields ranging from industrial and infrastructural sectors to biological and healthcare facilities. A considerable share of optical sensing technologies is developed using fiber-optic cables that form the veins of global communication. Various types of optical fibers have been fabricated using dielectric materials and are being used for static and dynamic physical fields monitoring [1–5], ultrasensitive and selective chemical sensing [6, 7], ultratrace gas detection [8], and biological samples' recognition [9, 10]. With the advancement in nanofabrication techniques, material development, broadband optical components, and signal processing methods, optical sensing has become more viable and has proliferated into various sectors. Such waveguided optical detection systems are compact, immune to electromagnetic interference, reconfigurable, integrable, and multiplexable. The waveguided optical systems have met with the demand of technical transfer from lab to industry at multiple levels. Based on features of the waveguide acting as sensor probe, several sensitive optical interrogation principles or techniques have been devised for monitoring external parameters. These principles predominantly involve modulation of transmitted wave properties, like intensity, wavelength, and phase, in the presence of external measureand field that can be efficiently observed by suitable means.

Owing to the versatile nature of optical sensors, wearable optical sensing systems are being developed for comprehensive health assessment. This has happened in view of increase in demand for point-of-care (POC) health diagnosis and real-time monitoring of different human activities. In fact, the market share of wearable

Optical and Electronic Fibers: Emerging Applications and Technological Innovations, First Edition.
Edited by Lei Wei.

sensing technology is forecasted to reach a value of US$ 5.5 billion by 2025 [11]. Wearable optical interrogation systems, based on waveguide or fiber configuration sensing mechanism, are being developed for physiological parameters monitoring, which facilitate detection of magnitude and nature of vital physiological parameters like joint movements, facial motions, heart and respiration rates, blood oxygen levels, and body temperature with high durability. To detect such vital parameters, optical sensing systems have been realized by fabricating flexible optical processed waveguides or by developing wearable patches with sensitive structured fibers. These systems are mounted on subject's body to collect the wealth of physiological information. The response of such systems is further analyzed computationally for health assessment and for making diagnosis decisions. In recent years, such sensing systems have been integrated with cloud servers and virtual interfaces for smart healthcare. Moreover, such sensing systems can be used to track public health for efficient evaluation of health metrics, identification of chronic ailments, and check the progress of epidemics and pandemics. Comprehensive integrated optical health monitoring systems facilitate the improvement of medical services and enable personalized health monitoring. Their productive utilization would empower physically challenged individuals and old-aged people, and enable faster response in emergency cases.

Processed optical fiber-based wearable sensing systems have the advantages, like they are more sensitive, noninvasive, corrosion resistant, energy efficient, and nonradiative in nature, unlike electronic systems. By realizing smart photonic textiles, an individual's all the vital health parameters can be interrogated simultaneously. Besides, flexible optical motion detection systems will form an important class of components for developing soft robots, tangible artificial intelligence (AI), and versatile human–machine interface. Further, advances in flexible optical sensing systems have motivated fabrication of 2D flexible broadband light sources and detectors for versatile utility. Also, newer computational techniques based on multivariable algorithms are being determined for handling the information collected by such sensitive systems. Thus, optical wearable systems are a promising class of device that would significantly improve the quality of human life and the health of society across the globe.

In this chapter, a brief idea about the important performance parameters of processed fiber-based wearable optical sensing systems have been discussed. Following this discussion, a general idea about the emerging techniques in wearable optical technology has been provided with their respective functionalities. Subsequently, an overview of recently developed flexible optical components and computational algorithms for data analytics is covered.

3.2 Performance Features of Wearable Sensors

The efficiency and viability of developed wearable sensors for physiological health monitoring is assessed by their performance features. These features are evaluated from experimental characterization of the system that include sensitivity, linearity and range of operation, response time and dynamic durability, biocompatibility, and integrability. Research on development of versatile wearable sensors is aimed at optimizing such features to achieve significant efficiency in monitoring health parameters. The following section includes brief description about the mentioned features inherent to wearable sensors.

3.2.1 Sensitivity

Sensitivity of the monitoring system is defined as the change of the observed quantity per unit change in measurand or external field parameter. For such systems, the value of sensitivity is determined from experimental data of the developed sensor. As in, a graphical representation of response of flexible optical sensor, such as change in propagating light intensity, or wavelength shift (dependent variable), or polarization, toward external parameters, such as mechanical strain, bending, and other physiological parameters (independent variable), is called its characteristics plot. The distribution of the data points in the characteristics plot is analyzed using statistical models, and a functional relationship is established that best predicts the system's response to external parameter. This functional relationship is called the response curve. The coefficient of the independent variable in the characteristic response curve determines the sensitivity of the system. The higher the value of sensitivity, the more likely it is to respond toward minute changes in external field and has higher resolution. Its value depends on sensing elements, mode of operation, encapsulation mobility, temporal stability, and noise level.

In case of intensity interrogation-based systems (Figure 3.1a), the magnitude of change in transmitted light intensity (ΔI) per unit change in physical parameter (ΔX), given as $S = \frac{\Delta I}{\Delta X}$, determines the sensitivity of the system from the response curve. The change in output intensity of sensing fiber depends on the working mechanism. In some cases, sensitivity is represented in terms of change in attenuation of transmitted power (in decibel [dB]) due to external field, while, in others, it is directly accounted as change in photodetector signal in units of voltage. In case of hetero-core fiber probes [12] and copolymer fiber- or polymer optical fiber (POF)-based sensing systems [13], the intensity attenuation is evaluated with change in bending angle. The attenuation (A) is calculated by cut back method [14] in dB using the formula, $A = 10\log_{10}\left(\frac{I}{I_0}\right)$, where, I and I_0 are the input and the output light intensities. The change in attenuation with strain (in percentage) or bending magnitude (in angle or curvature units) represents the sensitivity of such sensors. In certain systems [15], the change in output intensity is detected and represented in terms of voltage units or relative change in voltage units. Here, the sensitivity is popularly termed as optical gauge factor (G) and is given as $= \frac{\Delta P/P_0}{\Delta \varepsilon}$, wherein $\Delta P/P_0$ represents the relative transmitted power with respect to initial output power in absence of disturbance (P_0), and $\Delta\varepsilon$ is the change of external field magnitude. GF of microfiber structures is quite high due to their dimensions that

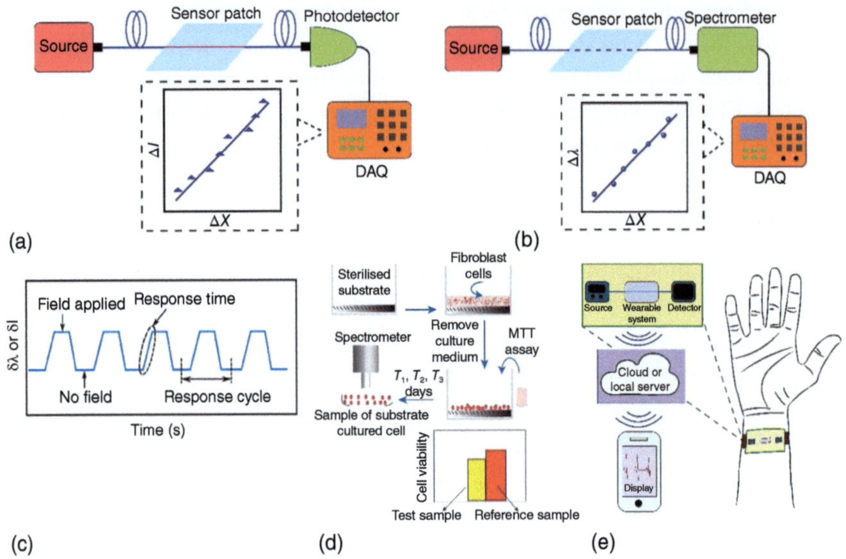

Figure 3.1 (*Source:* Rajan Jha): Performance features of optical wearable sensing system: (a, b) Schematic representation of basic setup for characterization of intensity-interrogated and wavelength-interrogated wearable sensing system, respectively. (insets) Graphical representation of response curve of each sensing system. (c) A graphical representation of real-time response of optical wearable sensor depicting the response time and cycle. (d) A stepwise representation of cytotoxicity test of flexible substrate by MTT assay protocol. (e) An illustration of an integrated optical wearable sensor for personalized healthcare. ΔI or $\Delta\lambda$: Change in intensity or wavelength; ΔX: Change in external field magnitude; $\Delta\lambda$: Change in Wavelength; ΔX: Change in external field magnitude; Field Applied: Response corresponding to applied external field; DAQ: Data Acquisition System; T1, T2, T3 represent time of observation in number of days like day 1, day 2, day 3; Test Sample: Cells cultured on bio-compatible sensor embedding substrate; Reference sample: Cells cultured on nutritious cell substrate.

lead to excitation of high evanescent field about the sensing region. The intensity of evanescent field is significantly susceptible to external disturbances and enhances light–matter interaction.

In case of wavelength interrogation-based systems (Figure 3.1b), the magnitude of shift in resonance wavelength with variation in external field magnitude determines the sensitivity and can be given as $S = \dfrac{\Delta\lambda}{\Delta X}$ from the response curve. Here, an optical spectrum analyzer (OSA) or spectrometer is used as detection unit that enables tracking of spectral distribution of transmitted light. Such systems capitalize on propagation resonance conditions and waveguide interferometric principles for detection. For fiber Bragg grating (FBG)-based sensing systems, the external parameter causes change in resonance wavelength (λ) condition, given as, $\lambda = 2n_{\text{eff}}\Lambda$, where neff is the effective refractive index, and Λ is the grating period. In this case, the relative change in resonance wavelength with applied field depends on the characteristics features of the fiber like, photoelastic coefficient (p_e), thermo-optic coefficient (a_m), and thermal expansion coefficient (b_n) of the fiber structure, given as [16] $\frac{\Delta\lambda}{\lambda} = (1 - p_e)\Delta\varepsilon + (a_m + b_n)\Delta T$, and p_e depends on effective refractive index of the grating structure. Thus, such sensors are sensitive to strain field and temperature variations with corresponding sensitivities that depend on fiber configurations. For interferometry-based sensing elements, like Mach–Zehnder- or Fabry–Pérot-type interferometers, the change in optical path length ($\Delta(nL)$) due to external strain/temperature variation causes shift in interference peak wavelength ($\Delta\lambda$), given as [17] $\Delta\lambda = \frac{\lambda}{nL}(\Delta nL + n\Delta L)$. The sensitivity is then determined from the slope of the plot between wavelength shift ($\Delta\lambda$) and external parameter magnitude. In interferometry systems, the sensitivity depends on the free spectral range (FSR) of pattern given as $\text{FSR} = \frac{\lambda^2}{nL}$, which can be modified by changing interferometer dimensions (L) and medium (n).

Besides, being dependent on the sensing mechanism, sensitivity of optical wearable sensors depends significantly on flexible substrates used for embedding the sensor probe. Woolen fabrics used as weaving material for optical fiber provide good stability and facilitates detection of minute motions, such as respiration, and can be sewn with clothes [18]. POFs sandwiched between plastic fabrics and bed linens show unaltered responsivity, as POFs already have configurable elastic modulus and refractive index for sensitive strain sensing and light confinement [19]. POFs have also been embroidered with moisture-wicking fabrics and are being used for monitoring health parameters [20]. Such fabrics provide strength to POFs and allow for interaction of light leaked from POFs to interact with tissues. For reduced graphene oxide-coated (RGO) polyester fibers that function by change in resistance of graphene coat under strain field, silicone gel substrates are used as base materials to mount the sensor probe on human body to realize wearability [21]. Among the elastomeric substrates, polydimethysiloxane (PDMS) has gained importance as one of the popular embedding materials. PDMS has high elastic modulus that can be altered by varying the cross-linker to elastomer ratio [22] and has considerable negative thermo-optic coefficient for detecting strain and temperature variations. Besides, it is transparent in optical region and has lower refractive index (~1.40 for elastomer: cross-linker ratio of 10 : 1) than silica that makes it suitable for embedding optical microfibers and other fiber configurations. PDMS films of low thickness (~100 μm) offer high flexibility to sense minute pressures and negate crosstalks between closely embedded similar sensors at optical frequencies [23]. It ensures stable performance and wearability of optical sensors, and affect the impact of external fields on embedded probes. Depending on orientation of optical probes within the PDMS membrane, the sensitivity tends to be different toward bending modes that aid determination of magnitude and direction of bending, and act as vector bending sensor [16]. The substrates used for embedding the optical sensing probe should have considerable elastic modulus to facilitate flexibility and be lightweight. Moreover, it should not interfere with light propagation along the waveguide but be able to transfer mechanical load to the waveguide.

For real-time monitoring of physiological parameters, the temporal stability and noise level of the sensing system under no-operation condition are vital features. The fluctuations in observed quantity like transmitted intensity or wavelength shift over time account for the noise level and can be quantified by $\frac{\Delta I}{\Delta t}$ or $\frac{\Delta \lambda}{\Delta t}$. This affects the signal-to-noise ratio (SNR in dB) of sensing system. For dynamic field monitoring like heart rate or respiratory rate, the SNR can be quantified as [24] $\text{SNR}(\text{dB}) = 10\log_{10}\left(\frac{E_s}{N_o}\right)$, where E_s is the energy of the signal, and N_o is the noise spectral density. The resolution of the sensing system is limited by such noise levels and need to be minimized.

For multicore fiber structures as sensing elements, the change in intensity due to bending of input fiber depends on the numerical aperture and orientation of cores that affect the transmitted intensity along each core [25].

3.2.2 Linearity and Range of Operation

As discussed in Section 3.2.1, the response curve predicted from statistical analysis represents the system's performance. The nature of the curve depends on sensor's characteristic response and can follow different relations. However, this nature should follow linear relationship to facilitate simple calibration and integrability with data processing units for an efficiently wearable optical sensor. The straight line equation is most prominently predicted by linear regression analysis of observed data, as only one external parameter is considered as the independent variable (Figure 3.1a,b (insets)). The fitted line, with corresponding coefficients and constants, is further analyzed statistically to test its predictive efficiency. In this regard, a quantity called coefficient of determination (R^2 value) is statistically evaluated that signifies the predictive efficiency of the obtained response curve with respect to the observed data. The value of R^2 varies between 0 and 1. Closer the value of R^2 to 1, better is the linear model. Thus, the linearity of a system's response signifies the degree of linear dependence of the change in observed quantity on variation in external parameter. Consequently, the range of external field variation for which linear response of the system is maintained is called the range of operation. It signifies the maximum value of external parameter, such as temperature, pressure, curvature, and strain, which can be detected without affecting the integrity and response stability of the system. When such a range is defined only for strain field, it is often referred to as stretchability of the system, as it indicates the maximum change in relative length of the embedded system. As most physiological movements are associated with strain field, stretchability of sensing system becomes an important feature to assess its performance.

Unlike electronic wearable systems, optical wearable systems do not involve intricate patterning and are free from fabrication defects. This reduces the chances of nonlinear responsivity and enable linear operation over specified range. For silica fiber-based sensing probes, the linear sensitivity and operational range is influenced by the embedding material and orientation of sensing probe in it. Fiber-optic probes embedded in PDMS elastomers have shown to have significant linear range toward curvature sensing (up to $40\,m^{-1}$) [26]. Microfiber-ring resonator embedded in PDMS-gold layers show high linear sensitivity toward pressure application (up to 600 Pa) and strain field responsivity up to 80% along longitudinal direction and up to 60% along lateral direction [27]. FBG-type sensors embedded in silicone gel have shown linearity response that depend on size [28] and orientation of the probe in the substrate [16] over a range that encompasses many body motions. However, as such systems have high sensitivity, the inherent trade-off between "broad linear range of operation" and "high sensitivity" limits the range of operation of system. On the other hand, POFs developed using PDMS have shown extensive linear response over broad range of strain field and have high stretchability. Dye-doped PDMS fiber demonstrated with high linear sensitivity over range of 0–100% strain field and graphene-mixed PDMS nanocomposite fiber showed linear strain sensitivity up to 150% strain. Additionally, thermoplastic elastomers processed into optical waveguides have been used to detect axial strains of up to 300% with significant reversibility [29]. Guo et al. have reported alginate–polyacrylamide hydrogel-flexible waveguides that have almost linear response for up to 700% applied axial strain [30]. However, such POFs function by intensity interrogation method and have lowered sensitivity. Thus, development of sensitive optical wearable systems that are operable over broad range is still a challenge.

3.2.3 Response Time and Dynamic Durability

Response time of an optical wearable system is the time duration taken for its response to reach a constant level under application of external field, as depicted in Figure 3.1c. For real-time monitoring of health parameters like breathing and temperature, low response time is a definite requirement. High sensitivity of optical systems has lowered the response time than electronic systems. In this regard, the hysteresis of optical flexible sensors is quite negligible as compared to electronic-based systems. Further, the response time depends on the detected parameter variation and nature of the substrate used for probe encapsulation. Parameters like temperature or humidity variation take substantial time to reach equilibrium, while pressure changes are almost instantaneous. Besides, substrate features like definite thermo-optic coefficient and viscoelastic properties of elastomers influence the response time. Among the reported systems, microfiber-based sensors embedded in PDMS or hybrid PDMS membrane have response time of order of microseconds (~10 μm) toward pressure sensing [15, 23]. Similarly, FBG-based systems have shown to have negligible response time toward pressure fields [31]. Wearable optical systems used for monitoring temperature variations have response time of few seconds (<100 seconds) due to definite thermo-optic coefficient of flexible substrate. Also, the response time of eccentric FBG system [32] and microknot resonator embedded in gelatin [33] for detecting relative humidity variation due to breathing are determined to be 92 and 84 ms, respectively. Furthermore, another feature of the wearable sensors that has been investigated is called the recovery time. Recovery time of the sensor is the time taken by the sensing system to return to its initial state after releasing the strain force. This recovery time for most sensors is either equal or a little greater than the response time. In case where the recovery time is higher than response time, the sensing system incurs certain degree of hysteresis that is attributed to certain elastomer-based substrates.

Dynamic durability is another crucial feature of a flexible sensor that establishes the repeatable response of system toward multiple loading–releasing cycles of external field (Figure 3.1c). This feature is essential while monitoring bending, strain field, and pressure fields in real time that indicate vital essential health parameters. It also signifies the degree of endurance of the skin mountable sensing patch toward regular stretching and relaxing motion for utilization in athlete health monitoring. While monitoring heart rate and respiratory rate in real time, each response cycle of the sensing system needs to be identical to facilitate proper diagnostics. Several optical sensors have been tested for multiple cycles of a particular magnitude. Microfiber-based sensing platforms have been reported to have high stability over 100 cycles of strain field [15], while FBG-based systems have been tested for 1000 cycles of stretching and releasing [34]. Hetero-core fiber-based sensor probes have been tested at different breathing frequencies for considerable time duration and show repeatable response [18, 35]. POFs have shown quite repeatable response over certain region of strain field [36]. In another work, macrobending-based sensing system has been tested for over 172 800 cycles for over 192 hours that showed almost stable response over the duration [37]. A durable optical sensor should possess high stability over long-term cyclic loadings for efficient utility. In fact, such repeatable response is expected toward each level of strain field over broad range.

3.2.4 Biocompatibility

Wearable optical sensors should be skin mountable, light weight, and compact to be placed on body or sewn in clothes. Skin-mountable optical sensors that do not affect the skin under continual usage are biocompatible in nature. The biocompatibility feature is assured by the presence of nonradiative components in the device, low heat generation, breathable, water resistant, and noncytotoxic to the skin. Unlike electronic wearable sensors, wherein microcontrollers and amplifiers are required, optical sensors use lesser components, as it requires a light source and photodetector/interrogator at its core that are powered by batteries. Processed optical fiber-based sensing probes are nonradiative in nature and do not generate any stray electromagnetic fields. Also, light propagation in any waveguide hardly releases any heat to the surroundings. Besides the optical sensing probe, the substrate used for embedding the probe needs to be compatible to the skin layers. Substrates that are flexible and hydrophobic in nature are preferred for sensors to be mounted on skin that allows for release of perspiration, corrosion resistance, and inhibit microbe growth. Additionally, these flexible substrates are required to be noncytotoxic so that they do not affect the epidermal layer of skin. The cytotoxicity nature is tested by following well-established protocol [38]. A schematic representation of cytotoxicity test using 3-[4,[5-dimethylthiazol-2-yl]-2,5-diphenyl tetrazolium bromide (MTT) assay is shown in Figure 3.1d. Here, the substrate acts as a biological medium or base for culturing fibroblasts that are important cells of the top skin layer. The cultured samples are analyzed by spectrometer to determine the cell viability and proliferation in comparison with controlled cultured samples. Further, to test the cell morphology, the cultured cells are immunostained and subsequently subjected to fluorescence examination. If the viability and optical density of samples are comparable, the flexible substrate is noncytotoxic in nature.

Woolen, cotton, and elastic-type fabrics used for encapsulating or holding optical sensor probes are usually sewn on clothes or cinched on body parts, and hardly affect the skin [18]. Fiber optic sensors embedded in polymer substrates have been referred to as smart textiles for health parameter monitoring. Primarily used flexible polymer substrates include silicone elastomers such as PDMS and Ecoflex [39] that have significant biocompatibility and hydrophobicity that make them corrosion resistant. Also, POFs have been developed with biocompatible polymers and elastomers to be used as flexible optical sensors. POFs made of polymeric materials like polymethyl methacrylate (PMMA), polystyrole, polycarbonate, and silicones are suitable to skin. PDMS-based POFs have shown to be highly biocompatible and lightweight to be used as a popular substrate for optical sensor probes. Besides, hydrogel waveguides have been used for extensive strain sensing for on-skin applications [30]. Recently, 2D materials have been developed that have high biocompatibility and sensing functionality called MXenes. These materials have been functionalized with polydopamine (PDA) for enhancing material interaction with elastic core fiber and are covered with corrosion-resistant PDMS layers to act as smart textiles for flexible sensors based on resistance variation [40]. However, their utility with optical sensing probes has remained elusive.

3.2.5 Integrability

Integrability of optical wearable sensors refer to its ability of being combined with miniaturized devices that can be linked with mobile data processing units for personalized noninvasive, continuous real-time health parameter tracking. A schematic representation of an integrated monitoring system can be visualized, as shown in Figure 3.1e. The most popular optical technique used as a wearable device is based on acquiring photoplethysmogram (PPG) signals from blood vessels in subcutaneous tissue. By observing the waveform, frequency, and amplitude of absorbed light signal, the heart rate and blood oxygenation level is determined. The accuracy of such device is a major challenge and lays the platform for competition among manufacturers. Popular tech companies like Microsoft, FitBit, Apple, Google, Sony, Motorola, and pharmaceutical companies have excelled in marketing such devices for public and personalized healthcare. To overcome the issues related with conventional PPG-based devices that utilize single point source and detector, organic light-emitting diode (OLED)-based devices adjoined with photodiodes have been reported to record PPG signals from blood vessels [41]. However, their widescale producibility is yet to be explored. Production of waveguide-based wearable optical devices for personalized use has remained an extensive challenge. In this regard, some prototypes based on intensity interrogation have been reported at laboratory level.

One such prototype of optical goniometer was reported that monitors bending activities over broad range and can be integrated with interfaces of mobile devices [42]. Most often, POFs-based sensing systems have been devised and mounted on gloves or clothes for monitoring [20, 30, 43, 44]. In one of the works, POFs-based photonic textiles were used for determining the blood oxygen level by observing the transmission of optical light intensity through the skin [45]. Among wavelength interrogation-based systems, FBG and microfibers embedded in polymeric substrates have been used for motion sensing and heart rate monitoring by mounting on skin and gloves [15, 34]. Although such sensor probes or patches are compact, lightweight, and biocompatible, yet their portability is limited due to non-availability of miniaturized high-resolution optical spectrometers. Besides fabricating skin-mountable devices, the signals from such sensing systems need to be compared with calibrated responses for real-time utility. Hence, the recorded response needs to be linear over the operational range and can then be analyzed by signal processing techniques to produce perceivable results. Moreover, the data can be stored in cloud servers and retrieved or shared for temporal health diagnostics. For processed optical fiber-based systems, the observed quantities are simple modulation of light signals with high SNR values that can be correlated with reference signal and relayed to local or cloud servers for smart analysis.

Another aspect of the integrable feature is to create a nexus of similar sensing devices for multipoint motion monitoring. This is technically referred to as the multiplexable feature of the sensors. Waveguide-based sensors offer the advantage of being cojoined along single waveguide channel that can illuminate the sensors along its length and collect their responses simultaneously. Multiplexed optical flexible sensors embedded in fabric have been reported for gait monitoring, wherein flexible LEDs are attached at certain separations along length of a single POF that transfers the response of each LED to a single detector or double detectors at both the ends [46–48]. As the sensor points are operated sequentially in time using microcontrollers, such systems are referred to as quasi-distributed sensing system. Hence, fiber-based flexible optical sensors' system offer the possibility of detecting motion at different points that can be used in medical, training, and corporate facilities to monitor several individuals in real time.

3.3 Processed Fiber-based Wearable Optical Sensors

Waveguide-based sensing systems employ processed waveguides as media for modulation of light due to physical, chemical, or biological disturbance and produce a detectable signal proportional to the magnitude and nature of the disturbance. When such sensor probes possess flexibility and skin-mountable property, or get embedded in flexible substrates, these are used as wearable sensors for physiological monitoring. The performance of such system depends on the configuration of the waveguide that control the modulation of light and offer wide scope to modify the configurations and develop extensive class of sensors. Depending on the light propagation mechanism within the waveguide, the intensity, wavelength or frequency, and phase of the propagating light gets modulated due to external perturbing field. This modulation is detected by a suitable detector for further analysis.

Here, we provide an overview about the two classes of wearable optical systems that have gained importance in recent years as an alternative to electronic wearable sensors. Each section is being categorized depending on the type of waveguide used as sensor probe.

3.3.1 Intensity Interrogation-based Sensing Mechanism

Processed optical fiber sensing systems that interact with external perturbing fields and produce a detectable change in the intensity of the propagating light are said to operate by intensity interrogation technique. The transmission losses and waveform of output signal, recorded mostly using photodetectors, provide a signature about the external field properties. Such systems are easy to integrate in flexible materials and link to analytic processors via wireless connectivity. Different types of processed waveguides act as sensor probe to develop intensity interrogation-based wearable optical sensing systems. Some of the popular techniques and their functionalities have been discussed below.

3.3.1.1 Micro-/Macro-bend Fiber-based Sensor Probe

This classical technique involves geometrical orientation or bending of optical fibers in various shapes and attaches it to wearable substrate for motion sensing. Application of strain or pressure on bent fibers cause loss in light intensity due to enhancement of evanescent field detected by a photodiode. Multiple turns of fiber are arranged on the flexible substrate to act as the sensing system. Using this technique, wearable sensors have been reported for lumbar spine posture monitoring by attaching pairs of fiber optic sensor along the spine for monitoring each region [49–51]. Similarly, a polychloreprene suit with attached multimode fiber (diameter = 1000 μm) sensors was used for 3D continuous monitoring of entire upper body that recorded the attenuation of propagating signal due to motion [52]. Macro-bending loss technique has been used for devising real-time respiration rate and heart rate monitoring wearable sensing system. A fiber-optic loop (Figure 3.2a(i)) attached to an elastic fabric is mounted on chest wall for respiration monitoring that records the light attenuation due to change in loop curvature under breathing motion [58]. In the similar way, a half-looped structured or U-shape bent fiber (Figure 3.2a(ii)) was used as sensor probe [54], while the authors used (Figure 3.2a(iii)) sinusoidal or wavy orientation of fiber [55] on flexible substrate to be mounted on abdominal region for respiration rate monitoring in another report. Yet, in another approach, the authors position the fiber in the form of figure of eight (Figure 3.2a(iv)) that showed enhanced sensitivity with lower noise level at limited region [49].

Instead of bending the fiber in different shapes, another approach was used in which the fiber is placed between corrugated surfaces that cause microbending of fiber [59]. This system is mounted on the body for respiration monitoring and respiratory-gated imaging of patient undergoing MRI scan. Such types of sensor probes need to have longer lengths to improve the sensitivity of the system, as the

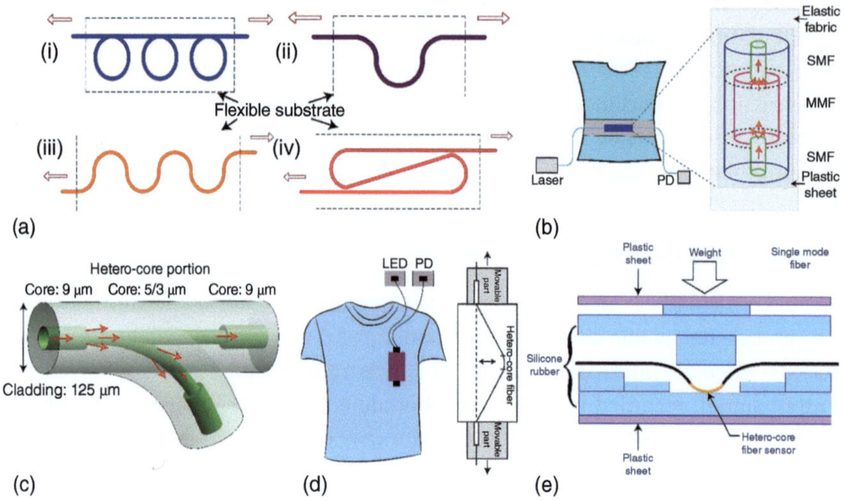

Figure 3.2 Macro-bend fiber and Hetero-core fiber-based sensing systems: (a) (i–iv) Configuration of fiber macro-bend structures for wearable optical sensing in shape of (i) loop bending (*Source:* Modified from [53]), (ii) half-loop bend (*Source:* Modified from [54]), (iii) wavy orientation (*Source:* Modified from [55]), and (iv) figure of eight (*Source:* Modified from [49]). (b) Single-mode fiber–multimode fiber–single-mode fiber configuration-based sensing probe for respiration and motion monitoring (*Source:* Modified from [56]). (c) Representation of hetero-core fiber structure used as sensor probe for intensity-interrogated optical sensing (*Source:* Rajan Jha). (d) Athlete's suit with affixed hetero-core fiber-based flexible sensor for body motion monitoring (*Source:* Modified from [57]). (e) Three-pin silicone-embedded hetero-core fiber configuration sensing system for respiration monitoring (*Source:* Modified from [12]).

bending loss along the length of the fiber would then be high. This makes the system bulky. In order to achieve higher sensitivity over broad range for motion and respiration monitoring by using a compact sensor working on microbending loss mechanism, Pang et al. used a combination of single mode–multimode–single mode (SMS) fibers on plastic substrate mounted on abdomen [56] (Figure 3.2b). Here, the curvature due to bending caused variation in effective refractive index exciting higher order MMF modes leading to loss in propagating light intensity. However, the sensitivity of such techniques is still low for sort fiber lengths and is affected by motion-induced noise. This raises issues against compactness and continuous wearability of the sensing system.

3.3.1.2 Hetero-core Fiber-based Sensor Probe

Hetero-core fiber-based structures are formed by splicing a section of smaller core diameter (~5/3 μm) fiber along the length of a larger core diameter (~9 μm) fiber (Figure 3.2c). Here, the total diameters of the fibers are similar to facilitate splicing of two fibers. As the light propagates from the larger core diameter to smaller core region or the hetero-core section due to mode mismatch, a significant power of light gets coupled to the higher order cladding mode as evanescent field about the smaller core section. This coupling of light to cladding region or the intensity of the evanescent field increases when the fiber bends about the hetero-core section. This causes decrease in output light intensity that is recorded in terms of optical loss or attenuation signal (dB), and provides a signature of the perturbing field. Such hetero-core structures are compact and have stable sensitivity over multiple cycles. It is observed that smaller the core size of hetero-core section, greater is the sensitivity and greater is the insertion loss [12]. Hence, by altering the core diameter and length of the hetero-core region, the system's sensitivity can be controlled. Multiple similar structures can be integrated along single fiber channel for multipoint monitoring.

The utility of hetero-core structures for developing wearable sensors has been pioneered by Nishiyama Michiko et al. Hetero-core structures embedded in smart clothing with virtual feedback interface has been demonstrated for full athlete's motion monitoring and guiding [57, 60, 61], as represented in Figure 3.2d. Also, such sensor probes have been installed in treadmill bases for gait monitoring [62] and smart mats to distinguish between walking and falling activities for washroom utility by old-aged groups [53]. For subtle pressure detection that enables respiration and heart rate monitoring, the hetero-core sensor probe is kept under tension by embedding it in a three-pin silicone framework [12] (Figure 3.2e). Small pressure fields due to arterial blood flow causes significant change in output light intensity. Similarly bent hetero-core structure has been used as human-like tactile sensor for object recognition [63]. By weaving a longer sensor probe along length of yarn piece and sewing it on clothes, heart, and respiratory rate has been monitored [18]. In the similar fashion, arm motion recognition is achieved by sewing such sensor probe on hand coverings [60]. Furthermore, Michiko et al. have integrated multiple similar sensing systems in tandem and distinguished each probe signal by time-differentiation of optical loss [64]. Additionally, the group has proposed that a Simple Management Network Protocol for remote management of such sensors' network is at par with standards and protocol of internet [65]. Although this class of sensing system shows promising results toward physiological parameters monitoring, yet its functionality is restricted to bending motion detection. Strain field detection that facilitates recognition of facial motion and voice recognition and temperature determination has not been achieved with such sensing systems.

3.3.1.3 Plastic Optical Fiber (POF)-based Sensor Probe

Although silica optical fiber-based sensor probes have high sensing capabilities, yet these have low flexibility and incur shattering issues creating a chance of skin injuries under breakage. In this regard, POFs have emerged as a promising substitute to develop highly sensitive, flexible, bio-compatible, and durable wearable sensors that operate over broad range of the measurand field. These can be reconfigured into various shapes, sizes, and be integrated and multiplexed to form a network of efficient sensors. These can be operated by intensity as well as wavelength interrogation technique for physiological monitoring. POFs are waveguides made of polymers or dielectric materials that have a higher indexed core/microstructured cladding and facilitate light propagation by the phenomenon of total internal reflection. Figure 3.3a depicts the basic structure of POF and its types. These are transparent to

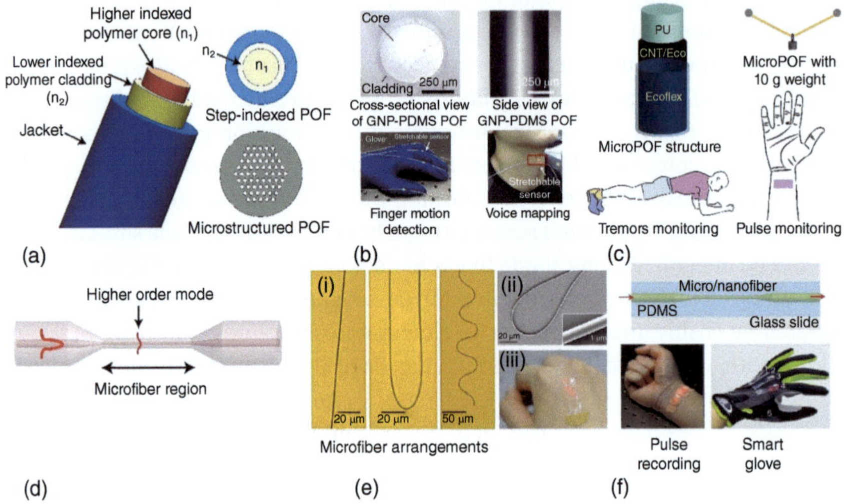

Figure 3.3 Polymer optical fiber and micro-/nanofiber-based wearable sensors: (a) Polymer optical fiber (POF) structure. (insets) Cross-sectional view of step-indexed and microstructured POF. (*Source:* Rajan Jha) (b) Fabricated PDMS POF for wide range strain sensing cross-sectional view and side view provided side by side. Images for tactile motion detection and voice recognition are included to depict its utility. (*Source:* Guo et al. [66] / Reproduced with permission of American Chemical Society). (c) Schematic representation of polyurethane (PU) microPOF for extensive strain sensing and high sensitivity. (inset) A 10 g weight being hanged with the fiber shows its tensile strength. It has been demonstrated for monitoring pressure associated with tremors during exercise and blood flow (*Source:* Modified from [67]). (d) Schematic representation of structure of optical microfiber (*Source:* Rajan Jha). (e) Different arrangements of embedded microfiber for optical sensing. (inset) Microscope image of microfiber and embedded microfiber mounted on subject's hand (*Source:* Zhang et.al. [23] / Institute of Optics and Electronics, Chinese Academy of Sciences). (f) Schematic representation of microfiber wearable sensor for multiparameter monitoring (*Source:* Rajan Jha). (bottom images) Sensor mounted on artery for pulse monitoring and on smart gloves for tracking finger motions (*Source:* Zhang et.al. [23] / Institute of Optics and Electronics, Chinese Academy of Sciences).

broad spectral range and can be formed in different shapes. In comparison with silica fibers, POFs have low Young's modulus (~an order lesser than silica waveguides) with high flexibility, higher elastic limits, greater ductility, lighter weight, higher sensitivity toward temperature (with negative thermos-optic coefficient), and higher fidelity for choice of polymer material. POFs have been used for various purposes. Popularly used biocompatible polymers for POF fabrication and healthcare monitoring are PMMA, plastic polycarbonate (PC), cyclic olefin copolymers (COC), pristine and doped PDMS and biodegradable polymers like hydrogels, and polylactic acid (PLA) variants. Various techniques have been devised for manufacturing POFs-like conventional polymer drawing for large-scale manufacturing and injecting it in designed preforms followed by thermal curing at laboratory level. Lately, with evolution of 3D printing technology, fabrication of customized POFs has become more feasible using various methods [68]. The developed POFs have been woven with fabrics or placed in specific configuration to facilitate strain/pressure sensing by intensity modulation of transmitted light. Here, we discuss a few of recent works for the reader's information.

POFs made of synthetic biodegradable polymers like PLA and PLGA are used as medical implants for *in-vivo* monitoring of neural signals by fluorescence photometry. Thin strands of such POFs have been used for deep tissue irradiation and phototherapy by fabricating microlens array [69] and bioabsorbable waveguide structures that probe deep tissues [70]. Another type of POFs made of hydrogels, prominently polymers, and co-polymers of poly(ethylene glycol) (PEG) for real-time monitoring of blood glucose level and strain sensing. Technically, a hydrogel-based glucose-sensitive waveguide was fabricated that enabled blood glucose measurements [71]. Hydrogel waveguides fabricated using biodegradable materials have been reported for *in-vivo* optogenetic measurements to assist neuropsychiatric treatments [72]. In another interesting application, electro-active polyacrylamide (PAAM) hydrogel fiber was used to sense electric potential variations of 2.2 μV due to modulation in propagating light intensity that could be used for electroencephalogram (EEG) signals detection. Also, hydrogels for strain sensing for up to 700% have been reported for monitoring multiple physiological parameters [30]. However, hydrogel-based sensors degrade in dry environments that affect their performance to great extent. Alternatively, POFs made of elastomers like PDMS, which has excellent light-guiding properties have been used for strain monitoring under different environmental conditions. No-core PDMS POF has shown high strain sensitivity of 3.507/ε over a range of 0–40% strain [73], while step-index PDMS POF (Figure 3.3b) has shown greater strain monitoring sensitivity and range of up to 80% facilitating pulse detection and temperature sensing ability [66, 74]. Researchers have successfully designed multifunctional hybrid POFs to act as brain implantable device for optogenetic, electrophysiological, and behavioral studies of neural circuits. Here, micro-optical POF made of PC (core) and COC (cladding) has been used as the optical channel, tin microwires encapsulated in poly(etherimide) (PEI) cladding as electrical channel, and the entire system of electrodes has been embedded in hydrogel POF as biocompatible brain mimicking for sensing brain micro-motions [75].

Besides, using different materials, different geometrical orientations of POFs have been reported for monitoring breathing and heart rate. PMMA POF arranged in wavy manner has been used for monitoring respiratory rate over different frequencies [76], while notched-shaped structure of POF showed higher sensitivity of 3.77/ε for respiration assessment [77]. To reduce the size or length of the sensing region without compromising the sensing ability, D-shaped structured POF [19], and fiber-cut configuration has been used for vital physiological parameter detection [78]. Recently, microfibers with cores made of polyurethane (PU) and coated with conductive (CNT/Eco) and compatible layers (Ecoflex) (Figure 3.3c) have shown to have high stretchability up to 400% strain field and high durability for up to 10 000 cycles of applied strain with low degradation [67]. The conductive layer provides higher tensile strength, and they have also been demonstrated for detection of body pressure (like pulse waveform) and vibration (like tremors during exercise).

3.3.1.4 Silica Micro-/Nanofiber (MNF)-based Sensor Probe

Silica micro-/nanofiber (MNF) have shown promising results for monitoring chemical, biological, and physical parameters, including time varying fields for real-time applications. These are fabricated by heating (~ up to melting point) and pulling an optical fiber such that its diameter of the fiber decreases to certain value adiabatically from both the sides, as shown in Figure 3.3d. The strong evanescent field about the microfiber region aids light–matter interaction that modulates the intensity of propagating light beam and makes such structure highly sensitive. Quite recently, MNFs have been used to fabricate wearable sensors for multiparameter monitoring ranging from limbs motion to pulse rate detection. By suitably embedding them in elastomers, prominently PDMS substrates, their mechanical strength and stretchability is increased for mounting on skin and detecting physical strain. Some of suitable configurations of microfiber in flexible substrate is shown in Figure 3.3e(i) with MNF microscope image in Figure 3.3e(ii), and its utility for joint motion detection is shown in Figure 3.3e(iii). In first of its kind, a straight MNF embedded in thin substrate of PDMS was demonstrated for detection of pressure and identifying loading/releasing events for over 10,000 cycles due to varying output light intensity [23]. It was used for monitoring pulse rate at wrist artery and tactile motion by installing multiple MNFs on a smart glove (Figure 3.3f). In the similar manner, the researchers developed a multifunctional skin-like wearable optical sensor (SLWOS) by embedding a bent MNF in PDMS substrate that increased its sensitivity and used it for monitoring, bending motion, pulse waveform, and body temperature over a range of from −20 to 130 °C [79]. A wavy optical microfiber in PDMS membrane was reported for precision pulse waveform and blood pressure monitoring [15]. This structure provided scope for wide reconfigurability in terms of altering the period of wavy structure in the membrane. To further enhance the sensitivity and reduce the loss in light intensity at the MNF–PDMS interface, researchers added a gold film and embedded a microknot resonator configuration of MNF in PDMS for vectorial strain monitoring in axial and lateral direction [27]. Due to shift in resonant wavelength under applied pressure, the intensity of output light changed, and was used for pressure (blood pressure) and strain detection over broad range with significant linearity. An alignment-free wearable structure, including a bent MNF in flexible liquid, was reported for cardiovascular health assessment of population of different age groups [80]. The designated microfiber-based sensor chip due to its compact size and patch-type shape could detect vital physiological signatures for personalized healthcare. Researchers went a step forward and encapsulated multiple MNFs in a single, flexible substrate and developed an interactive human–machine interface with the help of computational tools [81]. The sensing structure was put on a fabric (optical smart textile [OST]) and installed on a robotic arm for characterization that showed high durability for more than 30,000 cycles. By utilizing the sensor response data, machine learning based on super vector machine (SVM) algorithm was performed for high accuracy data classification and enabling a virtual interactive toy for interactive functionality. Such utilities of MNFs show a promising future for development of MNF-based wearable optical sensors and smart textiles for virtual reality. However, their mechanical stability and durability has remained a challenge that is most likely to be addressed soon.

3.3.2 Wavelength Interrogation-based Sensing Mechanism

Intensity interrogation-based wearable optical sensors facilitate miniaturization, integrability, and portability much like the electronic counterparts. However, their sensitivity is influenced by source power fluctuations, especially when the magnitude of measurand field, like EEG signals, is low. Moreover, for such sensing systems, it is difficult to monitor multiple parameters along the single fiber channel simultaneously, as all perturbations provide similar output of modulating the output light intensity. On the other hand, wavelength-interrogated sensing systems have high sensitivity, can operate at low source power, and facilitate simultaneous monitoring of different parameters due to wavelength selectivity. In this type of sensing systems, a spectrometer or wavelength interrogator is used for recording the output signal, and the observed quantity is usually wavelength or frequency shift of the received spectrum. In this section, we discuss few of waveguide-based sensing systems that are used for physiological monitoring in wavelength interrogation technique.

3.3.2.1 Processed Fiber Interferometers

The interferometry-based sensing systems are the classical and the most sensitive systems for sensing different parameters. These sensors' responses have been attributed to change in path length or phase difference between two propagating light beams of similar spectral composition, due to external perturbations. The observed quantity is shift in output spectrum peak or dip wavelength value depending on the magnitude of the external field variation. The prominent interferometer systems used in wavelength interrogation technique include Mach–Zehnder interferometer (MZI), Fabry–Pérot (FP), and Michelson types. For wearable optical sensing, these interferometers are developed within optical fibers as in-line modal interferometers and embedded in flexible substrates for physiological monitoring. These in-line interferometers are simpler to integrate and offer multiple degrees of reconfigurability to vary the sensing performance.

An in-line MZI was fabricated in which single-mode fiber (SMF) dual up-taper junctions acted as beam splitters, and a seven-core fiber (SCF) between the tapered junctions acted as a medium for excitation and recombination of core and cladding modes [82]. It was embedded in silicone sheet and demonstrated for tracking multiple fingers bending with high sensitivity of 2.744 nm/m and high linearity of 0.9967 (Figure 3.4a). A two-core fiber (TCF) concatenated asymmetrically along SMF, that is with core offset on output side, and used as an in-line Michelson interferometer (Figure 3.4b) for breathing and heart rate monitoring [83]. Although the intensity modulation of light is reported for real-time sensing, but its utility in wavelength interrogation technique is established due to variation in FSR with TCF bending. Furthermore, fiber-based Fabry–Pérot interferometer systems for monitoring vital physiological parameters have been reported. A fiber-based Fabry–Pérot interferometer encapsulated in capillary tube by ethyl alpha cyanoacrylate (EtCNA) binder was used for heart rate monitoring [88]. In such system, the power fluctuations of the cavity resonant wavelength are recorded for real-time monitoring, and its sensitivity can be varied by changing the cavity length and the corresponding resonant

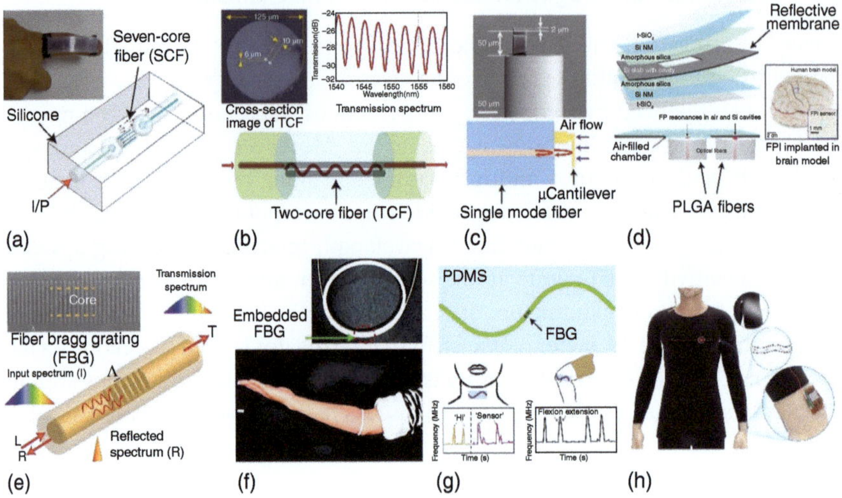

Figure 3.4 Wearable optical sensing systems based on fiber interferometers and FBG structures: (a) A seven-core fiber-based in-line Mach–Zehnder interferometer (MZI) for joint motion sensing (*Source:* [82]. Reproduced with permission of Elsevier). (b) A twin-core fiber-based fiber Michelson interferometer (MI) for respiration and heart rate monitoring (*Source:* Tan et al. [83] / Optical Society of America / CC BY 4.0). (c) A micro-cantilever-assisted fiber Fabry–Pérot interferometer (FPI) for respiration assessment (*Source:*Zhao et al. [84] / MDPI / CC BY 4.0). (d) A bioresobable fiber FPI for intracranial pressure monitoring with high sensitivity (*Source:* Shin et al. [85] / AAAS / CC BY 4.0). (e) Schematic representation of transmission and reflection spectra of FBG. (inset) Microscope image of FBG structure in fiber core (*Source:* Rajan Jha (Chapter Contributor)). (f) A wearable smart ring with embedded FBG for wide-angle-range joint monitoring (*Source:* Cheng et al. [86] / SAGE PUBLICATIONS, INC.) (g) S-shaped embedded FBG structure for multi-parameter monitoring ranging from voice recognition to joint movement (*Source:* Modified from [34]). (h) A fully photonic integrated FBG-based wearable optical system for vital parameters monitoring (*Source:* [87] Reproduced with permission of American Chemical Society).

wavelength. In similar interrogation approach, a microcantilever was attached to tip of an optical fiber (Figure 3.4c) for varied breathing rate monitoring and significant temperature tolerance for indoor conditions [84]. Further, researchers developed bioresorbable FPI sensors implanted in the brain for *in-vitro* sensing of intracranial pressure. This system interrogation relied on shift in resonant wavelength due to pressure-induced cavity length variation [85] (Figure 3.4d). Such promising systems open wide scope for development for extensive class of highly sensitive optical wearable sensors. However, most fiber interferometers still rely on intensity interrogation of spectral pattern that limit their sensitivity. This is due to its high sensitivity that entails difficulty in tracking wavelength shift in real time.

3.3.2.2 Fiber Bragg Grating (FBG) Structures

FBG structured sensing systems have become a viable option for realizing wavelength interrogation-based wearable optical sensors. As discussed earlier, the transmission and reflection spectra of in-line FBG structures are governed by the Bragg condition, $\lambda_r = 2n_{eff}\Lambda$, where n_{eff} is the effective refractive index, and Λ is the period of FBG. Figure 3.4e depicts a schematic representation of transmission and reflection spectra of a typical FBG. Change in n_{eff} value in presence of temperature field/chemical substances, or change in Λ, because of stretching or bending, leads to shifting of the resonant wavelength, λ_r. By recording the shift in resonant wavelength, the external perturbations such as temperature variations, chemical substances or strain fields are monitored. This functionality has been exploited to realize FBG-based wearable optical sensors for multitude of healthcare applications. Further, FBG-based sensing systems are simpler to integrate and can be monitored independently, as λ_r is specific to each FBG structure. Here, we discuss some of recent works related to it.

For joint monitoring, FBG-based wearable systems have shown promising results with high sensitivity and broad operational range. An FBG smart belt was reported with sensitivity of ~0.3 nm/degree of bending for knee joint monitoring [89], while a smart ring-type wearable system was used for elbow joint monitoring [86] (Figure 3.4f). A FBG sensor embedded in silicone substrate showed high linearity, and was used for identifying knee motions and carrying out gait analysis [28]. The utility of such systems for large angle bending was reported for joint monitoring that had detection range of 0°–200° with sensing resolution of 0.06° of bending [90]. Such sensors enable diagnosis of ailments like arthritis, osteoarthritis, and assessment of musculoskeletal disorders (MSD). Besides, highly accurate and linear responsive optical thermometer has been reported using FBG-based system for temperature detection over range of from −10 to 60 °C [91]. Recently, an implantable optical sensor was reported for *in-vivo* and *ex-vivo* cardiac monitoring with high efficiency [92]. Additionally, an array of FBG sensors along single fiber channel have been used for accurate cardiovascular diagnosis. By using an array of four FBGs, heart rate was monitored at different positions of a patient while sleeping to determine where the mean absolute error (MAE) is low that could be used for accurate monitoring of bed-ridden patients or during MRI scans [93]. In another report, the researchers positioned 12 FBGs at different locations on abdomen region to determine all breathing parameters, namely, respiratory rate, respiratory period, inspiratory, and expiratory period [94]. Likewise, by integrating multiple FBGs along a channel and attaching it to footwear soles, several motion monitoring systems have been reported [95, 96].

Further, by using a different interrogation technique for tracking frequency shift using dual-comb laser source, an FBG-based wearable sensor was reported for multiparameter monitoring like voice recognition and joint movement sensing [34] (Figure 3.4g). In fact, a completely integrated optical wearable sensor was reported for body temperature and heart rate monitoring [87] (Figure 3.4h). Such systems establish a definite scope for development of wearable optical sensor based on wavelength interrogation technique.

3.3.2.3 Polymer Optical Fiber Bragg Gratings (POFBGFs)

The Bragg grating technique of microinscription within fiber core has been extended to fabricate polymer optical fiber Bragg gratings (POFBGs) to realize wavelength interrogated flexible optical sensor probe for healthcare devising. POFBGs have higher sensitivity toward strain, curvature, and temperature over broader range as compared to equivalent silica fiber sensor probes due to their higher flexibility and elastic limits. Moreover, the extensive choice of polymers with significant biocompatibility makes them a better choice over silica FBGs for developing wearable optical sensors. A number of techniques has been reported for fabricating POFBGs of different forms, like long period gratings, tilted FBG, birefringent gratings, etc. [97].

In one of the early works, an eccentric core FBG inscribed in PMMA POF was fabricated, and its applications for strain, curvature, and temperature sensing were demonstrated [98]. The system facilitated vector bending monitoring due to different shift in wavelength in either direction, and showed high linearity toward strain and temperature sensing that could facilitate multiple physiological parameters monitoring. In another case, the researchers integrated multiple POFBGs in footwear sole (Figure 3.5a) for plantar pressure detection and gait analysis [99]. Its utility was extended for developing a force sensor platform with network of POFBGs and for monitoring body mass displacement depending on the subject's position on the sensing sole. Further, doped PMMA POFBGs have been fabricated using ultrafast inscription technique (in 7 ms) that is claimed to show 15 times more sensitivity than silica fiber FBGs [100]. The raw signal captured by a single POFBG, mounted on brachial artery of a person, was used to extract both cardiovascular and respiration signal with high sensitivity of 17 and 80 pm wavelength shifts, respectively (Figure 3.5b). In a similar work, POFBGs were used for detection of arterial pulse wave and blood pressure with enhanced SNR that is claimed to be about eight times

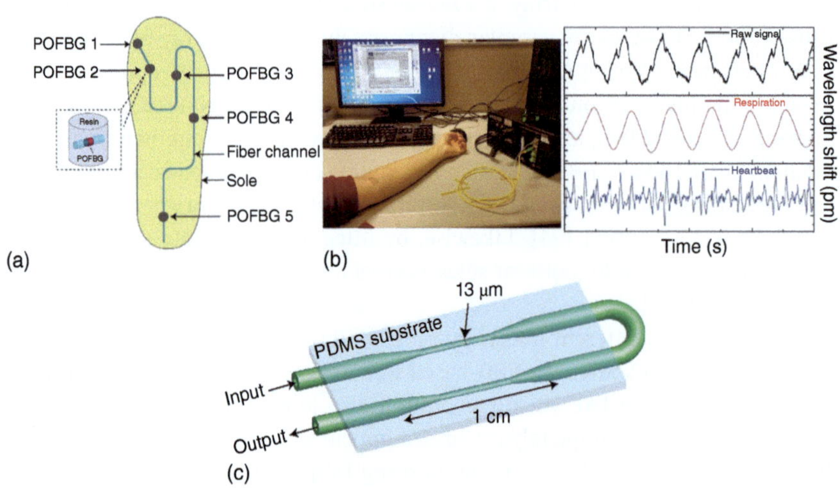

Figure 3.5 POFBG and Micro-/nanofiber-based wavelength-interrogated optical sensing systems: (a) Integrated POFBGs for plantar pressure monitoring and gait assessment (*Source:* Modified from [99]). (b) PMMA POFBG-based highly sensitive optical wearable sensor for monitoring respiration and heart rate simultaneously. The plot depicts the extraction of respiration and heart signal from single, raw data captured using POFBG (*Source:* Bonefacino et al. [100] / Springer Nature / CC BY 4.0). (c) U-shaped microfiber embedded in PDMS for multiparameter monitoring with high sensitivity (*Source:* Rajan Jha). $\Delta\lambda$: Wavelength shift; ΔT: Time axis.

higher than that for silica-based FBGs [101]. Such versatile utility of POFs has established their potential to be used for realizing smart wearable optical sensors for healthcare.

3.3.2.4 Micro/Nano Fiber Structures

Recently, microfiber-based sensing system operated by wavelength interrogation technique has gained importance, with few works reporting about their high sensitivity and multiparameter monitoring ability. These sensing systems have faster response and low hysteresis toward motion sensing. Bent microfiber embedded in PDMS was used to detect temperature variation by wavelength interrogation [79]. Recently, a U-shaped dual microfiber section was demonstrated for multiparameter monitoring with high sensitivity and fast response [102] (Figure 3.5c). The mechanical robustness of such sensing systems raises concerns for portability of the device. Besides, its high sensitivity is subject to motion-induced noises that lower the response SNR. However, such systems could find valuable application for monitoring serious patients in hospitals' special care units, where subtle motions depict patient's timely recovery.

3.4 Scope of Optical Wearable Sensors

The sensitivity and versatility of noninvasive optical wearable sensors guarantee their advanced functionality toward real-time all-round human health monitoring in the near future. As these are biocompatible, economical for mass production, energy efficient, customizable, scalable, and easy to fabricate that favor commercialization. However, the devisable aspect of such sensors is a challenge and needs to be addressed to counter the electronic counterparts in market. The prospects of optical wearable sensor to be marketized as tangible devices depend significantly on assembling the entire sensing system in a compact and skin-mountable housing for personalized utility. In this regard, besides developing a compact sensor probe, it is essential to have flexible and wearable compact light sources and detectors. In view of realizing miniaturized and wearable optical sensors, significant research has focused on developing flexible, broadband, integrable, compact, and energy-efficient light sources and photodetectors. Additionally, efficient computer algorithms and cloud-based servers need to be formulated for accurate health monitoring and diagnosis with the devices. Such developments facilitate realization of a comprehensive health monitoring system. Also, the utility of such optical sensors has been explored in detecting biochemicals, tracking applications in soft robotics and environmental surveillance that motivate further research in this area of optical sensing technology.

3.4.1 2D Materials for Miniaturized Wearable Sensors

With the advancement in thin film fabrication of semiconductor materials, flexible 2D broadband LEDs have been reported that emit in optical and near infrared (IR) regions. 2D excitonic LEDs fabricated using transition metal chalcogenides (TMDs) have been reported to exhibit significant efficiency at room temperatures. Among them are MoS_2-type LEDs emitting in optical to near IR wavelengths [103] and WeS_2-type junctions [104] emitting prominently in near IR range. Further, few layers of MoS_2 coupled with microdisk microsphere cavity act as 2D CW lasers in optical region at room temperatures [105]. Similarly, $MoTe_2$ layers on Si-photonic crystal nanocavity show efficient lasing action in the IR spectral region (~1132 nm) and have gained importance as on-chip nanolasers for Si-based nanophotonic devices [106].

On the similar grounds, compact semiconductors and complementary metal oxide (CMOS) photodetectors have been developed to be part of flexible devices that convert the radiation signal into photocurrent. Researchers have gone a step further to fabricate 2D flexible photodetectors for on-chip integrability. One such technique includes developing photoactive layers of nanomaterial or organic materials on flexible polymeric substrates to act as photosensitive component with high mechanical stretchability in optical and near IR regions. Crumpled structures of graphene–gold nanoparticles have been proposed with high optoelectronic performance and broad stretchability (~200%), while helical ZnO nanoparticles on flexible substrate have shown stretchability up to 600%. Paper substrates functionalized by ZnO and MoS_2 photoactive layers have also gained popularity as UV detectors due to their mechanical flexibility, low cost, biocompatibility, and biodegradability. The hygroscopicity and porous structure of paper enable efficient fabrication of photoactive layers on such substrates for flexible devices. Also, fiber-based substrates, like POFs, have been used to realize flexible photodetectors of varied shapes over broad spectral range. However, materializing a flexible broadband optical spectrometer for miniaturization of wavelength interrogation-based optical sensors has remained an unbeaten challenge.

3.4.2 Computational Modalities for Analytical Augmentation

With growing demand for advanced comprehensive healthcare facilities to combat pandemics and epidemics, regular and personalized monitoring of health parameters has become essential not only at individual level but public at large This requires regular data acquisition and their clinical analysis to diagnose health conditions that lays the foundation for extensive health informatics. Health informatics are strengthened by large sample-sized data pertaining to various health parameters like blood oxygenation levels, heart rate, joint movements that enable physicians to track chronic conditions, identify warnings for anomalies, and conduct accurate health metrics. Even at the individual level, devices that accurately track physical activities are in huge demand. Such data acquisition and their connectivity to user-friendly interfaces call for extensive computational modalities to satisfy the needs of an end-point user. Hence, robust computational algorithms have emerged as an integral part for data manipulation to extract vital information from complex and disturbance-infused data of optical wearable sensors. For instance, the discrete saturation transform (DST) signal extraction algorithm introduced by Masimo Corporation enabled more accurate determination of SpO_2 levels of blood under moving condition from pulse oximetry readings that revolutionized the oximeter industry [107].

Several electronic platforms have been launched for researchers to design customized hardware or signal processing systems (including CPU and GPU) that represent the detected signals into graphical forms for visualization of the user. Popular electronic prototyping platforms like Arduino have been used to design wearable sensor for bending monitoring [42]. These devices are supported with controllable graphical programming platforms, such as LabView, PyLabs, and other similar programming tools, for measurements, automation, and informative displays. In portable devices, the measured values are processed in hardware CPUs and GPUs to provide user-readable information. In the hardware units, the physiological information detected by the sensor is processed by algorithms for personalized diagnosis and health warnings. Computational processing like signal filtering and signal extraction enhances the recorded output signals, while analytical algorithms facilitate further diagnosis. In this regard, prominent computational algorithms based on machine learning techniques have opened new routes for realizing smart wearable devices for personalized diagnosis. For detected signals, like change in measured voltage or current due to light intensity variation, Artificial Neural Networks (ANN) algorithms have been used. Among the several ANNs, multilayered perceptron (MLP) neural network has been used for predictive analysis of intensity-interrogated fiber optic sensors output for force and absorptive sample detection [108–110]. In such methods, the MLP network is trained using different algorithms to adjust the weights of the neurons connectivity and produce a desired output. The choice of algorithm depends on the required accuracy and computational time for data analysis at each step. The output of the optimized network is tested for different input or measurand field values, and perform predictive analysis for smart diagnostics. Development of such software tools has enhanced the productibility and importance of wearable optical devices.

Besides personalized diagnosis, the programming platforms facilitate data linkage to local servers or cloud computers for further collective analysis. This process is often referred to as democratization of health information for extensive evaluation by stakeholders and public healthcare departments. Data democratization is followed by data mining, which has become an essential topic of research, especially in medical and other healthcare departments. Data mining is carried out on wearable sensors' data, such as echocardiograms (ECG), heart rate, respiration rate, blood pressure, blood glucose, and PPG signals. By proper handling of population health data, efficient healthcare aids can be realized for comprehensive healthcare of the community.

3.4.3 Additional Utilities of Processed Fiber Wearable Optical Sensors

Besides enabling personalized physiological monitoring utilities at domestic and public levels, flexible processed fiber optical sensors are being developed for biochemical sensing, medical treatments, interactive human–robot interfaces, environmental, and industrial utilities.

In view of biochemical sensing, an important parameter is pH of body fluids. Multiple silica FBGs have been demonstrated for pH detection [111], while doped

PMMA POFBG has also been used for highly sensitive pH detection with faster response time [112]. Further, POF-based glucose sensor has been reported with enhanced performance facilitated by parabolic concentrator tip for efficient light coupling [113]. Recent research activities have been focused on developing flexible optical biosensors for noninvasive probing of body fluids like sweat, saliva, and tears. For instance, nanoparticles-functionalized flexible substrates have been reported for realizing surface-enhanced Raman scattering (SERS) systems for tears and swab analysis [114]. Also, by growing metallic nanoparticles of polymeric substrates, flexible localized surface plasmon resonance system has been realized for higher refractive index sensitivity [115].

In the field of smart medical treatment, flexible optical sensors have been used at different fronts. For gait analysis and posture rehabilitation, advanced modular exoskeletons and orthotic devices have been reported that utilize POF-based sensing system [36]. These have been mounted on 3D supports for knee joint monitoring and hip joint analysis, while those encapsulated in shoe sole detected gait cycles. At another level, the performance of microstructured POFs for detection of ultrasounds in medical diagnosis has been investigated and was showed to be higher than silica-based optical sensors. Thus, POFs provide a safer and more biocompatible system for applications in ultrasonography, and can be extended for endoscopy [116]. Wearable optical sensors have the ability to be mounted on soft robots to monitor limb motions [34] and facilitate prosthetic interactions. Further, it would facilitate human–robot interaction via virtual interface [81] that would benefit physically challenged individuals. POFs arrangement operated in intensity interrogation has been demonstrated on active knee orthosis that facilitates human–robot interaction for motion assistance [117]. This would revolutionize the prosthetics industry with smart sensitive devices for interactive platforms.

Additionally, flexible optical sensors have been used to devise wearable environment monitoring system with commercial name as TZOA. The optical sensing system detects particulate matter in air, and assesses air pollution and UV exposure. The recorded data is linked to smartphone apps to display user readable figures. Similar devices can be developed for industrial parts surveillance in real time at different locations. Such flexible optical systems are mechanically robust and can be mounted on machinery parts.

3.5 Conclusions

This chapter provides a brief idea about the essential performance features of optical wearable sensing systems and their quantifications. Further, it includes a bird's eye view of the emerging optical wearable detection systems that capitalize on modulation of intensity and wavelength of light field for assessment of physiological parameters. Highly sensitive, durable, and broadly linear systems for monitoring body movements, voice, breathing and heart rate, temperature, and other parameters have been presented. Also, the chapter discusses about recent development of flexible 2D materials-based broadband light sources and photodetectors to realize all-optical unobtrusive personalized healthcare devices. Additionally, the emerging computational techniques that facilitate smart assessment for preventive care and handling of large data ensembles at the public level have been outlined. Their complimentary potential toward biosensing, and environmental and industrial monitoring have been stated. From such discussions, it is evident that integrated wearable optical systems will be pivotal in revolutionizing the healthcare industry in the near future. Such systems will become more ubiquitous in common man's lives much like the smartphones. Besides, the integrable feature would make them part of the Internet of Things (IoT) for efficient health metrics to track chronic diseases and pandemics.

Substantial research for development of wearable optical sensors is still underway to achieve higher fidelity in such systems. Challenges to develop durable skin-like ergonomic devices with efficient performance is yet to be beaten. Although POFs have facilitated realization of extensive class of flexible devices, their light-guiding properties need to be enhanced to tolerate stray mechanical fields and reduce error bars. In this regard, synthesizing more biocompatible and flexible doped or microstructured POFs with excellent waveguiding properties is still being explored. Such development would enable realization of multifunctional smart photonic textiles and will be overwhelmingly marketized. Feasibility of fast response and hysteresis-free sensing systems developed using highly sensitive sensor probes, like microfibers, has remained elusive. Moreover, more reliable data analytics for accurate data-driven diagnosis and personalized health assessment need to be developed and integrated with optical sensing systems.

Extensive research in the field of wearable optical sensing systems for healthcare has boosted the potential of optical sensors. Novel promising interrogation techniques have emerged in recent years to realize smart devices. It has opened the doors of more fundamental and technical research areas. Finally, their versatile potential lays the foundation for growth of new group of industries dedicated to commercializing public healthcare facilities.

References

1 Ma, G., Wang, Y., Qin, W. et al. (2021). *High Voltage* 6: 367–561.
2 Chatterjee, K., Pal, S.K., and Jha, R. (2021). *Adv. Opt. Mater.* 9: 2001574.
3 Dass, S., Chatterjee, K., Kachhap, S., and Jha, R. (2021). *J. Lightwave Technol.* 39: 3974.
4 Chatterjee, K., Arumuru, V., Patil, D., and Jha, R. (2022). *Sci. Rep.* 12: 3798.
5 Min, R., Liu, Z., Pereira, L. et al. (2021). *Opt. Laser Technol.* 140: 107082.
6 Gorai, P. and Jha, R. (2021). *Adv. Photonics Res.* 2: 2100044.
7 Leibl, N., Haupt, K., Gonzato, C., and Duma, L. (2021). *Chemosensors* 9: 123.
8 Hodgkinson, J. and Tatam, R.P. (2013). *Meas. Sci. Technol.* 24: 012004.
9 Butt, M.A., Khonina, S.N., and Kazanskiy, N.L. (2021). *Fiber Integr. Opt.* 40: 14.

10 Li, X., Chen, N., Zhou, X. et al. (2021). *J. Biophotonics* 14: e202100068.

11 MarketResearchReports.Biz (2015). Wearable sensors 2015–2025: the market for wearable sensors will reach $5.5bn by 2025: MarketResearchReports.Biz. https://www.globenewswire.com/news-release/2015/11/18/788287/10156627/en/Wearable-Sensors-2015-2025-The-Market-For-Wearable-Sensors-Will-Reach-5-5bn-by-2025-MarketResearchReports-Biz.html (accessed 31 July 2022).

12 Nishyama, M., Miyamoto, M., and Watanabe, K. (2011). *J. Biomed. Opt.* 16: 017002.

13 Krehel, M., Rossi, R.M., Bona, G.L., and Scherer, L.J. (2013). *Sensors* 13: 11956.

14 Ziemann, O., Krauser, J., Zamzow, P.E., and Daum, W. (2008). *POF Handbook: Optical Short Range Transmission Systems*. Berlin, Heidelberg: Springer.

15 Zhu, H.T., Zhan, L.W., Dai, Q. et al. (2021). *Adv. Opt. Mater.* 9: 2002206.

16 Ge, J., Xu, L., Fok, M.P. et al. (2018). *Opt. Lett.* 43 (11): 2503–2506.

17 Zhu, C., Zhuang, Y., Zhang, B. et al. (2019). *IEEE Photon. Technol. Lett.* 31: 35.

18 Koyama, Y., Nishiyama, M., and Watanabe, K. (2018). *IEEE Sens. J.* 18: 6175.

19 Wang, Y.L., Liu, B., Pang, Y.N. et al. (2021). *IEEE Trans. Instrum. Meas.* 70: 1–8.

20 Quandt, B.M., Braun, F., Ferrario, D. et al. (2017). *J. R. Soc. Interface* 14: 0060.

21 Zhang, J., Cao, Y., Qiao, M. et al. (2018). *Sens. Actuators, A* 274: 132.

22 Wang, Z., Volinsky, A.A., and Gallant, N.D. (2014). *J. Appl. Polym. Sci.* 131: 41050.

23 Zhang, L., Pan, J., Zhang, Z. et al. (2020). *Opto-Electron. Adv.* 3: 1.

24 Zaunseder, S., Vehkaoja, A., Fleischhauer, V., and Antink, C.H. (2022). *Biomed. Signal Process. Control* 74: 103538.

25 Zawawi, M.A., O'Keeffe, S., and Lewis, E. (2015). *J. Lightwave Technol.* 33: 2492.

26 Li, S., Zhao, H., and Shepherd, R.F. (2017). *MRS Bull.* 42: 138.

27 Li, J.-H., Chen, J.-H., and Xu, F. (2018). *Adv. Mater. Technol.* 3: 1800296.

28 Resta, P., Lo Presti, D., Schena, E. et al. (2020). A wearable system for knee flexion/extension monitoring: design and assessment. *2022 IEEE International Workshop on Metrology for Industry 4.0 & IoT (MetroInd4.0&IoT 2020)*, pp. 273–277 (7–9 June 2022), IEEE.

29 Leber, A., Cholst, B., Sandt, J. et al. (2019). A wearable system for knee flexion/extension monitoring: design and assessment. *Adv. Funct. Mater.* 29: 1970030.

30 Guo, J., Liu, X., Jiang, N. et al. (2016). *Adv. Mater.* 28: 10244.

31 Kanellos, G.T., Papaioannou, G., Tsiokos, D. et al. (2010). *Opt. Express* 18: 179.

32 Bao, W., Chen, F., Lai, H. et al. (2021). *Sens. Actuators, B* 349: 130794.

33 Yi, Y., Jiang, Y., Zhao, H. et al. (2020). *ACS Sens.* 5: 3404.

34 Guo, J., Zhao, K., Zhou, B. et al. (2019). *Adv. Opt. Mater.* 7: 1900086.

35 Nishiyama, M., Sonobe, M., and Watanabe, K. (2016). *Biomed. Opt. Express* 7: 3675.

36 Leal-Junior, A.G., Frizera, A., Vargas-Valencia, L. et al. (2018). *IEEE Sens. J.* 18: 7085.

37 Witt, J., Narbonneau, F., Schukar, M. et al. (2012). *IEEE Sens. J.* 12: 246.

38 Azadi, S., Peng, S., Moshizi, S.A. et al. (2020). *Adv. Mater. Technol.* 5: 2000426.

39 Massaroni, C., Saccomandi, P., and Schena, E. (2015). *J. Funct. Biomater.* 6: 204.

40 Luo, J., Gao, S., Luo, H. et al. (2021). *Chem. Eng. J.* 406: 126898.

41 Smith, J., Bawolek, E., Lee, Y.K. et al. (2015). *Electron. Lett.* 51: 1312.

42 Lim, C.K., Luo, Z., Chen, I.M., and Yeo, S.H. (2011). *Front. Mech. Eng.* 6: 13.

43 Yang, C., Guo, J., and Niu, M. (2017). *Optica* 4 (10): 1285–1288.

44 Wang, D., Sheng, B., Peng, L. et al. (2019). *Polymers* 11: 1433.

45 Rothmaier, M., Selm, B., Spichtig, S. et al. (2008). *Opt. Express* 16: 12973.

46 Avellar, L.M., Leal-Junior, A.G., Diaz, C.A.R. et al. (2019). *Sensors* 19: 3356.

47 Leal-Junior, A., Avellar, L., Frizera, A., and Marques, C. (2020). *Sci. Rep.* 10: 13867.

48 Leal-Junior, A.G., Díaz, C.R., Leitão, C.P. et al. (2019). Polymer optical fiber-based sensor for simultaneous measurement of breath and heart rate under dynamic movements. *Opt. Laser Technol.* 109: 429–436.

49 Augousti, A.T., Maletras, F.X., and Mason, J. (2005). *Opt. Fiber Technol.* 11: 346.

50 Dunne, L.E., Walsh, P., Smyth, B., and Caulfield, B. (2007). Design and evaluation of a wearable optical sensor for monitoring seated spinal posture. In: *2006 10th IEEE International Symposium on Wearable Computers, ISWC*, 65–68. IEEE.

51 Williams, J.M., Haq, I., and Lee, R.Y. (2010). *Med. Eng. Phys.* 32: 1043.

52 Munoz, R., Leija, L., Diaz, F., and Alvarez, J. (1995). 3D continuous monitoring system for localization of upper limb based on optical fiber. In: *Proceedings of 17th International Conference of the Engineering in Medicine and Biology Society*, vol. 17, 1595–1596. IEEE.

53 Nishiyama, M., Sasaki, H., Nose, S. et al. (2010). *Mater. Manuf. Process.* 25: 264.

54 Grillet, A., Kinet, D., Witt, J. et al. (2008). *IEEE Sens. J.* 8: 1215–1222.

55 De Jonckheere, J., Narbonneau, F., Jeanne, M. et al. (2009). OFSETH: Smart medical textile for continuous monitoring of respiratory motions under magnetic resonance imaging. *Proceedings of the 31st Annual International Conference of the IEEE Engineering in Medicine and Biology Society: Engineering the Future of Biomedicine, EMBC 2009*, pp. 1473–1476. IEEE.

56 Pang, Y.N., Liu, B., Liu, J. et al. (2021). *IEEE Sens. J.* 21: 4610.

57 Koyama, Y., Nishiyama, M., and Watanabe, K. (2013). *IEEE Trans. Instrum. Meas.* 62: 828.

58 Davis, C., Mazzolini, A., Mills, J., and Dargaville, P. (1999). *Med. Eng. Phys.* 21: 619.

59 Lau, D., Chen, Z., Teo, J.T. et al. (2013). *IEEE Trans. Biomed. Eng.* 60: 2655.

60 Nishiyama, M., Sasaki, H., and Watanabe, K. (2006). Wearable sensing clothes embedding a hetero-core optic fiber for recognizing arm segment posture and motion. *Proc. IEEE Sensors* 1519–1522.

61 Nishiyama, M., Sasaki, H., and Watanabe, K. (2007). *Sensors and Smart Structures Technologies for Civil, Mechanical, and Aerospace Systems 2007*, vol. 6529, 65291Y. SPIE.

62 Otsuka, Y., Koyama, Y., and Watanabe, K. (2014). *Procedia Eng.* 87: 1465.

63 Yamazaki, H., Nishiyama, M., Watanabe, K., and Sokolov, M. (2016). *Sens. Actuators, A* 247: 98.

64 Nishiyama, M., Sasaki, H., and Watanabe, K. (2008). *IEEE Sens. J.* 8: 1055.

65 Goh, L.S., Anoda, Y., Kazuhiro, W., and Shinomiya, N. (2014). *Sensors* 14: 468–477.

66 Guo, J., Zhou, B., Yang, C. et al. (2019). *Adv. Funct. Mater.* 29: 1902898.

67 Gao, J., Fan, Y., Zhang, Q. et al. (2022). *Adv. Mater.* 34: 2107511.

68 Luo, Y., Canning, J., Zhang, J., and Peng, G.D. (2020). *Opt. Fiber Technol.* 58: 102299.

69 Kim, M., An, J., Kim, K.S. et al. (2016). *Biomed. Opt. Express* 7: 4220.
70 Nizamoglu, S., Gather, M.C., Humar, M. et al. (2016). *Nat. Commun.* 7: 10374.
71 Yetisen, A.K., Jiang, N., Fallahi, A. et al. (2017). *Adv. Mater.* 29: 1606380.
72 Wang, L., Zhong, C., Ke, D. et al. (2018). *Adv. Opt. Mater.* 6: 1800427.
73 Shentu, Z., Kang, J., Zhu, Z. et al. (2021). *Opt. Fiber Technol.* 63: 102531.
74 Guo, J., Zhou, B., Zong, R. et al. (2019). *ACS Appl. Mater. Interfaces* 11: 33589.
75 Park, S., Yuk, H., Zhao, R. et al. (2021). *Nat. Commun.* 12: 3435.
76 Leal-Junior, A.G., Díaz, C.R., Leitão, C. et al. (2019). *Opt. Laser Technol.* 109: 429.
77 Zheng, W., Tao, X., Zhu, B. et al. (2014). Fabrication and evaluation of a notched polymer optical fiber fabric strain sensor and its application in human respiration monitoring. *Text. Res. J.* 84: 1791.
78 Sartiano, D. and Sales, S. (2017). *Sensors* 17: 2900.
79 Pan, J., Zhang, Z., Jiang, C. et al. (2020). *Nanoscale* 12: 17538.
80 Li, L., Liu, Y., Song, C. et al. (2022). *Adv. Fiber Mater.* 4: 475.
81 Ma, S., Wang, X., Li, P. et al. (2022). *Adv. Fiber Mater.* 4: 1108.
82 Li, H., Li, H., Lou, X.P. et al. (2019). *Optik* 179: 298.
83 Tan, F., Chen, S., Lyu, W. et al. (2019). *Biomed. Opt. Express* 10: 5940.
84 Zhao, C., Liu, D., Cai, Z. et al. (2022). *Biosensors* 12: 168.
85 Shin, J., Liu, Z., Bai, W. et al. (2019). *Sci. Adv.* 5: eaaw1899.
86 Cheng-Yu, H., Abro, Z.A., Yi-Fan, Z., and Lakho, R.A. (2021). *J. Ind. Text.* 50: 1660.
87 Li, H., An, Z., Zhang, S. et al. (2021). *ACS Photonics* 8: 3607.
88 Li, Y., Dong, B., Chen, E. et al. (2021). *IEEE J. Sel. Top. Quantum Electron.* 27: 5600206.
89 Abro, Z.A., Hong, C., Chen, N. et al. (2020). *Text. Res. J.* 90: 386.
90 Umesh, S., Padma, S., Srinivas, T., and Asokan, S. (2018). *IEEE Sens. J.* 18: 216.
91 Rosli, E.B. and Noor, U.M. (2018). *IOP Conf. Ser.: Mater. Sci. Eng.* 340: 012020.
92 Ferraro, D., D'Alesio, G., Camboni, D. et al. (2021). *IEEE Sens. J.* 21: 14051.
93 Lo Presti, D., Santucci, F., Massaroni, C. et al. (2021). *Sci. Rep.* 11: 21162.
94 Lo Presti, D., Massaroni, C., Formica, D. et al. (2017). *IEEE Sens. J.* 17: 6037.
95 Ravi, T., Ranganathan, R., Ramesh, S.P., and Dandotiya, D.S. (2021). 3D printed personalized orthotic inserts using photogrammetry and FDM technology. In: *Fused Deposition Modeling Based 3D Printing* (ed. H.K. Dave and J.P. Davim), 349–361. Springer Cham.
96 Tavares, C., Domingues, M.F., Frizera-Neto, A. et al. (2018). *Sensors* 18: 1334.
97 Luo, Y., Yan, B., Zhang, Q. et al. (2017). *Sensors* 17: 511.
98 Chen, X., Zhang, C., Webb, D.J. et al. (2010). *Meas. Sci. Technol.* 21: 094005.
99 Vilarinho, D., Theodosiou, A., Leitão, C. et al. (2017). *Sensors* 17: 2924.
100 Bonefacino, J., Tam, H.Y., Glen, T.S. et al. (2018). *Light Sci. Appl.* 73: 17161.
101 Haseda, Y., Bonefacino, J., Tam, H.Y. et al. (2019). *Sensors* 19: 5088.
102 Mishra, P., Kumar, H., Sahu, S., and Jha, R. (2023). Flexible and wearable optical system based on U-shaped cascaded microfiber interferometer. *Adv. Mater. Technol.* 8 (3): 2200661.
103 Sundaram, R.S., Engel, M., Lombardo, A. et al. (2013). *Nano Lett.* 13: 1416.
104 Pospischil, A., Furchi, M.M., and Mueller, T. (2014). *Nat. Nanotechnol.* 9: 257.
105 Salehzadeh, O., Djavid, M., Tran, N.H. et al. (2015). *Nano Lett.* 15: 5302.

106 Shang, J., Cong, C., Wu, L. et al. (2018). *Small Methods* 2: 1800019.

107 Goldman, J.M., Petterson, M.T., Kopotic, R.J., and Barker, S.J. (2000). *J. Clin. Monit. Comput.* 16: 475.

108 Saracoglu, Ö.G. (2008). *Sensors* 8: 1585.

109 Efendioglu, H.S., Sahin, A.K., Yildirim, T., and Fidanboylu, K. (2011). Design of hetero-core smart fiber optic macrobend sensors. *ELECO 2011 – 7th International Conference on Electrical and Electronics Engineering*, Bursa, Turkey (1–4 December 2011). IEEE.

110 Efendioglu, H.S., Yildirim, T., and Fidanboylu, K. (2009). *Sensors* 9: 7167.

111 Aldaba, A.L., González-Vila, A., Debliquy, M. et al. (2018). *Sens. Actuators, B* 254: 1087.

112 Cheng, X., Bonefacino, J., Guan, B.O., and Tam, H.Y. (2018). *Opt. Express* 26: 14610.

113 Hassan, H.U., Nielsen, K., Aasmul, S., and Bang, O. (2015). *Biomed. Opt. Express* 6: 5008.

114 Xu, K., Zhou, R., Takei, K., and Hong, M. (2019). *Adv. Sci.* 6: 1900925.

115 Ghosh, A.K., Sarkar, S., Nebel, L.J. et al. (2021). *Adv. Funct. Mater.* 31: 2101959.

116 Gallego, D., Sáez-Rodríguez, D., Webb, D. et al. (2014). Interferometric microstructured polymer optical fiber ultrasound sensor for optoacoustic endoscopic imaging in biomedical applications. In: *23rd International Conference on Optical Fibre Sensors*, vol. 9157 (ed. J.S. Santos, J.M. López-Higuera, D. Julian, and M. López-Amo), 91574X. SPIE.

117 Leal-Junior, A.G., Frizera, A., Marques, C. et al. (2018). Polymer optical fiber strain gauge for humanrobot interaction forces assessment on an active knee orthosis. *Opt. Fiber Technol.* 41: 205.

4

Electrochemical Plasmonic Fibers for Operando Monitoring of Renewable Energy

Xiaobin Xue, Xile Han, Fu Liu and Tuan Guo

Jinan University, Institute of Photonics Technology, Xingye East 855, Guangzhou, 510632, China

4.1 Introduction

The drive toward renewable energy technologies, such as solar, wind and ocean, to help fight climate change puts a premium on ways to generate, store, and deliver power from these noncontinuous energy sources. One intriguing piece in this power management puzzle lies in energy storage devices (lithium-ion batteries, hydrogen fuel cells, supercapacitors as examples)—ultrahigh-capacitance structures that can sop up, store, and deliver significant amounts of charge, and that can withstand hundreds of thousands of charge–discharge cycles. Such electrochemical energy devices have already found use in a wide variety of clean energy systems at scales ranging from mobile medical devices to electric cars to power plants. But how can these closed devices be monitored in real time to ensure that they are working at peak efficiency and to prevent potentially catastrophic failures in the renewable-power grid? *In-situ* and continuous monitoring of the activities of clean energy devices over the whole power transfer chain (energy generation, storage, and consumption) is necessary.

Right now, assessing the charge state of an energy storage device (lithium-ion batteries as an example) typically involves measuring externally accessible electrical parameters like voltage and current. However, the lack of internal cell information hampers the accuracy and predictive capability of current battery management system (BMS) algorithms and models. There is a growing interest in spatiotemporal imaging of the thermal flows within a cell using temperature sensors [1, 2]. Typically, they are employed in electronic sensing devices, such as thermocouples (TC) [3], thermistors [4], infrared thermography [5], and resistance temperature detectors (RTDs) [6]. Additionally, batteries exhibit a breathing effect where they expand and contract in volume during cycling. These changes, along with electrode

Optical and Electronic Fibers: Emerging Applications and Technological Innovations, First Edition.
Edited by Lei Wei.

volume expansion associated with solid electrolyte interphase (SEI) growth, result in significant mechanical stress inside the battery materials (like cracks), which is detrimental to their performance. Sensing internal strain and pressure is often accomplished using strain gauges [7]. However, the aforementioned temperature and stress sensors are unsuitable for embedding within batteries due to their short resolution and accuracy, huge measurement apparatus, or higher size, as well as their inability to withstand the harsh electrochemical environment.

Technologies such as ultrasonic detection [8, 9], optical color contrast [10, 11], and electrochemical window infrared detection [12, 13] have been applied to certain extent at the laboratory level to obtain internal battery state information. However, these technologies involve large detector components, are not suitable for embedded installation, focus primarily on point measurement, or require opening a transparent window into the battery and utilizing professional color recording and analysis software. While these methods have a certain value for individual batteries, they present significant challenges for application at the module level.

A more direct method for assessing the charge state of energy storage device is to open the storage device and peer at it with an electron microscope. The transmission electron microscopy (TEM) has been widely used to observe the dynamic process of the batteries during charging or discharging status [14–17]. Recently, the cryo-electron microscopy (cryo-EM) method has been used to monitor the atomic structure and interfaces of battery materials [18], which has made a progress in understanding of the mechanisms of energy storages. However, these techniques have limitations such as the requirement for expensive instrumentation, challenges in sample preparation, and limited real-time monitoring capabilities. However, in addition to requiring expensive instrumentation, these techniques suffer from certain limitations, such as sample preparation challenges and limited real-time monitoring capabilities.

With the progress of technology, optical fiber sensors represent an ideal solution for battery testing (Figure 4.1a) due to their flexibility, small size, light weight, high temperature and pressure survivability, chemical resistance, and absence of conductivity and immunity to electromagnetic interference [19–21], while providing superior measurement performance compared to other technologies. Optical fiber sensors operate by correlating some of the many characteristic parameters (such as wavelength, intensity, phase, polarization state, etc.) of the interrogating light signals with factors such as local temperature, stress, pressure, and refractive index, which are all crucial parameters for battery monitoring [22]. Integrating optical fiber sensors inside batteries for real-time temperature measurements was first reported in 2013 [23]. Subsequently, the researchers further proposed advanced sensing schemes for cells internal strain and temperature discrimination by using a polarization-maintaining fiber Bragg grating (FBG) [24] and a Fabry–Perot interferometer (FPI) [25]. This was followed by multipoint measurements in another group [26].

The key problem of implanted optical fiber sensors for battery detection is not the basic measurement of stress and temperature, but the perception of the phase transition process, electrochemical behavior, interface dynamic response characteristics, and safety evolution characteristics of key materials in the battery.

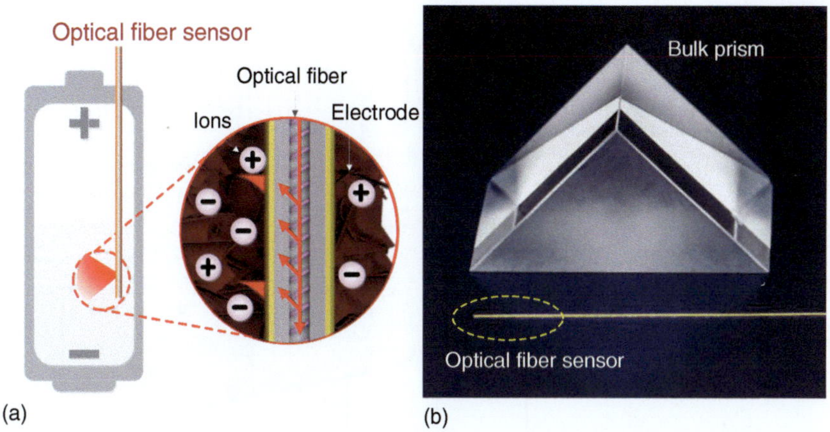

Figure 4.1 (a) Schematic diagram of in-situ monitoring of optical fiber battery. (b) Comparison between bulk prism-based and optical fiber-based SPR sensor.

Fiber-optic-based methods offer remotely operated "in-cell" solutions, enabling real-time monitoring of electrochemical processes and dynamic changes with cell operation [27–30]. By combining different sensing mechanisms, such as evanescent field, surface plasmon resonance (SPR) field, Raman spectroscopy, fluorescence spectroscopy, or light absorption, fiber-optic probes can decode relevant internal parameters, improve temporal and spatial resolution, and provide information on chemical composition, structural changes, and comprehensive information on electrochemical performance.

In this chapter, the opportunity to exploit electrochemical surface plasmon resonance (EC-SPR) method for operando monitoring energy storage devices is demonstrated. When light strikes the interface between a metal and a dielectric, part of the light beam's energy is transferred into a subwavelength-scale electromagnetic surface wave in a phenomenon known as SPR. It turns out that the characteristics of the resonance are highly sensitive to chemical or physical changes at the metal–dielectric boundary. The increased sensitivity achieved with plasmon waves arises because of the large localization of electromagnetic energy in the layer immediately adjacent to the metal surface [31]. Any perturbation in that layer, such as the bonding of analytes on receptor molecules modifies the local refractive index of the dielectric and the plasmon phase velocity. It offers a promising way to simultaneously explore optical and electrochemical properties of chemical matrix accompanying redox reactions. However, all above experiments are carried out using a bulky prism, it may be hard to be employed in hard-to-reach space or long-distance energy monitoring. Plasmonic optical fiber sensor is cost-effective and much simpler to interrogate compared to the well-established bulky prism configurations discussed above, as sketched in Figure 4.2. Its compact size makes it possible to be inserted into various hard-to-reach environments for in-situ detection either as a handheld probe or as a set of remotely operated devices fixed at various locations along a fiber-optic cable. The optical fiber sensor proposed here consists of a gold-coated tilted fiber

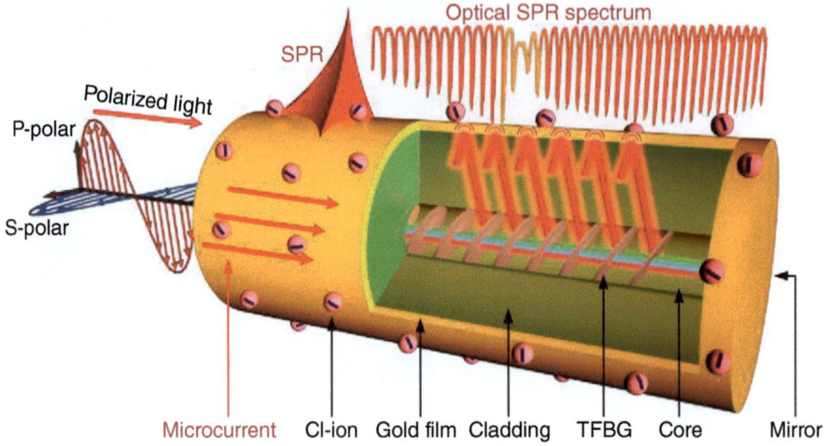

Figure 4.2 Setup of plasmonic fiber-optic-embedded sensing system for real-time monitoring renewable energy storage and alarming the potential safety hazards.

Bragg grating (TFBG) imprinted in a commercial single-mode fiber core, in which the nanometric-scale gold coating over the fiber surface offers the simultaneous detection of both the comprehensive monitoring the electrochemical information (current, potential, and stored charge) and the optical information (surface plasmon wave). This chapter focuses on the principles and characterization of in-situ monitoring of electrochemical activities of renewable energy devices using EC-SPR optical fiber sensing technology, and presents recent representative work on clean energy resource and storage device monitoring.

4.2 Sensing Principle

4.2.1 TFBG-assisted Plasmonic Excitation by Thin Metal Film Coating

The surface plasmon polariton (SPP) is an electromagnetic wave that propagates at the boundary between a metal and a dielectric. In a metal-coated fiber, hybrid-guided SPP-claddings mode can be excited successfully by a TFBG when the effective index of the cladding mode is phase matched to that of the SPP [32], expressed as

$$N_{\mathrm{eff}}^{\mathrm{SPP}} = \sqrt{\frac{\varepsilon_{\mathrm{m}}\varepsilon_{1}}{\varepsilon_{\mathrm{m}} + \varepsilon_{1}}} \tag{4.1}$$

where ε_{m}, ε_1 are the relative permittivities of the metal and external dielectric, respectively. For instance, the $N_{\mathrm{eff}}^{\mathrm{SPP}} = 1.328$ if we consider Au and water at a wavelength of 1550 nm wavelength and a cladding mode with such effective index will be hybridized with the SPP at a wavelength determined by the period of the TFBG. In the transmission spectra shown in Figure 4.3 (a) (2 cm length, 556.015 nm period, and 12° titled FBG with 45 nm thickness of Au on the cladding surface), the amplitudes of cladding modes near 1545 nm attenuate and "disappear" close to SPP phase-matched wavelength with P-polarized incident light where radially polarized cladding modes (i.e. TM-like) are excited. There is no SPP excitation for azimuthally polarized cladding modes (i.e. TE-like).

More details of the plasmonic hybrid-guided cladding mode fields are shown in Figure 4.3b. The simulation was done by using FIMMWAVE with an 8.2 μm core diameter and a refractive index of 1.449311, a 125 μm cladding diameter and a refractive index of 1.444078, a 45 nm thick Au coating with refractive index of $0.5240 + 10.7427i$, and a 20 μm layer of water above the gold (water refractive index = 1.3154). As for the mode field on the left of Figure 4.3b, the SPP is excited with an effective mode index $n = 1.3287 + 0.00129i$, and the energy density at the interface between Au and water is enhanced. The mode power fraction in water increases from 0.68% to 25.7% and in the metal layer from 0.035% to 0.736%, compared to the S-pol guided mode without SPP excitation but at a very similar wavelength (and real part of the effective index) shown on the right of figure 4.3b. Because of the large dispersion of SPP modes, a small permittivity changes near the metal surface can lead to large

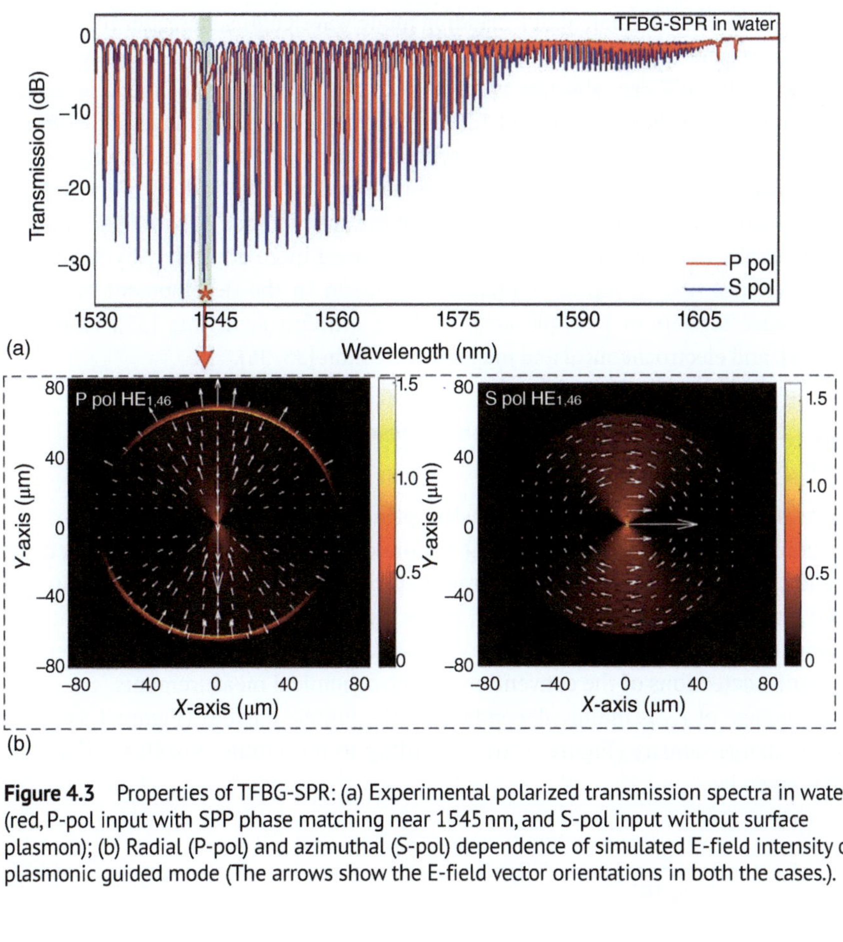

Figure 4.3 Properties of TFBG-SPR: (a) Experimental polarized transmission spectra in water (red, P-pol input with SPP phase matching near 1545 nm, and S-pol input without surface plasmon); (b) Radial (P-pol) and azimuthal (S-pol) dependence of simulated E-field intensity of plasmonic guided mode (The arrows show the E-field vector orientations in both the cases.).

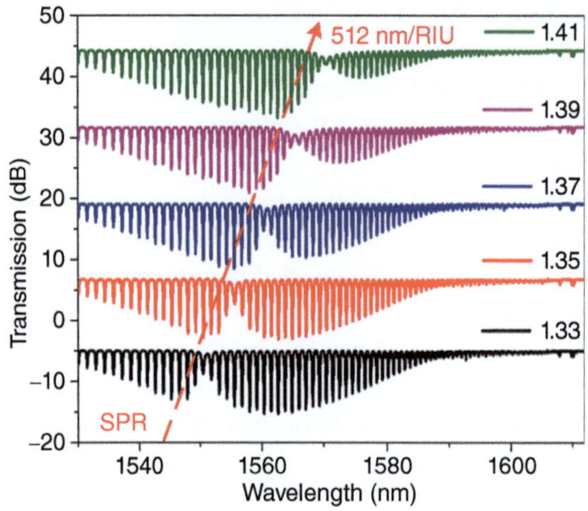

Figure 4.4 Experimental spectra of TFBG-SPR as a function of the refractive index of external dielectric layer. (a series of transmission spectra with offset to clarify the linearity of the SPR shift response with external index).

changes in the transmission at SPP cladding mode wavelengths, hereafter referred to as surface plasmon resonance (SPR resonance). Figure 4.4 shows a highly linear response of the SPR (i.e. absolute value of real part permittivity of metal is much larger than that of dielectric around 1550 nm wavelength) over a range of SRI from 1.33 to 1.41, where the guided cladding modes become too lossy to build strong transmission resonances along the grating. The corresponding RI sensitivity of the SPR is around 512 nm/RIU, which is similar to that of the cut-off mode of bare fiber TFBGs, but potentially more accurate as it can be determined more accurately by following the amplitudes of a group of resonances as shown in the development of high-performance sensors in multiple areas such as ultrafast switching [33], biomedicine [34], and electrochemical and magnetic detection [35, 36].

4.2.2 Electrochemical Surface Plasmon Resonance (EC-SPR) Sensing Method

Figure 4.5 indicates the comprehensive principles of electrochemical surface Plasmon resonance (EC-SPR) sensing technologies, which is directly related to faradic current and double-layer charging current. It changes the permittivity of metal film and thus the SPR frequency, and is a powerful tool to identify the electrochemical activity adjacent to the electrode interface, which is hard to be achieved by the "bulk" detections of the conventional electrochemical measurements.

The change of ε_m is mainly dependent on the faradic current (Figure 4.5a) and surface charge density (Figure 4.5d). According to the Drude model, ε_m of metal film is given by

$$\varepsilon_{\mathrm{m}} = 1 - \frac{n_{\mathrm{e}} e^2}{\varepsilon_0 m_{\mathrm{e}} 4\pi^2 f^2} \tag{4.2}$$

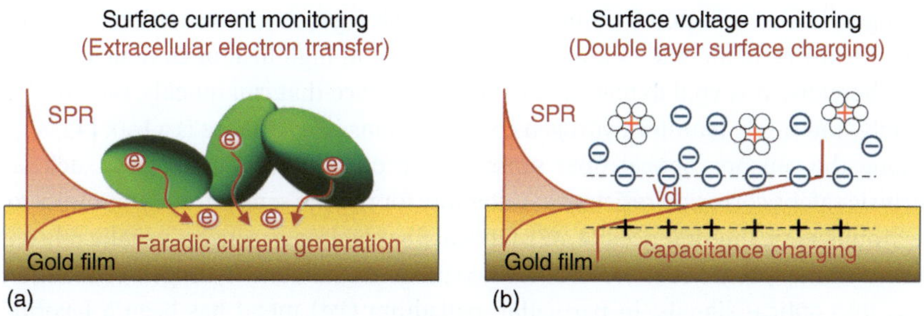

Figure 4.5 Principle of electrochemical sensing using surface plasmon resonance: (a) extracellular electron transfer process-induced faradic current; (b) capacitance charging-induced dielectric property changes.

where e, m_e, and n_e are the electron charge, electron mass, and electron density, respectively; f is light frequency, and ε_0 is vacuum dielectric constant of 8.85×10^{-12} F/m.

The change of ε_m is mainly originated from electron transfer process-induced faradic current $\Delta\varepsilon_{m(fc)}$ and the surface charging-induced dielectric property changes $\Delta\varepsilon_{m(dl)}$ of the metal film.

$$\Delta\varepsilon_{\mathrm{m(fc)}} = \left(\alpha_R D_R^{-1/2} - \alpha_0 D_0^{-1/2}\right)\left(nF\pi^{1/2}\right)^{-1} \times \int_0^t i(t')(t-t')^{-1/2}\,dt' \tag{4.3}$$

where α_0 and α_R are the changes in the local refractive index per unit concentration for the oxidized and reduced molecules, respectively; D_0 and D_R are the diffusion coefficients of the reaction species; n is number of electrons transferred per reaction; and F is the Faraday constant.

$$\Delta\varepsilon_{\mathrm{m(dl)}} = -\frac{\left(\varepsilon_{\mathrm{m}} - 1\right)c_{\mathrm{dl}}\Delta V_{\mathrm{dl}}}{ed_{\mathrm{m}}n_e} \tag{4.4}$$

where d_m is the thickness of thin metal film; and c_{dl} is interfacial capacitance density (capacitance per unit area).

The above equations establish simple relations between SPR signal modulation versus the value of localized faradic current, interfacial capacitance, and potential. When a positive or negative potential is applied on the gold film, electrons flow in or out of the gold film, resulting in a change of electron density. This will influence the value of ε_m and thus SPR response.

4.3 Recent Progress of Operando Monitoring of Renewable Energy

4.3.1 Ultrafast and Repeatable Hydrogen Monitoring

Hydrogen as a promising energy carrier is a perfect candidate to supply the energy demand of the world and concomitantly reduce toxic emissions [37–39]. However, hydrogen is also a highly flammable and explosive gas at concentrations ranging from 4% to 75% in air, with a low ignition energy and high heat of combustion [40, 41]. Therefore, it is vital to have an inexpensive device that can quickly, sensitively, reliably, and safely monitor hydrogen concentrations in case there is a leak [42, 43]. Among the sensors, optical fiber sensors have outstanding advantages, such as intrinsic safety, corrosion resistance, suitability for remote sensing, and immunity to electromagnetic interference [44, 45]. Optical fiber hydrogen sensors usually rely on sensitive and specific reactive thin film coatings to translate hydrogen concentrations into optical signals: in particular, palladium (Pd) metal has been a favorite coating as it absorbs hydrogen in proportion to its relative concentration, thereby forming either palladium hydride (PdH*x* where *x* is the atomic ratio of H:Pd) or a mixture of molecular hydrogen in Pd. This chemical change is associated with volume changes as well as modifications to the real and imaginary part of the metal permittivity [46].

Figure 4.6a shows the experimental setup for the measurement of the complex refractive index of thin Pd films exposed to various concentrations of H_2 based on an Au–Pd-coated 23° TFBG with a ~200 nm thick gold mirror deposited on the cleaved end by direct current (DC) magnetron sputtering followed by a protective layer of 500 nm of SiO_2 using radiofrequency (RF) sputtering [47]. The Pd coatings used were also deposited by magnetron sputtering. Light was launched in the core of the plasmonic TFBG from a broadband source (BBS) with wavelength range of 1280–1560 nm, and the reflection spectrum was monitored by an optical spectrum analyzer (OSA) with a wavelength resolution of 0.02 nm. A linear polarizer and a polarization controller (PC) were placed upstream of the coupler to adjust and orient the state of polarization of light so as to provide the strongest SPR excitation. Figure 4.6b shows a typical experimental transmission spectrum of an 18 mm long, 23° tilt FBG (internal angle relative to the fiber cross section) coated with 30 nm of gold and measured in air with input light p-polarized [48]. The fine comb of cladding mode resonances is more complex than that of TFBGs with smaller tilt angles, but a clear SPR signature is observed near 1312 nm. In fact, the tilt angle was chosen so that strong coupling occurs to high-order cladding modes with a real part of their effective index close to 1.01, i.e. synchronous with the effective index of the surface plasmon of a gold surface in air at these wavelengths. In Figure 4.6c, the main outcomes of this analysis are that the identified resonance redshifts rapidly with increasing Pd thickness, and that its amplitude first decreases and then increases again in simulations. These changes appear to saturate for thicknesses exceeding 60 nm, indicating that the combined metal layers become thicker than the field penetration. For thicknesses between 3 and 10 nm approximately, the amplitude of the simulated resonance becomes quite insensitive to thickness changes while the wavelength shifts rapidly in the sense that the reaction of Pd with H_2 leads to volume changes in addition to chemical changes [46]; so the optical response that is measured by the TFBG will only be impacted by the chemical change (i.e. the change in permittivity) and not by thickness changes if a thickness in the aforementioned range is used. As shown in Figure 4.6d, the measured amplitude of the 1312 nm resonance is seen to increase when the ambient medium changes from pure N_2 to 3% H_2 in N_2, while the wavelength does not appear to

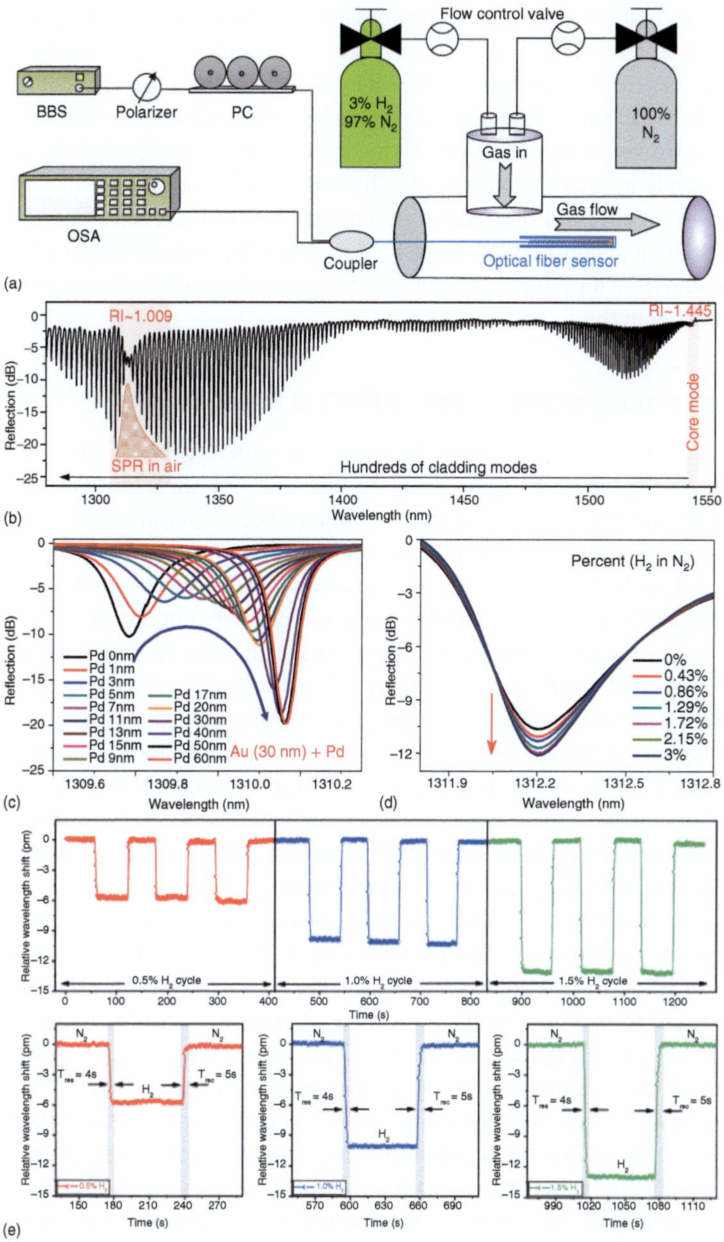

Figure 4.6 Schematic of a plasmonic fiber-optic sensing system for in-situ H_2 monitoring: (a) the experimental setup to measure the coated TFBG response to various concentrations of H_2. (b) experimental transmission spectrum in air for a 23°-tilt TFBG in standard single-mode optical fiber with 30 nm of gold deposited on the cladding surface and the input light polarized linearly along the vertical axis, i.e. p-polarized relative to the tilt direction; (c) simulated results show how the selected mode resonance shifts in spectrum with increasing thickness of Pd; (d) measured reflected spectrum near the 1312 nm resonance as a function of H_2 concentration; (e) response and recovery characteristics of TFBG sensor at specific hydrogen concentrations of 0.5%, 1.0%, and 1.5% in nitrogen environment. *Source:* Copyright (2018), with permission from Royal Society of Chemistry.

change. This means that either the thickness does not change (unlikely) or that the intensity response of the TFBG is indeed insensitive to small thickness changes when its initial thickness is near 7 nm, as shown by the simulations in Figure 4.6c. Figure 4.6e details the ultrafast response and recovery characteristics of TFBG sensor at different hydrogen concentrations. To be specific, the response time is as small as four seconds needed for "seeing" the hydrogen molecules after entering the nitrogen environment, and the recovery time involved in the reverse process is down to about five seconds. This large angle-tilted FBG sensor whose SPR generates in air could be a perfect candidate for the promising green energy gas–hydrogen monitoring in the future.

4.3.2 *In-situ* Monitoring of State of Charge (SOC) of Battery

Energy storage devices (ESDs), including batteries and supercapacitors [49], are efficient ways to solving the challenge of using the unstable renewable energy (sun, wind, or ocean) at any time [50, 51]. It is of great importance to monitor the health of these ESDs precisely and timely for stably storing the discontinuous energy [52]. Usually, traditional methods for detecting the performance of energy storage devices mainly rely on the electrochemical testing, including cyclic voltammetry (CV) and galvanostatic charge–discharge (GCD) measurements. However, these measurements cannot be operated when the devices are simultaneously at work, that is, they are ex-situ processes. Several mainstream methods for monitoring the inner reaction process of ESDs (mainly Li-ion batteries) under working state have also been proposed. For example, the *in-situ* transmission electron microscope (TEM) was used to reveal the dynamic variation process in Li-ion batteries [53, 54]. However, this kind of *in-situ* test can only be performed in special battery with nanoscale, which is far away from the true device in actual application.

An effective tool for observing the state of charge of energy storage is to integrate optical fiber sensors. Among them, the plasmonic optical sensors consisting of a TFBG and a layer of decorated gold coating in nanoscale for providing plasmon waves were designed to precisely monitor the state of charge for typical supercapacitors [29]. As shown in Figure 4.7a, the TFBG plasmonic optical fiber sensor was attached firmly to one of electrodes (MnO_2 nanosheets coated on carbon fabric) of the supercapacitor, and both the sensor and the supercapacitor were soaked into liquid LiCl electrolyte. The active material MnO_2 possesses a pseudocapacitive effect, that is, it can store energy through not only electric double-layer capacitive function, but also some redox reaction and fast intercalation/deintercalation on its surface. When the supercapacitors were under CV measurements, the whole process was divided into two parts: charging segment from −0.8 to 0.8 V and discharging segment from 0.8 V back to −0.8 V, as presented in Figure 4.7b. Three characteristic points (red points in Figure 4.7b) were chosen for explaining the monitoring mechanism. At −0.8 V, the surface of the electrode was full of positive charges so that negative electronic polarization happened on the gold layer. At 0 V, the supercapacitor was neither charged nor discharged, thus resulting in the steady state of the gold film. While at 0.8 V, negative charges were found on the surface of electrode, and the opposite polarized situation would appear on the gold coating. If the supercapacitors were performed a reversible charge and discharge operation from 0 to 0.8 V, the optical SPR spectra collected at the same time demonstrates an obvious correspondence to the intensity of the electronic polarization on the gold layer, as shown in the blue line of the Figure 4.7c. Besides, the core mode (red line in Figure 4.7c) was not influenced by the electrochemical energy storage process, which can be used as an in-situ thermometer. The repeated charging/discharging measurements under CV, GCD, and galvanostatic charge and voltage holding (GCVH) methods presented excellent stability (Figure 4.7d), indicating the great potential in actual application of this sensing system.

4.3.3 *In-situ* Monitoring of Ion Activities in Battery

The critical factors in battery performance and the maximum energy storage capacity are determined by the charging and discharging processes, which, in turn, depend on ion intercalation and deintercalation kinetics between the electrodes and the electrolytes. In particular, understanding ion transport kinetics and electrolyte–electrode interactions at electrode surfaces of batteries in operation is essential to determine their performance. However, it remains a challenging task to capture in real time the details of surface-localized and rapid ion transport at the microscale. The reason for this is that existing methods to quantify charge transport near electrodes are predominantly restricted to inoperative batteries and to time scales that are much longer than the fastest processes occurring real-time operation [52, 55]. To address this, an electrochemical SPR fiber-optic sensor was incorporated into batteries for in-situ detection of chemical and electrochemical events, without perturbing battery operation [27].

Figure 4.8a shows the all-fiber-coupled electrochemical SPR fiber-optic sensing system. The optical fiber sensing probe was tightly attached to the surface of working electrode. This configuration also ensures strain-free operation of the sensor to eliminate the effect of cross sensitivity to strain of the higher order cladding and plasmonic modes when the sensor is fixed at one end. The three-electrode system was driven by an electrochemical workstation. Both the optical signal and electrochemical signal were recorded and analyzed by a computer in real time. The dash–dot–line inset presents the voltage of the battery (top) and corresponding optical output of the SPR fiber sensor (bottom), simultaneously. As shown in Figure 4.8b, the attenuation band of the TFBG spectrum due to the SPR effect is highlighted in red. As the attenuation band of the SPR shifts, the amplitude of the plasmonic resonance changes (located on the short wavelength side of the SPR envelope and identified by the red asterisk "*".) and allows an indirect, but precise, quantitative measurement of very-small SPR changes. To further verify the correlation between electrochemical behavior and optical signal, the differential of the optical power is obtained, which is obtained by deriving the SPR power level with respect to time, which represents the rate of change of the optical power. Unlike voltage change, the SPR power level change is due to the local ion concentration decrease at the diffusion layer caused by H^+/Zn^{2+} intercalation. For better understanding, two groups of

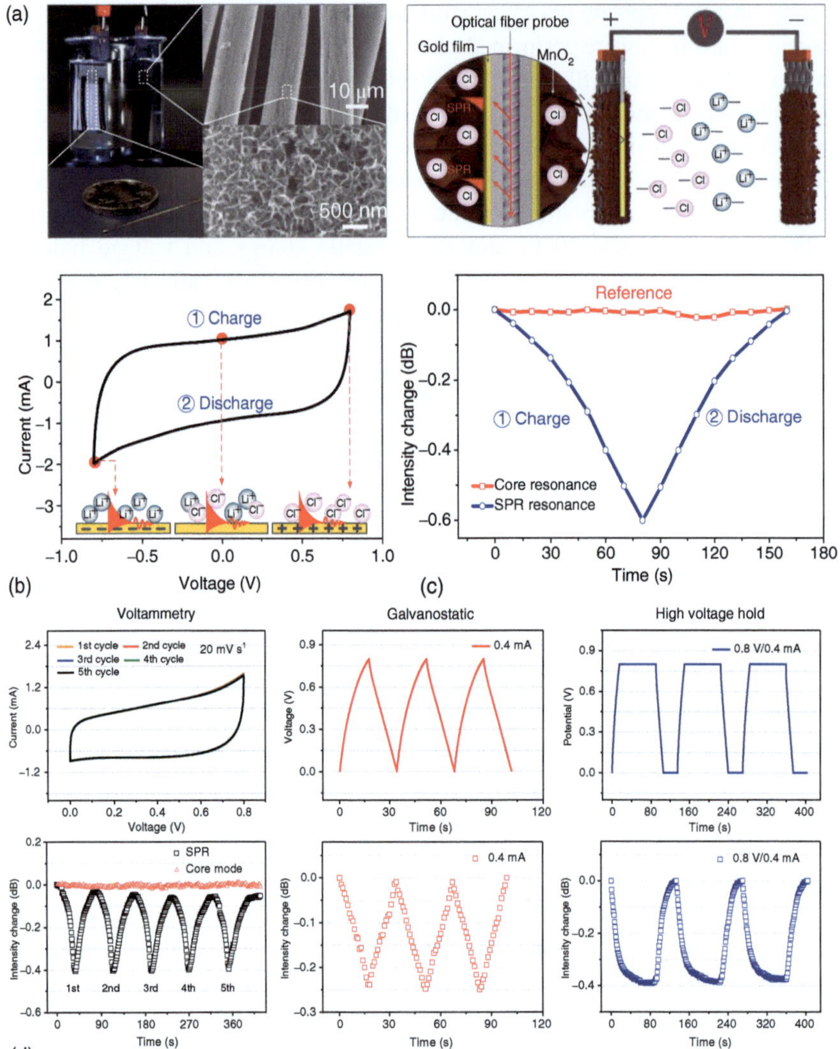

Figure 4.7 (a) The scheme of the electrochemical SPR sensing device, and the corresponding optical and microcosmic images of the experimental equipments. (b) CV curve of the supercapacitor during a polarizing cycle from −0.8 to 0.8 V. Insets show the corresponding polarizing states for the supercapacitor under a voltage of −0.8, 0, and 0.8 V. (c) Intensity change of the selected SPR resonance (blue dot) and the core Bragg resonance (red block) following one cycle of the CV curve of the supercapacitors with voltage window ranging from 0 to 0.8 V. (d) The responses of the SPR sensor under repeating charging/discharging tests with different operated methods (CV, GCD, and GVCH measurements). *Source:* Copyright (2018), with permission from *Springer Nature*.

aqueous electrolytes were set to compare the difference between H^+/Zn^{2+} two-step intercalation and only H^+ intercalation. As shown in Figure 4.8c, the intercalation mechanism and polarization properties of H^+/Zn^{2+} ions are verified by the joint

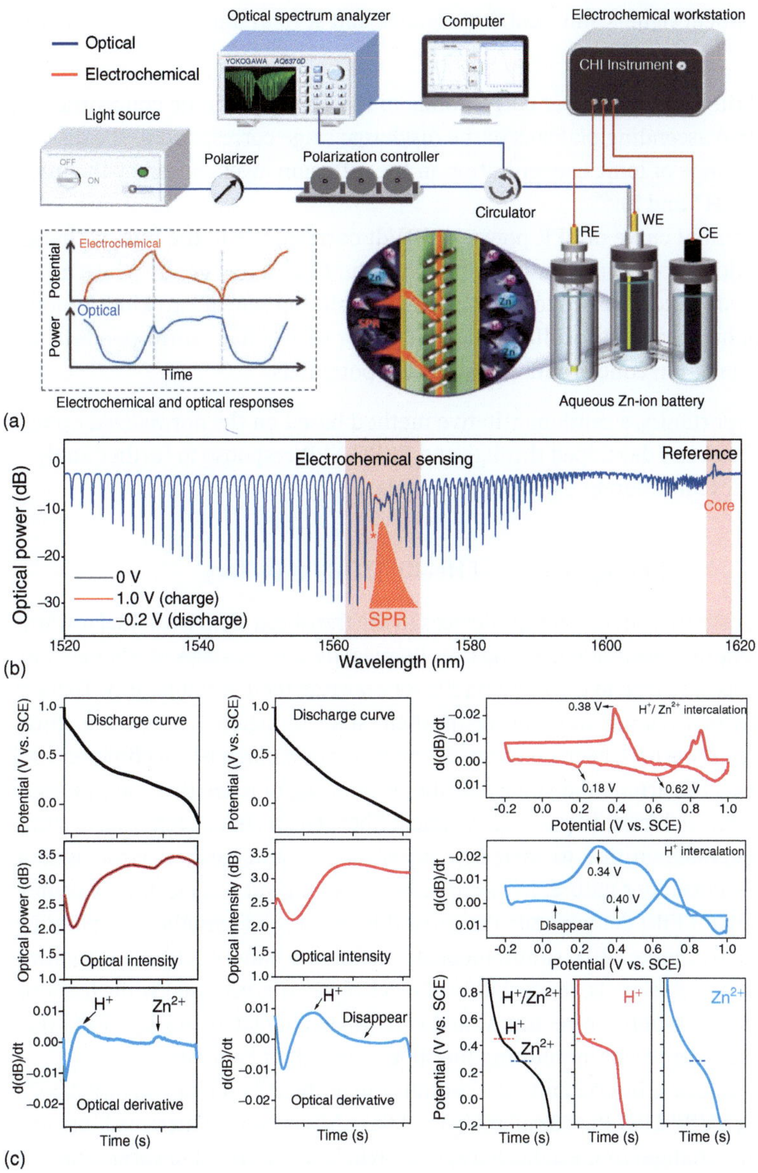

Figure 4.8 (a) Experimental setup of a plasmonic fiber-optic sensing system and an electrochemical calibration system for Zn-ion battery measurement. RE: reference electrode, WE: working electrode, and CE: counter electrode. The dash–dot–line inset presents the voltage of the battery (top) and corresponding optical output of the plasmonic fiber sensor (bottom), simultaneously. (b) Reflection spectral response of a mirror-ended TFBG optical fiber (tilt angle of 12°) for the different electronic polarization states of the electrode; the spectral change of the most sensitive SPR-coupled cladding mode (marked by a red asterisk "*") is used for electrochemical monitoring; other cladding mode resonances at longer wavelengths and the core mode resonance near 1610 nm can be used for reference. (c) Electrochemical curve (black), SPR power curve (red), and differential of light power (blue) versus the discharge stages at 1 M $ZnSO_4$ + 0.4 M $MnSO_4$ and 0.4 M $MnSO_4$ aqueous electrolyte. *Source:* Copyright (2022), with permission from *Springer Nature*.

analysis of optical signals and electrical signals, and several important information is obtained:

(1) The evolution of SPR power corresponds to the change of ionic concentration, and the two ascending plateaus at the discharge stage correspond to the discontinuous change of ionic concentration in the diffusion layer caused by the intercalation of H^+ and Zn^{2+}.
(2) The differential curve of SPR power d(dB)/dt corresponds to the rate of change of the ionic concentration in real time, and the two characteristic peaks at the discharge stage correspond to the maximum evolution rate of two kinds of ions.
(3) The mapping curve of the differential SPR power with voltage corresponds to the change rate of ion concentration at different potentials.

And most importantly, a semiquantitative method based on the normalized optical rate curves has been developed through the TFBG-SPR response to further analyze the ion kinetics for different electrode materials.

4.3.4 *In-situ* Monitoring of State of Health (SOH) of Battery

Batteries are essential energy storage devices in the transition from fossil fuel towards sustainable energy. Operando tracking the chemical dynamics/states of a battery and its components is critical to their second life for boosting their sustainability. To meet such objectives, accurate monitoring of their state of health becomes a must. Considerable efforts have been placed in the quest for new diagnostic techniques [56] beyond the use of current and voltage or the positioning of a few thermal probes in the latest electric vehicles. In particular, optical fiber sensors have been used to study electrolyte degradation due to their low invasiveness, smallness, electromagnetic immunity, and insulator nature. In 18650 cells containing various electrolyte combinations, a single TFBG demonstrate the capability to operando monitor temperature (T) and refractive index (RI), simultaneously, inside the battery [28]. By exploiting the time-resolved RI signals, the turbidity of electrolytes was accessed while tracking electrolyte deterioration. These unravelled electrolyte characteristics by TFBG provide additional metrics for monitoring the state of health of batteries.

The TFBG sensor is inserted into the central void of the jelly roll of $Na_3V_2(PO_4)_2F_3$/hard carbon (NVPF/HC) 18,650 cells (Figure 4.9a), thus removing the sensibility to ε. Consequently, changes in the collected spectra will be on ascribed to variations of T and RI. In order to assess the diffusion/reaction kinetics of electrolyte, galvanostatic intermittent titration technique (GITT) is performed during the formation of cells. Figure 4.9b shows the temporal voltage profile and the TFBG spectra during the galvanostatic intermittent titration technique (GITT) test. Such a dramatic change in the TFBG spectrum contours after 30 hours (Figure 4.9b, bottom) implies copious electrochemically driven decompositions [57]. Looking into the spectra every 10 hours (Figure 4.9c), the amplitudes of most of the cladding resonances simultaneously decrease after 30 hours. This response is related to the excessive depletion of electrolyte and the subsequent precipitation of solid particles such as byproducts, which could then enter the evanescent field of TFBG or even be deposited on the cladding. This causes a dramatic light scattering effect of the TFBG and leads to a

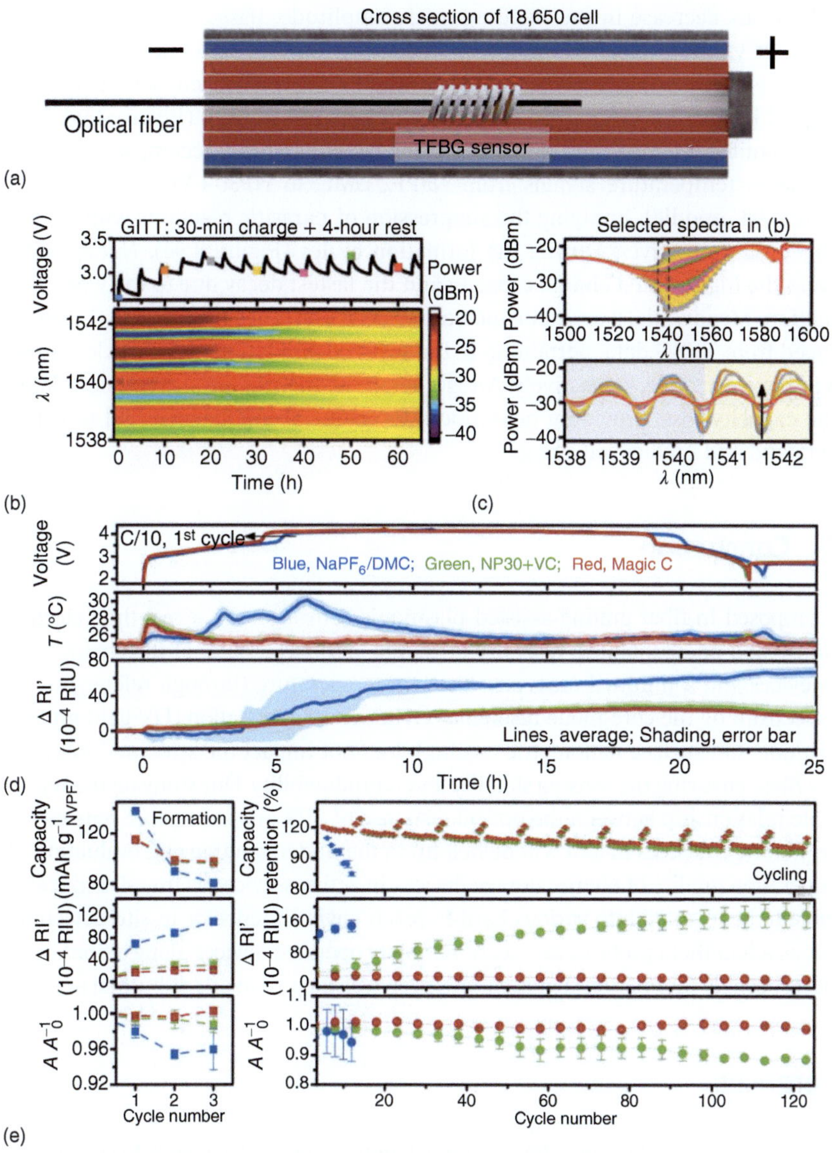

Figure 4.9 (a) A schematic showing the integration of TFBG into the central void of the jelly roll of an 18,650 cell. (b) The temporal voltage (top) and enlarged TFBG spectrum contour (bottom) during the GITT (30 min charge followed by four hour rest) of an NVPF/HC cell containing $NaPF_6$/DMC electrolyte at a current of C/10. The voltage rises and decreases during the charge and the rest, respectively. (c) Full and enlarged spectra of TFBG at the selected points in (b) in the same colors. The arrow in (c) indicates the amplitude suppression of a guided cladding resonance after 30 hours. (d) The temporal voltage (top), T (middle), and $\Delta RI'$ (bottom) of two cells for each electrolyte, during the first formation cycle at C/10 with six hour rest between charge and discharge. (e) The charge capacity, charge capacity retention (relative to the 5th cycle), $\Delta RI'$ (middle), and $A\,A_0^{-1}$ (bottom) as a function of cycle number during the formation at C/10 and the cycling test at 1 C. *Source:* Copyright (2021), with permission from Royal Society of Chemistry.

simultaneous decrease in the cladding mode amplitude. Based on this, TFBG are injected in various 18,650 Na-ion cells filled with other electrolyte formulae (as shown in Figure 4.9d) to analyze the implication for the cell capacity loss. The slight voltage shifts in the time scale between different electrolytes (Figure 4.9d, top, arrow) manifest the presence of copious side reactions. This is accompanied by the decrease in temperature signals from $NaPF_6$/DMC to NP30+VC and Magic C (Figure 4.9d, middle), implying the suppression of parasitic reactions with proper cosolvent and additives. During the formation cycles (Figure 4.9e), $NaPF_6$/DMC delivers the highest first charge capacity and the fastest decay due to the worst SEI formation. Meanwhile, a dramatic increase in $\Delta RI'$ and a decrease in $A\ A_0^{-1}$ can be deduced from TFBG data, pertaining to the severe degradation and particle saturation in the electrolyte, respectively. Altogether, these metrics are strongly correlated to the capacity loss, providing new inputs for artificial intelligence [58] to better understand and predict cell aging.

4.4 Conclusion

The proposed in-fiber grating-assisted plasmonic sensor shows a real-time electrochemical and photochemical response to charge–discharge states of the energy storage devices and a minimal cross-sensitivity to temperature (through referencing of the spectrum by the core-mode resonance). The sensor fabrication (UV-light-grating inscription and surface nanometric coating) does not impact the structural integrity of the fiber, ensuring the sensing stability and reproducibility. Our study on the reproducible relationship between the optical response of SPR and the charge state of the energy storage devices in-situ will be helpful to the understanding and evaluation of the operation quality of energy storage devices in active service. Its tiny size makes it feasible to be inserted into various hard-to-reach environments for in-situ detection, either as a handheld probe or as a set of remotely operated devices along a fiber-optic cable, especially for environmental monitoring over the urban and suburban areas.

References

1 Lee, J.H., Lee, H.M., and Ahn, S. (2003). Battery dimensional changes occurring during charge/discharge cycles-thin rectangular lithium ion and polymer cells. *J. Power Sources* 119–121: 833–837.

2 Bandhauer, T.M., Garimella, S., and Fuller, T.F. (2011). A critical review of thermal issues in lithium-ion batteries. *J. Electrochem. Soc.* 158: 3.

3 Mutyala, M.S.K., Zhao, J.Z., Li, J.Y. et al. (2014). In-situ temperature measurement in lithium ion battery by transferable flexible thin film thermocouples. *J. Power Sources* 260: 43–49.

4 Bolsinger, C. and Birke, K.P. (2019). Effect of different cooling configurations on thermal gradients inside cylindrical battery cells. *J. Energy Storage* 21: 222–230.

5 Goutam, S., Timmermans, J.M., Omar, N. et al. (2015). Comparative study of surface temperature behavior of commercial LI-ion pouch cells of different chemistries and capacities by infrared thermography. *Energies* 8: 8175–8192.

6 Wang, P.D., Zhang, X.Y., Yang, L. et al. (2016). Real-time monitoring of internal temperature evolution of the lithium-ion coin cell battery during the charge and discharge process. *Extreme Mech. Lett.* 9: 459–466.

7 Leung, P.K., Moreno, C., Master, I. et al. (2014). Real-time displacement and strain mappings of lithium-ion batteries using three-dimensional digital image correlation. *J. Power Sources* 271: 82–86.

8 Hsieh, A.G., Bhadra, S., Hertzberg, B.J. et al. (2015). Electrochemical-acoustic time of flight: in operando correlation of physical dynamics with battery charge and health. *Energy Environ. Sci.* 8: 1569–1577.

9 Gold, L., Bach, T., Virsik, W. et al. (2017). Probing lithium-ion batteries' state-of-charge using ultrasonic transmission – Concept and laboratory testing. *J. Power Sources* 343: 536–544.

10 Villemin, E. and Raccurt, O. (2021). Optical lithium sensors. *Coord. Chem. Rev.* 435: 213801.

11 Yang, L., Chen, H.S., Song, W.L., and Fang, D.N. (2018). Effect of defects on diffusion behaviors of lithium-ion battery electrodes: in-situ optical observation and simulation. *ACS Appl. Mater. Interfaces* 10: 43623–43630.

12 Zhu, S.X., Han, J.D., An, H.Y. et al. (2020). A novel embedded method for in-situ measuring internal multi-point temperatures of lithium ion batteries. *J. Power Sources* 456: 227981.

13 Rudolph, S., Schröder, U., Bayanov, I.M. et al. (2013). High resolution state of charge monitoring of vanadium electrolytes with IR optical sensor. *J. Electroanal. Chem.* 694: 17–22.

14 Gu, M., Parent, L.R., Mehdi, B.L. et al. (2013). Demonstration of an electrochemical liquid cell for operando transmission electron microscopy observation of the lithiation/delithiation behavior of Si nanowire battery anodes. *Nano Lett.* 13: 6106–6112.

15 Liu, X.H. and Huang, J.Y. (2011). In-situ TEM electrochemistry of anode materials in lithium ion batteries. *Energy Environ. Sci.* 4: 3844–3860.

16 Gu, M., Belharouak, I., Zheng, J.M. et al. (2013). Formation of the spinel phase in the layered composite cathode used in Li-ion batteries. *ACS Nano* 7: 760–767.

17 Zeng, Z.Y., Liang, W.-I., Liao, H.G. et al. (2014). Visualization of electrode-electrolyte interfaces in $LiPF_6$/EC/DEC electrolyte for lithium ion batteries via in-situ TEM. *Nano Lett.* 14: 1745–1750.

18 Li, Y.Z., Li, Y.B., Pei, A. et al. (2017). Atomic structure of sensitive battery materials and interfaces revealed by cryo-electron microscopy. *Science* 358: 506–510.

19 Culshaw, B. (2004). Optical fibre sensor technologies: Opportunities and-perhaps-pitfalls. *J. Lightwave Technol.* 22: 39–50.

20 Tosi, D., Macchi, E.G., Braschi, G. et al. (2014). Fibre-optic combined FPI/FBG sensors for monitoring of radiofrequency thermal ablation of liver tumors: *ex vivo* experiments. *Appl. Opt.* 53: 2136–2144.

21 Madan, A., Yap, S.H.K., Paulose, V. et al. (2020). Investigation of a Bragg grating-based Fabry-Perot structure inscribed using femtosecond laser micromachining in an adiabatic fibre taper. *Appl. Sci.* 10: 1069.

22 Raghavan, A., Kiesel, P., Sommer, L.W. et al. (2017). Embedded fibre-optic sensing for accurate internal monitoring of cell state in advanced battery management systems part 1: Cell embedding method and performance. *J. Power Sources* 341: 466–473.

23 Yang, G., Leitão, C., Li, Y.H. et al. (2013). Real-time temperature measurement with fibre Bragg sensors in lithium batteries for safety usage. *Measurement* 46: 3166–3172.

24 Matuck, L., Pinto, J.L., Marques, C., and Nascimento, M. (2022). Simultaneous strain and temperature discrimination in 18650 Li-ion batteries using polarization-maintaining fibre Bragg gratings. *Batteries* 8: 233.

25 Nascimentoa, M., Novaisa, S., Ding, M.S. et al. (2019). Internal strain and temperature discrimination with optical fibre hybrid sensors in Li-ion batteries. *J. Power Sources* 410-411: 1–9.

26 Jia, T., Zhang, Y., Ma, C.Y. et al. (2023). The early warning for overcharge thermal runaway of lithium-ion batteries based on a composite parameter. *J. Power Sources* 555: 232393.

27 Wang, R.L., Zhang, H.Z., Liu, Q.Y. et al. (2022). Operando monitoring of ion activities in aqueous batteries with plasmonic fibre-optic sensors. *Nat. Commun.* 13: 547.

28 Huang, J.Q., Han, X.L., Liu, F. et al. (2021). Monitoring battery electrolyte chemistry via in-operando tilted fibre Bragg grating sensors. *Energy Environ. Sci.* 14: 6464–6475.

29 Lao, J.J., Sun, P., Liu, F. et al. (2018). In-situ plasmonic optical fibre detection of the state of charge of supercapacitors for renewable energy storage. *Light: Sci. Appl.* 7: 34.

30 Cai, S.S., Liu, F., Wang, R.L. et al. (2020). Narrow bandwidth fibre-optic spectral combs for renewable hydrogen detection. *Sci. China Inf. Sci.* 63: 222401.

31 Wang, S.P., Huang, X.P., Shan, X.N. et al. (2010). Electrochemical surface plasmon resonance: basic formalism and experimental validation. *Anal. Chem.* 82: 935–941.

32 Liu, F., Zhang, X., Li, K. et al. (2021). Discrimination of bulk and surface refractive index change in plasmonic sensors with narrow bandwidth resonance combs. *ACS Sens.* 6: 3013–3023.

33 Liu, F. and Albert, J. (2019). 40 GHz-rate all-optical cross-modulation of core-guided near infrared light in single mode fibre by surface plasmons on gold-coated tilted fibre Bragg gratings. *APL Photon.* 4: 126104.

34 Guo, T., Liu, F., Liang, X. et al. (2016). Highly sensitive detection of urinary protein variations using tilted fibre grating sensors with plasmonic nanocoatings. *Biosens. Bioelectron.* 78: 221–228.

35 Yuan, Y., Guo, T., Qiu, X.H. et al. (2016). Electrochemical surface plasmon resonance fibre-optic sensor: In-situ detection of electroactive biofilms. *Anal. Chem.* 88: 7609–7616.

36 Zhang, Z.C., Guo, T., Zhang, X.J. et al. (2016). Plasmonic fibre-optic vector magnetometer. *Appl. Phys. Lett.* 108: 101105.

37 Amirante, R., Cassone, E., Distaso, E., and Tamburrano, P. (2017). Overview on recent developments in energy storage: mechanical, electrochemical and hydrogen technologies. *Energy Convers. Manage.* 132: 372–387.

38 Dimitriou, P. and Tsujimura, T. (2017). A review of hydrogen as a compression ignition engine fuel. *Int. J. Hydrogen Energy* 42: 24470–24486.

39 Hosseini, S.E. and Wahid, M.A. (2016). Hydrogen production from renewable and sustainable energy resources: promising green energy carrier for clean development. *Renewable Sustainable Energy Rev.* 57: 850–866.

40 Abdalla, A.M., Hossain, S., Nisfindy, O.B. et al. (2018). Hydrogen production, storage, transportation and key challenges with applications: a review. *Energy Convers. Manage.* 165: 602–627.

41 Nakayama, J., Kasai, N., Shibutani, T., and Miyake, A. (2019). Security risk analysis of a hydrogen fueling station with an on-site hydrogen production system involving methylcyclohexane. *Int. J. Hydrogen Energy* 44: 9110–9119.

42 Huebert, T., Boon-Brett, L., Black, G., and Banach, U. (2011). Hydrogen sensors-a review. *Sens. Actuators, B* 157: 329–352.

43 Buttner, W.J., Post, M.B., Burgess, R., and Rivkin, C. (2011). An overview of hydrogen safety sensors and requirements. *Int. J. Hydrogen Energy* 36: 2462–2470.

44 Zhang, Y.N., Peng, H.J., Qian, X.L. et al. (2017). Recent advancements in optical fibre hydrogen sensors. *Sens. Actuators, B* 244: 393–416.

45 Joe, H.E., Yun, H., Jo, S.H. et al. (2018). A review on optical fibre sensors for environmental monitoring. *Int. J. Pr. Eng. Man.-GT.* 5: 173–191.

46 Lewis, F. a. (1960). The hydrides of palladium and palladium alloys. *Platinum Met. Rev.* 4: 132–137.

47 Zhang, X.J., Cai, S.S., Liu, F. et al. (2018). In-situ determination of the complex permittivity of ultrathin H_2-infused palladium coatings for plasmonic fibre optic sensors in the near infrared. *J. Mater. Chem. C* 6: 5161–5170.

48 Caucheteur, C., Guo, T., Liu, F. et al. (2016). Ultrasensitive plasmonic sensing in air using optical fibre spectral combs. *Nat. Commun.* 7: 13371.

49 Wang, X.F., Lu, X.H., Liu, B. et al. (2014). Flexible energy-storage devices: design consideration and recent progress. *Adv. Mater.* 26: 4763–4782.

50 Mei, A., Li, X., Liu, L. et al. (2014). A hole-conductor-free, fully printable mesoscopic perovskite solar cell with high stability. *Science* 345: 295–298.

51 Wang, Z.L. (2017). New wave power. *Nature* 542: 159–160.

52 Grey, C.P. and Tarascon, J.M. (2017). Sustainability and in-situ monitoring in battery development. *Nat. Mater.* 16: 45–56.

53 Huang, J.Y., Zhong, L., Wang, C.M. et al. (2010). In-situ observation of the electrochemical lithiation of a single SnO_2 nanowire electrode. *Science* 330: 1515–1520.

54 Liu, X.H., Liu, Y., Kushima, A. et al. (2012). In-situ TEM experiments of electrochemical lithiation and delithiation of individual nanostructures. *Adv. Energy Mater.* 2: 722–741.

55 Liu, D.Q., Shadike, Z., Lin, R.Q. et al. (2019). Review of recent development of in-situ/operando characterization techniques for lithium battery research. *Adv. Mater.* 31: 1806620.

56 Huang, J.Q., Blanquer, L.A., Bonefacino, J. et al. (2020). Operando decoding of chemical and thermal events in commercial Na(Li)-ion cells via optical sensors. *Nat. Energy* 5: 674–683.

57 Yan, G.C., Corte, D.A.D., Yin, W. et al. (2018). Assessment of the electrochemical stability of carbonate-based electrolytes in Na-ion batteries. *J. Electrochem. Soc.* 165: A1222–A1230.

58 Vegge, T., Tarascon, J.M., and Edstrom, K. (2021). Toward better and smarter batteries by combining AI with multisensory and self-healing approaches. *Adv. Energy Mater.* 11: 2100362.

5

Fiber Optofluidic Microlasers Toward High-performance Biochemical Sensing

Yiling Liu[1], *Xi Yang*[2], *Yanqiong Wang*[1] *and Yuan Gong*[1]

[1] *University of Electronic Science and Technology of China, School of Information and Communication Engineering, Key Laboratory of Optical Fiber Sensing and Communications (Ministry of Education of China), No. 2006, Xiyuan Ave., High-Tech Zone (West), Chengdu, Sichuan, 611731, People's Republic of China*
[2] *Peking University, School of Physics, State Key Laboratory for Mesoscopic Physics and Frontiers Science Center for Nano-optoelectronics, No. 5, Yiheyuan Road, Haidian District, Beijing, 100871, People's Republic of China*

5.1 Introduction

Optical fiber sensors have attracted considerable interest in biochemical sensing due to their compact size, relatively low cost, and biocompatibility. With the development of novel optical fiber structures, materials, and fabrication technologies, optical fiber biosensors based on grating, interferometers, surface plasmonic resonance (SPR), and fluorescence have been explored to monitor biochemical information of refractive index (RI), concentration, molecular weight, and interaction between molecules. To enhance the light–matter interaction, fiber optofluidic lasers (FOFLs) were proposed by integrating optical fiber, laser, and microfluidics to achieve high sensitivity and high signal-to-noise ratio.

In FOFLs, optical fibers simultaneously serve as optical microcavities, microfluidic channels, and sometimes optical transmission components. As an optical microcavity, optical fiber can support standing wave in Fabry–Pérot (FP) cavities or traveling wave in whispering-gallery mode (WGM) cavities to achieve optical resonance in the axial, angular, or radial directions. In addition to serving as laser resonators, optical fibers can also be employed as microfluidic channels. The liquid gain medium is withdrawn into the optical fiber to incorporate with the microcavity, enabling population inversion and optical amplification of lasing. Many types of materials, including laser dyes, quantum dots (QDs), perovskite materials, and fluorescent proteins can be employed as liquid gain medium. Besides, the optical transmission properties of optical fiber make it an excellent candidate for remote pumping or *in vivo* imaging for clinical diagnosis, such as optical fiber endoscope.

Optical and Electronic Fibers: Emerging Applications and Technological Innovations, First Edition.
Edited by Lei Wei.

Optical fiber provides many unique properties for lasing and sensing applications. First, high Q-factor and small mode volume of optical fiber microcavity enables strong light–matter interaction and narrow laser linewidth to achieve high sensitivity and wavelength multiplexing. Second, thanks to the high aspect ratio of optical fiber, optical microcavities could be continuously distributed along the optical fiber to perform high throughput biodetection through spatial multiplexing. Multiplexed and multifunctional analysis will be enabled by the highly integrated hollow structures along the axial of optical fibers. Third, optical fibers are commercially available and can be mass-produced by a fiber drawing tower with ultralow cost and high reproducibility. It makes FOFL an ideal candidate for disposable biosensors, avoiding cross-contamination and recalibration in clinical diagnosis. Finally, the abundant optical fiber materials like silica, polymer, hydrogel, and spider silk, show excellent biocompatibility and ease of surface modifications, which makes FOFLs a promising biosensing platform. Based on these key features, FOFLs have made great progress in biosensing, especially in the applications of biomolecule and cell analysis.

In this chapter, it starts with a brief introduction of basic theory and sensing mechanism of optical microcavity and FOFL, followed by a detailed discussion of different optical fiber resonators explored in FOFLs. The typical biochemical sensing applications of FOFLs toward sensitive, fast, high throughput, and disposable biodetection are also concluded.

5.2 Theory

5.2.1 Optical Microcavity and Its Sensing Principle

In recent years, optical microcavities have achieved significant development in terms of physical principles and technical breakthroughs. Thanks to their properties of small volume, compact structure, and strong light–matter interaction, optical microcavities played an important role in various research fields, including optical frequency combs, ultranarrow linewidth optical filters, low-threshold lasers, and ultrasensitive sensors. Besides, optical microcavities are also involved in the fundamental physics domains of cavity quantum electrodynamics (QED), quantum information processing, and quantum optics. Toward different sensing applications, the light distribution in optical microcavities can be flexibly tuned by the design of structures and materials.

5.2.1.1 The Principle of Optical Microcavity

Optical microcavities confine light in a small volume by resonant recirculation, leading to a high optical energy density and thus strong light–matter interaction. It can be classified into optical cavities working with standing wave or traveling wave, according to the optical resonance mechanism. The standing wave mainly occurs in Fabry–Pérot (FP) cavities, photonic crystal (PC) microcavities, and distributed Bragg reflection (DBR) cavities. Taking the FP cavity as an example, it usually consists of two reflective mirrors, and the light is bounced back and forth between them. When the round-trip optical path length is equal to an integer number of the wavelength, constructive light interference takes place inside the FP cavity. This is known as the resonant condition, which can be expressed as

$$2nL = q\lambda \tag{5.1}$$

where n is the effective RI of the microcavity mode, L is the cavity length, q is an integer, and λ is the resonant wavelength. The medium inside the optical cavity could be air, liquid, or solid optical material.

The traveling wave in the optical cavity is formed in the microring, like microdisks, microspheres, microdroplets, and microbottles. In these resonators, light propagates along the curved boundary via total internal reflection (TIR), enabling a WGM. The resonant condition is approximated as

$$\pi nd = m\lambda \tag{5.2}$$

where d is the resonator diameter, and m is an integer.

5.2.1.2 The Sensing Mechanism of Optical Microcavities

Optical microcavities confine photons into a structure at the micrometer scale, which greatly enhances the light–matter interaction and leads to ultrahigh sensitivity for sensing. The light in optical microcavities can be coupled in and out by tapered optical fibers (Figure 5.1a), or integrated waveguides for on-chip optical microcavities. Many publications have reported the optical microresonators and their sensing applications, particularly for the detection of individual molecules, ions, viruses, and nanoparticles, as well as gyroscopes, magnetic fields, and temperatures.

There are four types of representative sensing mechanisms, including resonance wavelength shift, mode splitting, broadening, and dissipative induced mode broadening [1]. Detecting the resonance wavelength shift is the most frequently used

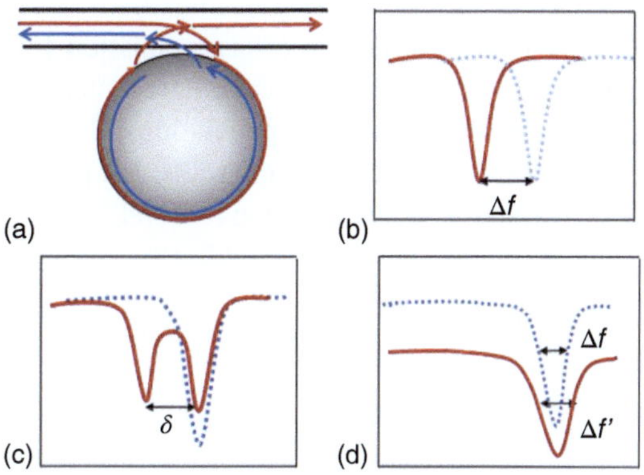

Figure 5.1 The sensing mechanisms of optical microcavities. (a) Single nanoparticle-microcavity-tapered fiber coupling system, (b) resonance wavelength shift, (c) mode splitting, and (d) mode broadening. *Source:* Yiling Liu.

method. As shown in Figure 5.1b, the resonance wavelength varies with changes in the surroundings. It can be utilized not only to quantify the physical parameters around the microcavities, such as temperature, humidity, or electric field, but also to detect biochemical parameters like particle size and concentration, or even single molecules. Mode splitting is observed by adsorbing nanoparticles or biomolecules onto the microcavity surface, which is corresponding to the mode coupling of clockwise and counter-clockwise modes. Hence, the ratio between splitting (δ) and the linewidth is employed as the sensing signal to characterize the information of analytes (Figure 5.1c). The ratio of splitting must be larger than the mode linewidth to make the mode splitting detectable. The particle scattering can also induce an increase of the mode linewidth in the transmission spectrum, which is defined as mode broadening. This mechanism is naturally resistant to ambient thermal noise and laser frequency noise compared to wavelength shift, and has a lower limit of detection than the mode splitting (Figure 5.1d). The above three mechanisms are dependent on particle scattering, which is related to the real part of the polarizability. The dissipative sensing mechanism is primarily based on the imaginary part of the polarizability, which will induce the mode linewidth change via the absorption loss. This method is appropriate for detecting analytes with high absorption coefficients, such as metal nanoparticles and carbon nanotubes [2].

5.2.2 Optofluidic Laser and Its Sensing Principle

5.2.2.1 The Principle of Laser Emission

Laser is light amplification by stimulated emission of radiation with the characteristics of coherent, unidirectional, monochromatic, and high intensity. Stimulation emission occurs in the gain medium. When the gain medium absorbs a photon, its electrons are pumped to an upper energy level and relax to a lower energy state quickly to emit two identical photons. Those photons can further stimulate more atoms in a chain reaction to achieve population inversion, leading to light amplification. Optical microcavities could select the resonant modes and amplify the light with optical feedback.

A four-level model is briefly introduced to the principle of laser emission [3]. Figure 5.2 depicts a typical energy level model of a dye molecule. Rate equations describe the time evolution of the dye molecule energy level populations. It starts with assuming that the laser pumping makes a stimulated pump transition probability of $W_{03} = W_{30} = W_P$ between level 0 and 3. The rate equation for level 3 is

$$\begin{aligned}\frac{dN_3}{dt} &= W_P\left(N_0 - N_3\right) - \left(\gamma_{32} + \gamma_{31} + \gamma_{30}\right)N_3 \\ &= W_P\left(N_0 - N_3\right) - \frac{N_3}{\tau_3}\end{aligned} \tag{5.3}$$

where τ_3 is the lifetime for decay of level 3 to all of the lower levels and denotes $1/\tau_3 = \gamma_3 = \gamma_{32} + \gamma_{31} + \gamma_{30}$. Ni denotes the populations at the ith level at time t ($i = 0$, 1, 2, 3), W_P is the normalized pump intensity, and $\gamma_3 j$ denotes the relaxation

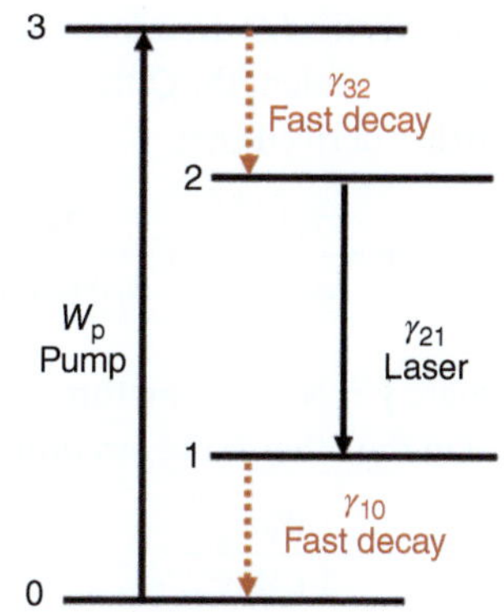

Figure 5.2 Schematic diagram for optofluidic laser with a four-level model. *Source*: Yiling Liu.

probability of populations from level 3 to lower levels (j = 0, 1, 2). The steady-state population of level 3, $dN_3/dt = 0$, is written as

$$N_3 = \frac{W_P\tau_3}{1 + W_P\tau_3}N_0 \tag{5.4}$$

Generally, the direct transitions from ground level 0 to upper level 2 can be negligible. Therefore, the rate equation for the levels 2 and 1 can be expressed as

$$\frac{dN_2}{dt} = \gamma_{32}N_3 - (\gamma_{21} + \gamma_{20})N_2 = \frac{N_3}{\tau_{32}} - \frac{N_2}{\tau_2} \tag{5.5}$$

and

$$\frac{dN_1}{dt} = \gamma_{31}N_3 + \gamma_{21}N_2 - \gamma_{10}N_1 = \frac{N_3}{\tau_{31}} + \frac{N_2}{\tau_{21}} - \frac{N_1}{\tau_{10}} \tag{5.6}$$

where τ_{ab} denotes the relaxation time from level a to level b. Similarly, the steady-state population of level 2 and level 1 is

$$N_2 = \frac{\tau_2}{\tau_{32}}N_3 \tag{5.7}$$

and

$$N_1 = \left(\frac{\tau_{10}}{\tau_{21}} + \frac{\tau_{32}\tau_{10}}{\tau_{31}\tau_2}\right)N_2 \tag{5.8}$$

in which the total populations denote $N = N_0 + N_1 + N_2 + N_3$.

Considering that the level 3 to 2 relaxation rate is very fast, and the level 2 has a long lifetime, τ_2 is significantly greater than τ_{32}, leading to the ignorable populations at level 3 and level 1. Therefore, $N \approx N_0 + N_2$. According to the laser threshold condition,

$$N_2\sigma_e(\lambda) = N_0\sigma_a(\lambda) + L_Q \tag{5.9}$$

where N_2 and N_0 are the concentrations of gain molecule at levels 2 and 0, respectively. σ_e and σ_a are the emission and absorption cross sections at lasing wavelength, respectively. LQ describes the loss of cavity and is expressed as $2\pi n/\lambda_L Q$ with λ_L the lasing wavelength, Q the Q-factor, and n the effective RI of laser mode. It can be further derived that

$$\gamma = \frac{\sigma_a(\lambda)}{\sigma_e(\lambda)}\left(1 + \frac{2\pi n}{\lambda_L N Q \sigma_a(\lambda)}\right) \tag{5.10}$$

where $\gamma = N_2/N$ is the threshold ratio of the excited molecule population for lasing.

On the other hand, according to the rate equation, γ can be approximated as:

$$\gamma = \frac{W_P \tau_{rad}}{1 + W_P \tau_{rad}} \tag{5.11}$$

$$W_P = \frac{I_{th}}{E_0 \Delta t}\sigma_a \tag{5.12}$$

where τ_{rad} is the lifetime of upper level, I_{th} is the laser threshold, Δt is the pulse width of pump laser, $E_0 = hv = hc/\lambda_P$ is the energy of single pump photon with λ_P the wavelength of pump laser, h the Planck constant, and c the speed of light in vacuum. Above the threshold, the laser output can be expressed as

$$I_{laser} = \left(\frac{I_P}{I_{th}} - 1\right) \times \frac{\delta_e I_{sat}}{2} \tag{5.13}$$

where I_{laser}, I_P, and I_{sat} are the intensity of laser output, pump laser, and the saturated intensity, and δ_e is the external coupling factor.

5.2.2.2 The Sensing Mechanism of the Optofluidic Laser

According to the threshold condition Eq. (5.9), the emission cross-section, the concentration of gain molecules in the excited state, the absorption cross-section, and the nature of the cavity are all related to the laser's output characteristics. These parameters in sensing applications are affected by the characteristics of the analyte, including analyte concentration, molecular mass, intermolecular interactions, and molecular conformational changes, causing changes in the laser output [4]. Therefore, changes in laser output parameters, such as laser intensity, spectra, and thresholds, can be monitored to reflect tiny changes in biological processes. The more typical way of sensing is to monitor the optofluidic laser intensity changes or wavelength shifts. The biochemical sensing approach based on the laser intensity has the advantages of a simple detection device and method as well as a lower cost. According to Eqs. (5.11)–(5.13), the laser output intensity is related to the absorption cross-section, emission cross-section, refractive index, gain molecule number, and Q value of the cavity. The laser output intensity will change if the analyte causes a change in any of these parameters. Using the number of gain molecules as an example:

$$I_{laser} \propto \frac{1}{\gamma} \tag{5.14}$$

where the laser output intensity is inversely proportional to γ. In Eq. (5.10), the γ increases as the number of gain molecules decreases, causing the laser output intensity drops as well.

The other method of biochemical sensing is based on laser emission wavelength demodulation. The laser emission wavelength is related to the gain medium's emission spectrum, absorption spectrum, and microcavity features, according to laser theory. For example, in Eq. (5.2), it shows that the smaller the effective refractive index (n), the smaller the resonant wavelength (λ), and therefore the laser emission wavelength is blue-shifted to match the resonance condition.

In addition to changes in laser intensity and wavelength shift, sensing can be achieved by measuring the spatial distribution of light, polarization properties, etc. Additionally, the sensing involving biochemical reactions can be performed by converting the pump laser threshold into threshold concentration or threshold reaction times.

5.3 Optical Fiber Microresonators for Optofluidic Lasing

Optical fibers could support optical resonance for lasing in different ways. In this section, structures, resonance mechanisms, and unique properties of optical fiber microresonators are introduced, including fiber microring resonator, photonic bandgap (PBG) fiber, fiber Fabry–Pérot (FFP) cavity, and random scattering.

5.3.1 Fiber Microring Resonator

Various microstructures, including microspheres, microdroplets, microdisks, microtubes, microbottles, optical fibers, and on-chip microdisks, could serve as microring resonators. Different from other types of microring resonators, optical fiber microring resonators are based on the microring structure embedded in its cross-section, and can be integrated with microfluidic channel, as well as be continuously distributed along its axis. Thanks to the development of advanced fabrication technologies, various types of fiber structures are prepared and used in fiber microring resonators, such as common optical fiber (COF), hollow optical fiber (HOF), microstructured optical fiber (MOF), and micro-/nano-optical fibers.

5.3.1.1 Common Optical Fibers

Common single-mode fiber (SMF) or multimode fiber (MMF) with a diameter of 125 μm can be commercially mass-produced (~50 km in length) with low cost (~US$ 0.01/m), owing to its wide application in optical fiber communications. Its round cross-section serves as a microring resonator with a Q-factor of about 10^7 to support WGMs and offers optical feedback for COF-OFLs. The COF is surrounded by the gain molecules in the case of COF-OFLs and usually interacts with the gain through evanescent waves (Figure 5.3a).

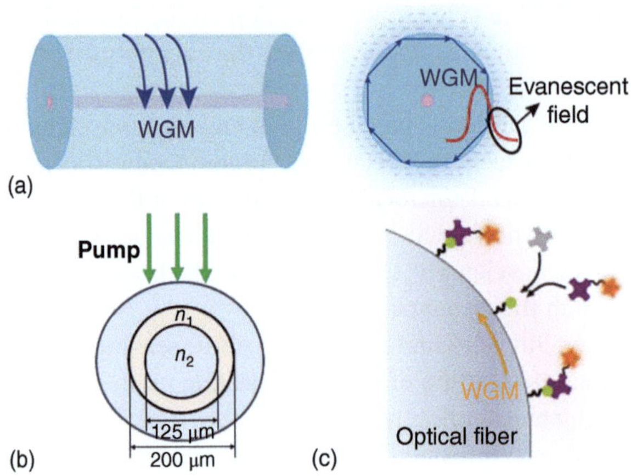

Figure 5.3 FOFLs with COF. (a) Resonance mechanism in COFs [5]. Schematic diagrams of FOFLs based on (b) fiber-in-capillary structure and (c) single layer of gain molecules. *Source*: (a) Reproduced with permission from Yang et al. [5] / John Wiley & Sons. (b) Yiling Liu. (c) Yuan Gong.

The fiber-in-capillary structures were frequently used for lasing with the surrounding bulk solution of gain medium (Figure 5.3b), in which the COF and glass capillary are used as optical microcavity and microfluidic channel, respectively [6]. The laser emission can be directionally coupled out through a fiber prism in contact with the outer capillary wall. In addition, the capillary is also important to reduce the lateral volume of the bulk gain and thus suppresses the fluorescent background resulting from the gain medium beyond the evanescent waves.

Instead of utilizing a bulk solution of gain medium, a single layer of gain molecules can be conjugated on the COF surface (Figure 5.3c) [7], which holds two impressive characteristics. First, the numbers of gain molecules engaged in lasing are much fewer than that in the case of using bulk solutions. Second, all of the gain molecules could interact with the evanescent wave and contribute to the lasing action. These provide COF-OFLs with distinctive features of weak fluorescence background, low-volume sample and reagent consumption, and high sensitivity for bioassay.

5.3.1.2 Hollow Optical Fibers

As shown in Figure 5.4, HOFs can serve as a liquid-core optical ring resonator (LCORR) or optofluidic ring resonator (OFRR) with an inner diameter of a and a wall thickness of b. In contrast to the COF, HOF is more compact to integrate the optical microcavity with microfluidic channel, which allows the gain medium to flow through the hollow core and interact with the high Q-factor WGMs (~10^6), offering optical feedback for low-threshold lasing. The light distribution in the cross-section of HOF is decided by its wall thickness and the refractive index (RI) of liquid gain.

In a thin-walled HOF (b is a few micrometers; the specific thickness is related to the RI difference $\Delta n = |n_2 - n_1|$.), lasing is observed when the RI of dye solution is lower ($n_1 < n_2$, $n_2 > n_0$) or higher ($n_1 > n_2 > n_0$) than silica. The optical resonance in the

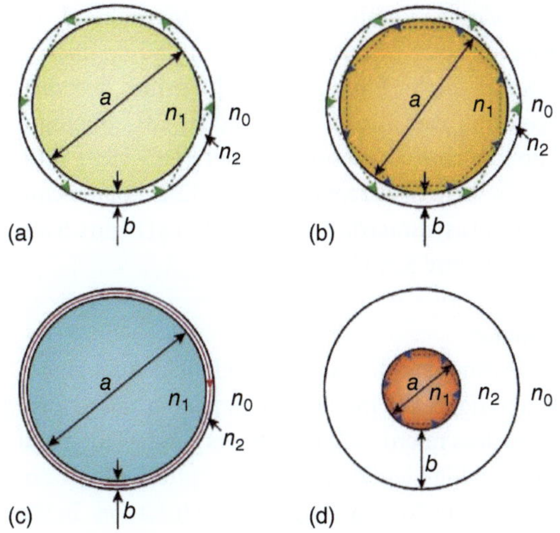

Figure 5.4 Schematics of FOFLs based on HOF with a thin wall of several micrometers and with a RI distribution of (a) $n_1 < n_2, n_2 > n_0$ and (b) $n_1 > n_2 > n_0$, (c) with an ultrathin thickness around 1–2 μm; and (d) HOF-OFL with a thick wall and RI distribution of $n_1 > n_2 > n_0$. n_0, n_1, and n_2 are the RI of air, liquid gain, and silica, respectively [5]. *Source:* Reproduced with permission Yang et al. [5] / John Wiley & Sons.

former case mainly occurs on the silica–air interface (Figure 5.4a) and, in the latter case, may happen on both the silica–air and silica–liquid interfaces (Figure 5.4b) [5]. The interaction of the gain medium and the evanescent wave can be enhanced by reducing the HOF thickness down to the wavelength scale (Figure 5.4c). When the wall thickness is around 1–2 μm, a high *Q*-factor of $>10^7$ and a low threshold of $<1\,\mu J/mm^2$ for FOFL can be achieved.

Lasing with thick-walled HOF often requires a higher RI of gain solution than that of silica ($n_1 > n_2 > n_0$). The TIR occurs on the liquid–silica interface, and the liquid microcavity supports WGMs for lasing (Figure 5.4d). High *Q*-factor lasing modes could exist only around the inner silica–liquid boundary, and only a negligible portion of the lasing could leak out to the outer surface. Thanks to the strong overlap between the resonant mode and the gain medium, a threshold lower than $0.5\,W/cm^2$ was achievable by using R6G in quinoline ($n_1 \approx 1.626$, $n_2 \approx 1.458$) as the gain medium.

5.3.1.3 Microstructured Optical Fiber

Microstructured optical fiber (MOF), which periodically arranges plenty of air voids running along the length of the fiber with either a solid core (suspended core MOF) or a hollow core (air-core MOF), can provide efficient optical feedback through WGMs or PBGs. MOFs often integrate a fiber microring resonator with a built-in microfluidic channel to address stability and overcome fragility issues. The optical resonators embedded in MOF allow for the high-efficiency light–matter interaction, easy surface functionalization, immune to external contamination, and ultralow consumption of reagents/analytes for microanalysis. Recently, a low-threshold microcavity dye laser based on a hollow core MOF (HC-MOF) was demonstrated.

Furthermore, a single longitudinal-mode laser emission is also achieved, and the factors influencing single-mode emission were discussed, including pump energy, pump incident angle, and laser detection angle (Figure 5.5a) [8]. Besides, a stable, single longitudinal mode lasing operation can also be designed by introducing a microfluid control device to the hollow-core antiresonant fiber (HC-ARF) [10]. FOFLs based on other types of MOFs were also reported, including hollow-core negative-curvature MOF, symmetrical and asymmetric twin-hole MOFs (Figure 5.5b) [11], solid-core PCF, and Kagome PCF (Figure 5.5c) [9].

5.3.1.4 Optical Microfiber Ring Resonator

Optical microfiber, fabricated by the heating-and-drawing method, has a diameter of 0.5–3 μm and can be bent, wrapped, and tightened into a knot, loop, or a coil to form a ring resonator supporting WGM. Low-loss micro- and nanofibers can be fabricated due to the surface tension of fused silica at high temperatures, and optical microresonators with high Q-factor can be formed in the form of loop or knot structures. In the thin optical microfibers, only low-order transverse modes with strong evanescent waves are excited. The strong light–matter interaction between the gain molecules and the evanescent wave greatly reduces the lasing threshold. Single-mode lasing can also be achieved by reducing the diameter of the μFRRs. In 2006, a rare-earth-doped microfiber was tightened into a knot to support optical resonance and laser

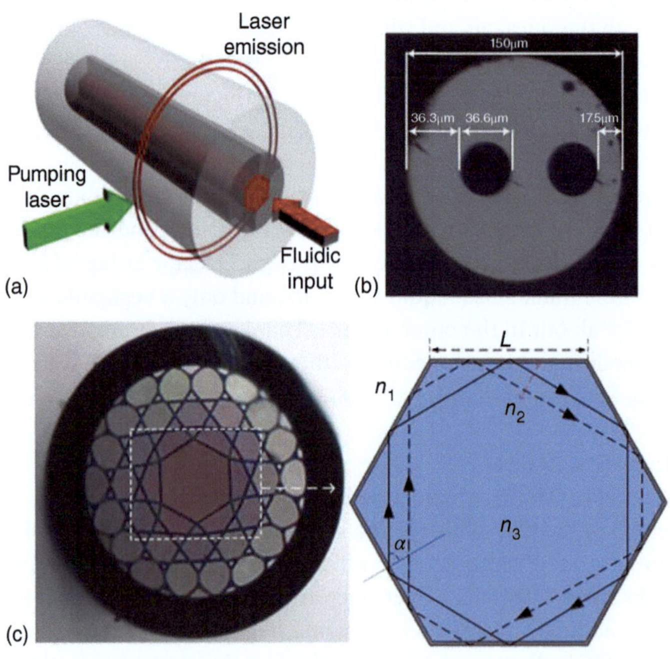

Figure 5.5 HC-MOFs for FOFLs. The structures of microresonators based on (a) simplified HC-MOF [8], (b) twin hole and (c) Kagome PCF MOFs [9]. *Source*: (a) Reproduced with permission from Yu et al. [8] / IEEE. (b) Yuan Gong (Chapter Contributor). (c) Reproduced with permission from Yan et al. [9] / IEEE.

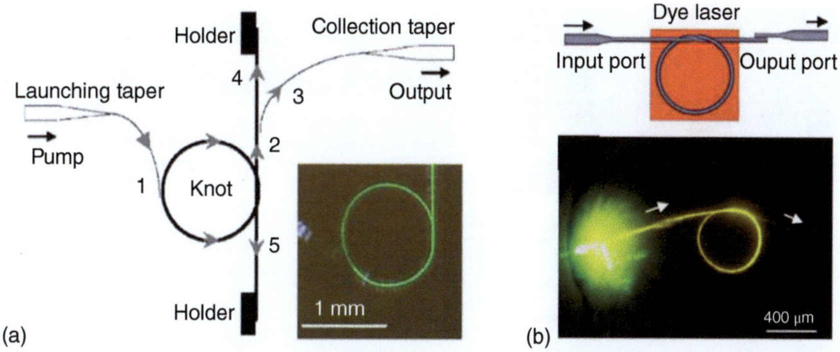

Figure 5.6 FOFLs with optical microfiber ring resonator. (a) Schematic diagram of the structure of a microfiber knot laser [12]. *Source*: Reproduced with permission from Jiang et al. [12] / AIP Publishing. (b) Schematic diagram and image of the μFRR with in-fiber coupling in and out [13]. *Source*: Reproduced with permission from Jiang et. al. [13] / AIP Publishing.

emission (Figure 5.6a) [12]. Further, by taking the advantage of waveguide effect, all-fiber coupling of the pump and the laser emission can be achieved with the bi-tapers of the microfiber (Figure 5.6b) [13].

5.3.1.5 Other Resonant Microstructures

To further enhance the *Q*-factor, optical fiber microcavities with 3D optical confinement, including microbottles, microdisks, and microspheres, are designed and fabricated by heating the optical fiber with flame, electric arc, or CO_2 laser. Because of the surface tension, the melted SiO_2 can be self-assembled into 3D microresonators. The remaining part of the optical fiber can serve as a stem to conveniently manipulate the microresonator. All the 3D optical fiber microresonators support WGMs for lasing. Thanks to the strong light confinement, an ultrahigh *Q*-factor of 10^8 can be achieved, enabling a low-threshold lasing. Single-mode WGM lasing was observed in a polymeric bottle microresonator by spatial pump modulation (Figure 5.7a) [14]. Using high-efficiency nanoheater [16], a tunable resonance polymer-coated silica microsphere was demonstrated (Figure 5.7b). In addition, aggregation-induced emission (AIE) materials were used to create a microbottle laser with high sensitivity (0.34 pm/με), high resolution, and real-time response (Figure 5.7c) [15].

5.3.2 Photonic Bandgap Fiber Microcavity

The structure of photonic bandgap fiber (PBG fiber, also named Bragg fiber) is similar to that of the antiresonant reflecting optical waveguide (ARROW) or the film stack of interferometric coatings (Figure 5.8a), in which the high- and low-index materials are alternately stacked around a solid or hollow core to achieve a PBG [17]. The PBG fibers can be fabricated by the same fiber preform to obtain identical multiple reflections, which is preferable for single-use sensors. Different from the WGM resonators, the multireflections in the cross-section of the PBG fiber enable almost all the gain molecules to contribute to lasing.

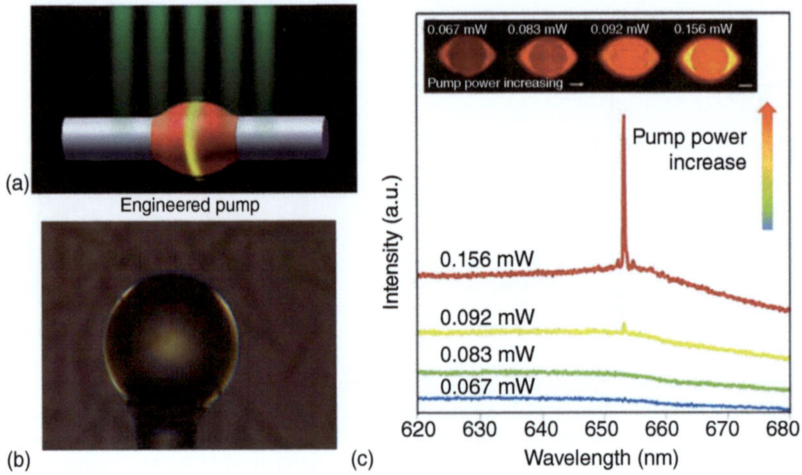

Figure 5.7 Illustration of 3D-resonant microstructures. (a) A microbottle cavity [14] and (b) an optical fiber microsphere cavity. (c) AIE-based microbottle single-mode laser [15]. *Source*: (a) [14]. Reproduced with permission. © 2017, Springer Nature. (b) Yiling Liu (Chapter Contributor). (c) Reproduced with permission from Guo et al. [15] / The Royal Society of Chemistry.

The hollow-core PBG fibers have many unique properties for optofluidic lasing. The reflection band in PBG fiber can be broader than 100 nm in theory, which makes it flexible to employ different kinds of gain mediums for a multicolor laser. The diameter of the hollow core, as large as 250 μm, can be designed with high flexibility. It can serve as a microfluidic channel to conveniently deliver the liquid gain or analytes via capillary action. When filled with aqueous or colloidal gain materials, such as organic dye in the water/copolymer matrix or the quantum dots (QDs) in hexane solvent, azimuthally polarized radial-lasing emission and light amplification were observed. For example, zero-angular-momentum laser with purely radial emission was demonstrated by filling PBG with laser dye. In the electrode integrated PBG fiber, the laser intensity and direction can be modulated by contacting two opposite-facing liquid–crystal microchannels [19]. Excellent spatial and angular selectivity makes this laser promising for medical therapeutic applications. In addition, the hollow core of PBG fiber can also minimize the dependence of light transmission characters on the materials of the fiber, which is helpful to pump the gain medium axially. Light can be transmitted along the fiber even with steep angles to the fiber axis, which is different from telecom optical fibers to satisfy a critical angle. Due to the excellent optical transmission of PBG fiber, lasing over a continuous length of fiber was achieved (Figure 5.8b) [18].

5.3.3 Fiber Fabry–Pérot Cavity

FP cavities are the most commonly used resonators due to their simple structure and ease of fabrication. The key technology for constructing FP cavities is the alignment of two reflective surfaces. Because optical fibers can be sliced to create a flat

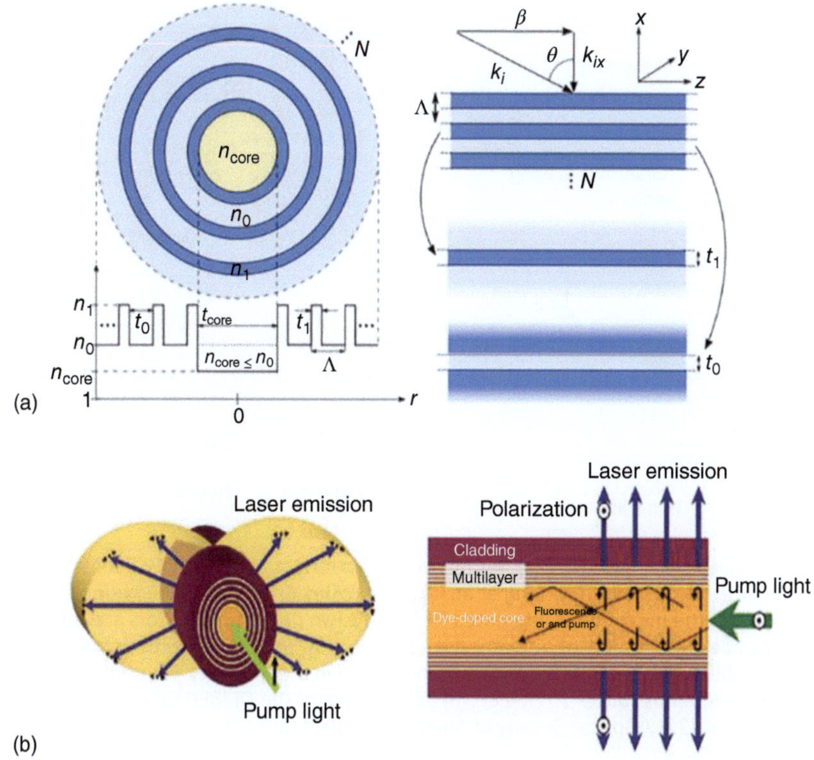

Figure 5.8 FOFLs with PBG cavity. (a) Resonance mechanism of PBG fibers and planar ARROWs. *n*, refractive index; *t*, thickness; subscript 0 for the hollow core; subscripts 1 and 2 for high- and low-RI layers, respectively; *N* denotes how many periods of layer pairs. *Source:* Rowland et al. [17] / Optical Society of America / CC BY 4.0. (b) Schematic of the PBG-FOFLs. Source: Shapira et al. [18] / Optical Society of America / CC BY 4.0.

surface and coated with reflective films in a batch, photons are bounced back and forth in the FP cavity, forming a standing wave at the resonance frequency (Figure 5.9a). High *Q*-factor is easily available from highly reflective coatings, as the state-of-the-art thin film deposition technology enables a reflectivity as high as 99.999%.

The FFP cavity is convenient to integrate with microfluidic channel. The continuous flow of liquid gain significantly decreases the influence of photobleaching, and enhances the stability and sustainability of the laser output. It has been demonstrated that a digital microfluidic dye laser was fabricated by the incorporation of the FFP cavity and microfluidic channel. Controlling the droplet flow allows for quick altering of the laser wavelength and intensity (Figure 5.9b) [20]. Recently, MOFs were sandwiched between two mirrors to achieve special properties. By utilizing the filtering effect of twin-hole MOF, a single-mode laser emission in FFP (Figure 5.9c) was demonstrated [21]. Besides, optical microbubbles can be placed in the FP cavity to enhance the stability and *Q*-factor of microcavity (Figure 5.9d) [22].

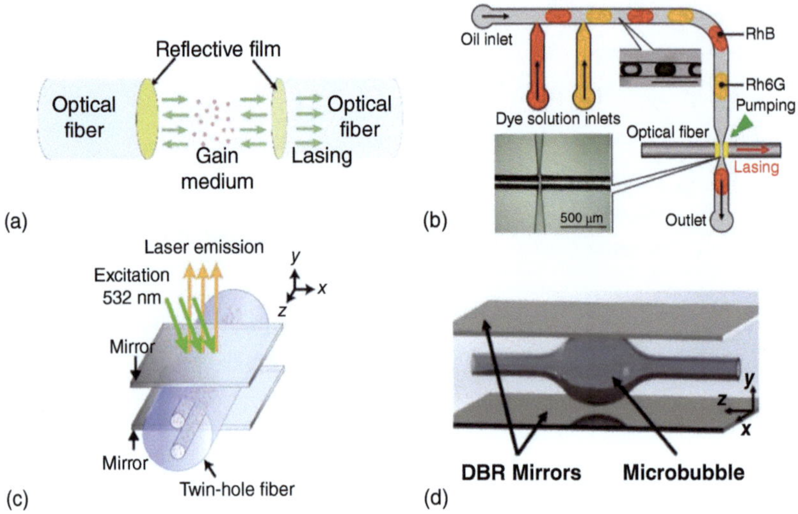

Figure 5.9 FOFLs with FFP cavity. (a) Resonance mechanism of an FFP cavity. Source: Reproduced with permission from Yang et al. [5] / John Wiley & Sons. (b) Schematic representation of the microfluidic integrated FFP cavity. Two droplet generators are set in series and can deliver droplets of different dyes. *Source:* [20]. Reproduced with permission. © 2011, AIP Publishing LLC. (c) FOFL with twin-hole MOF in the FP cavity. *Source:* [21]. Reproduced with permission. © 2017, IEEE. (d) 3D structure of the OMBFP cavity. *Source:* Chen et al. [22] / Optical Society of America / CC BY 4.0.

The FFP cavity has two additional advantages for sensing applications. First, the large overlap between light and liquid enables the strong light–matter interaction and minimizes the requirement of Q-factor for lasing. It is beneficial to the low-threshold lasing and sensitive biochemical sensing. The Fresnel reflection from optical fiber end faces (less than 4% reflection) was used to provide optical feedback for optofluidic lasing [23]. Second, single-mode laser emission is available due to the filtering effect of FFP. The number of longitudinal modes can be considerably reduced by employing a SMF instead of a MMF to transmit the pump laser. Shortening the FP cavity length or exploiting the in-cavity filtering effect have been performed to achieve single-mode laser emission.

5.3.4 Random Scattering

Random lasers (RLs), with optical feedback from random scattering rather than an optical resonator, have the advantages of easy fabrication and low cost, which have been investigated in the past decade for applications in imaging and biomedicine (e.g. cell analysis and tumor tissue detection). Because the scattering provides the optical feedback, RLs usually have comparatively low Q-factor but don't rely on any external cavities. In general, RLs can be divided into two different feedback mechanisms, i.e. incoherent feedback (red, Figure 5.10a) and coherent feedback (green,

(a)
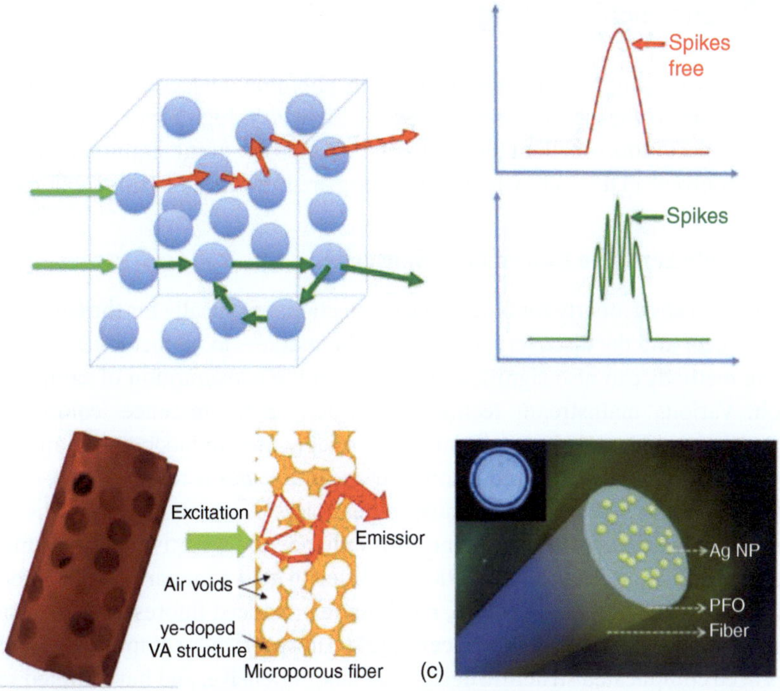

(b) (c)

Figure 5.10 FOFLs with random scattering. (a) Resonance mechanisms of random FOFLs with incoherent feedback (red) and coherent feedback (green). Corresponding laser spectra are shown. *Source:* Reproduced with permission from Luan et al. [24] / Elsevier. (b) Schematic diagram of the microporous fibers (inverse structure) and light amplification via multiple scattering in a microporous fiber. *Source:* Reproduced with permission from Ta et al. [25] / The Royal Society of Chemistry. (c) Schematic diagram of plasmonic random laser on fiber facet. *Source:* Shi et al. [26] / Optical Society of America / CC BY 4.0.

Figure 5.10a) [24]. The main feature of a coherent random laser is the spikes in the laser spectra, which shows a fundamental difference from that of the incoherent RLs.

For optical fiber-based random laser, scatterer can combine with optical fiber in several ways. The simplest one is by filling the optical fiber with colloid nanoparticles doped dye-like liquid crystal, semiconductors, and dielectric particles. To further control the lasing property, the disordered dielectric film can be coated on the inner surface of a HOF to achieve random lasing, in which the scatterers are separated from the gain medium. (Figure 5.10b) [25]. On the other hand, scatterers can be deposited on the fiber facet. Based on plasmonic scattering, Shi et al. achieved a low-cost RL using self-assembled Ag nanoparticles and polymer film coated on the fiber facet, demonstrating the detection of refractive index and IgG (Figure 5.10c) [26]. Additionally, a random lasing network made of dye-doped polymeric nanofibers is capable of amplifying light along its branches and trapping it at its nodes due to multiple scattering. It has also shown that amyloid proteins can fold into higher order structures to support random laser emission.

5.4 Biochemical Sensing Based on FOFLs

FOFLs are promising for wide applications in photonic devices and biochemical sensors, thanks to their high *Q*-factor, narrow linewidth, high reproducibility, low cost, high aspect ratio, excellent biocompatibility, and ease of integration. Four typical applications of FOFL in biochemical sensing are introduced in this section.

5.4.1 Highly Sensitive Biochemical Sensors

Sensitivity is a key property for biochemical detections, especially for the early diagnosis of cancer and the discovery of new trace biomarkers in biomedicine research. Sensitive methods can also significantly cut down the consumption of samples by dilution. Various mainstream technologies, such as fluorescence, colorimetry, chemiluminescence, and electrochemical, are widely used in biochemical analysis. The sensitivity of fluorescent and colorimetric technologies is relatively low due to the low signal intensity and strong background noises. A variety of approaches are implemented to amplify the signal and thus enhance the sensitivity. For example, local surface plasmon resonance (LSPR) supported in gold nanoparticles can modulate the fluorescent emission process, resulting in enhanced fluorescence emission to achieve plasmon-enhanced fluorescence [27]. However, those amplification techniques need complicated structure design, are labor intensive, and have limited sensitivity improvement.

Optofluidic laser employs lasing emission as the sensing signal, which is fundamentally different from fluorescence-based detection. It has strong light–matter interaction that can be explored for broad applications in sensitive biochemical sensing. Enzyme-linked immunosorbent assay (ELISA) is the gold standard for immunoassay, which is incorporated in the FP cavity to form optofluidic laser ELISA for detecting interleukin-6 with an exceptional limit of detection of 1 pg/l and a dynamic range of six orders of magnitude [28]. The detection of ions, explosives, uric acid, and the DNA high-resolution melting analysis has shown that optofluidic laser is more sensitive than the traditional methods. Besides biomolecules, cellular detection can also be implemented by cell-in-cavity or microcavity-in-cell structures.

Optical fiber can serve as an optical cavity with high *Q*-factor and also as microfluidic channels, enabling it as an ideal candidate to incorporate with optofluidic laser for sensitive biodetection. The typical applications of FOFLs for sensitive biochemical analysis are presented in Figure 5.11. In 2012, A bio-switchable optofluidic laser based on DNA Holliday junctions was implemented to detect Mg^{2+}, showing a sensitivity improvement of over one order of magnitude than the conventional fluorescence (Figure 5.11a) [29]. Different from the bulk solution, the gain molecules could be cross-linked to the fiber surface to maximize the light–matter interaction. Almost 100% of gain molecules participate in the lasing process, leading to the weak fluorescence background and ultrahigh sensitivity to the number of the gain molecules. Digital DNA detection was accomplished by Wonsuk Lee et al., binding a monolayer of DNA molecules on the surface of COF (Figure 5.11b) [30]. To explore the

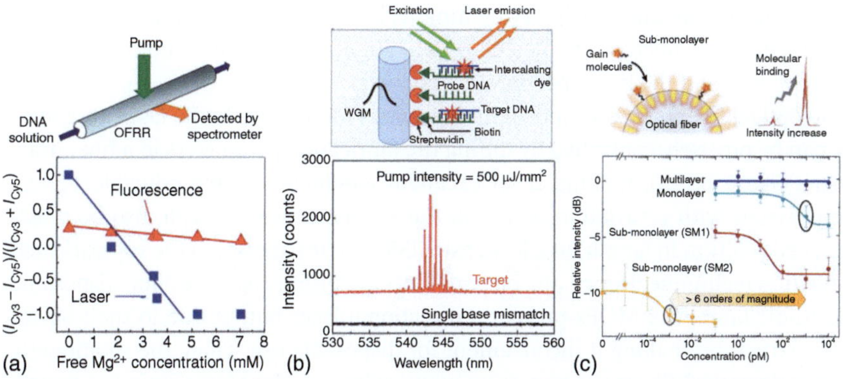

Figure 5.11 Sensitive biodetection based on FOFLs. (a) Schematic of the bio-switchable laser based on OFRR, and the comparison of laser and fluorescence for the detection of free Mg^{2+} ions. *Source:* Reproduced with permission from Zhang et al. [29] / The Royal Society of Chemistry. (b) Schematic of the optofluidic laser with a monolayer of DNA molecules. *Source:* Lee et al. [30] / Royal Society of Chemistry / CC BY 3.0. (c) Illustration of submonolayer laser. *Source:* Yuan Gong.

sensitivity limit, gain molecules should be further reduced to approach threshold. Recently, a submonolayer laser was developed by immobilizing the Cy3 molecules on the outer surface of telecom optical fiber (Figure 5.11c) [7]. Based on this scheme, the ultimate sensitivity shows six orders of magnitude enhancement compared to the monolayer biolaser. It was utilized to detect a Parkinson's disease biomarker, alpha-synuclein, with a lower limit of detection of 0.32 pM in serum.

5.4.2 Disposable Biochemical Sensors

The development of affordable and portable devices for biomedical test has attracted significant attention in recent years, particularly for disposable sensors. Disposable sensors can avoid cross-contamination and eliminate the unstable wash process. It requires the sensors to be massively produced with low cost and high reproducibility. The most successful applications of disposable sensors are the commercially available paper microfluidics, notably in reproductive health (pregnancy tests, ovulation tests), diabetes care (glucose sensing strips), and pH monitoring (pH paper). However, it is mainly used for qualitative evaluation, which can provide limited information on analytes. Consequently, quantitative disposable sensors are in demand.

Optofluidic laser also shows great potential for disposable applications. The major challenge for optofluidic laser achieving disposability is to massively produce optical microcavities with low cost and high reproducibility. Reproducible microcavities or microlasers are strongly dependent on the development of materials and fabrication technologies. Through soft lithography, a reproducible optofluidic laser ELISA was developed, in which four SU-8 reaction wells were microfabricated directly on the dielectric mirrors to form a planoconcave FP cavity [31]. However, the cost was relatively high, making it difficult to achieve disposability. More recently, Zhang et al. employed an anisotropic dry etching technique to mass produce the high-Q microcavities and microlasers on a chip [32]. High repeatability of both the microdisk geometry and the laser spectra was obtained.

The optical fibers serve as optical microcavities for an optofluidic laser. Its geometry can be precisely controlled by tuning the fabrication parameters in a fiber draw tower (Figure 5.12a), making it an excellent candidate for reproducible optical microcavities with a variation of 1% on the kilometer scale. It can also be mass-produced (~50 km in length) at a low cost (US$ 0.01/m), thanks to the sophisticated fabrication technique. Hence, the optical fibers are promising to develop disposable optofluidic laser. The MOFs present a directional laser output due to their asymmetric structure, making it inconvenient for disposable use. Then, the rotationally symmetric optical fibers can ensure an angular independent laser emission to provide high reproducibility. Recently, the thin-walled hollow optical fibers are mass-produced using fiber drawing tower and further demonstrate a disposable optofluidic laser immunosensor for protein detection (Figure 5.12b) [33]. A low coefficient of variation of 3.3% confirms the high reproducibility of the disposable FOFL sensors. The disposable manner enables a large number of tests, allowing the statistical analysis of laser emission and thus protein concentrations. Similar properties can also be demonstrated using telecom optical fibers. Except for silica optical fiber, polymer optical fiber can also be mass-produced to explore the potential of disposability. By physical drawing, the polymer microfibers can be readily mass-produced by

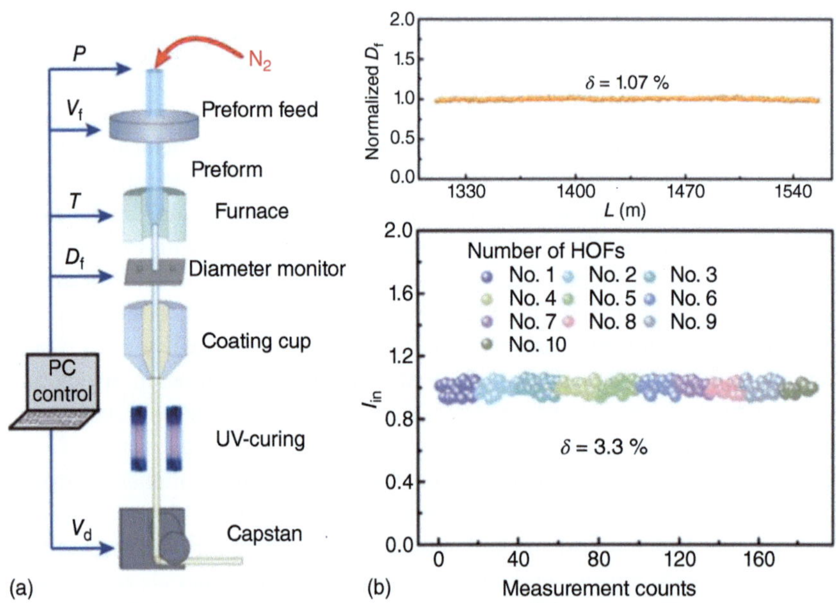

Figure 5.12 Disposable biosensors based on FOFLs. (a) Schematic diagram of a fiber draw tower to precisely control the geometry of optical fiber. (b) The reproducibility of the HOF geometry and the laser emission. *Source:* [31]. Reproduced with permission. © 2020, The Royal Society of Chemistry.

drawing the composite solution on the substrate using a metal tip [34]. Electrospinning is another scalable strategy to massively fabricate polymer microfibers. Thanks to the excellent fabric compatibility, the microfibers can be woven into the clothes to fabricate a wearable sensor [35].

5.4.3 Fast, High-throughput Biochemical Sensors

Analytical efficiency is a key factor to assess the potential of biosensor. Test efficiency can be improved by reducing assay time and increasing test throughput, which is valuable for biomedical applications of DNA sequencing, biodiagnostics, and drug screening. Thanks to the wavelength multiplexing and fast response of photons, optical sensors show great potential for highly efficient bioanalysis.

Benefiting from the small sample volumes, ease of integration, and rapid turnaround times of microfluidics, optofluidic lasers are one of the most striking technologies for highly efficient bioanalysis. Compared to fluorescence spectra with a linewidth of about 30–50 nm, laser spectrum exhibits a narrow linewidth (less than a few nanometers), leading to a higher capacity for wavelength multiplexing. Single-mode laser emission with subnanometer linewidths allows large-scale spectral multiplexing [36]. In terms of detection speed, miniaturized sensors have a high surface-to-volume ratio, making them suitable for rapid binding of molecules and thus for fast bioanalysis [37].

The test efficiency of the optofluidic laser can be further enhanced by integrating with optical fiber, which provides spatial multiplexing in addition to the wavelength multiplexing due to the small cross-section and large aspect ratio of optical fiber. For example, a lasing display project was performed by carefully controlling the size, location, and composition of dye-doped segments along a PBG fiber (Figure 5.13a) [18]. In addition, a distributed FOFL based on the optical microcavities quasi-continuously distributed along the fiber axis was developed. Taking advantage of the distributed optical microcavities, sample channels are arrayed along the fiber on the chip scale to achieve colorimetric detection. A 2D chip can be obtained through wavelength and spatial multiplexing to enhance the multi-functionality, enabling up to 2500 sensing elements. The high reproducibility of optical fiber microcavities makes it possible to construct FOFL arrays. On the other hand, due to the high surface-to-volume ratio, the MOF possesses a short diffusion distance of analytes and thus a short assay time, enabling rapid detection to enhance the test efficiency. Figure 5.13b shows an example of a fast assay based on FOFL, in which the molecules are immobilized on the inner surface of the HOF. And a sequential bioconjugation method is proposed to overcome the difficulty of washing residual reagents out of the miniature structures [38].

5.4.4 Cell and Organism Analysis

Analysis of cells and tissues enables a deep understanding of their structure and function, as well as the responses to external stimulations, including strain, temperature, or drugs, which provide in turn the information for health monitoring,

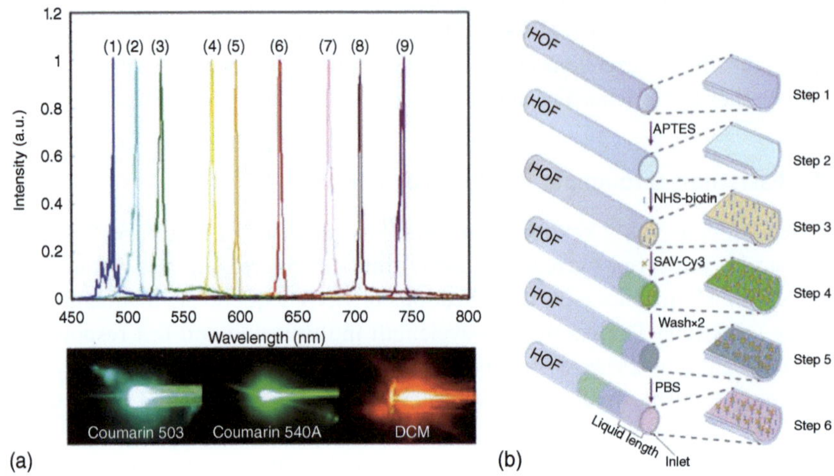

Figure 5.13 Fast and high-throughput bioanalysis based on FOFLs. (a) Laser emission spectra from fibers doped with nine different dyes. The lasers producing emission spectra 1–3 are constructed using the same hollow-core PBG fibers having a fundamental reflection band gap at 500 nm and were pumped at 355 nm. *Source:* Shapira et al. [18] / Optical Society of America / CC BY 4.0. (b) Conceptual illustration of sequential bio-conjugated FOFL for fast and wash-out-free bioassay. *Source:* Reproduced with permission from Yang et al. [38] / The Royal Society of Chemistry.

disease diagnosis, surgery guidance, and therapy. Optofluidic biolasers have been demonstrated for labeling, tracking, and imaging of cells and tissues, owing to their excellent biocompatibility, spectral multiplexing capability, high contrast ratio, and strong background suppression. Various biocompatible materials have been employed in biolasers, such as gelatin, amyloids, polysaccharides, starch, lipids, DNA, silk, liquid crystal, wood, and hydrogels.

A few examples of FOFLs for the analysis of cells and organisms are presented. FOFL technology is suitable for *in vivo* detection with little invasion due to the optical transmission characteristics and high aspect ratio of optical fibers. M. Humar and S.H. Yun demonstrated a cellular laser in subcutaneous adipose tissue where collagenase and lipophilic Nile red dye were injected [39]. To explore the deep tissue, optical fiber was utilized to guide the pulsed pump laser and collect the laser output. In contrast, when using free-space pumping and light collection, the tissue significantly reduces the penetration depth of the light due to severe light scattering and absorption. On the other hand, micro-/nanoscale biolasers can be implanted or integrated into living cells due to the miniaturization of the structure and the good biocompatibility of the material. Micro-/nanoscale structures, such as nanowires, were implanted into cells as nanolaser probes for cell biology exploration and biomedical research (Figure 5.14b) [41]. By using polymeric optical fibers as flexible building blocks, FOFLs have been developed that can be integrated with soft materials such as PMMA, PS, hydrogels, DNA, nanowires, and liquid crystals. In 2020, a

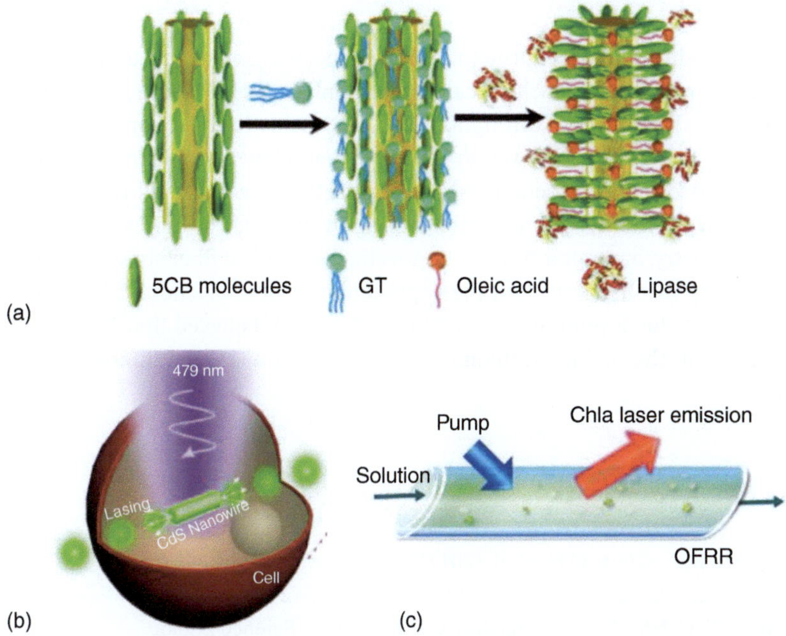

Figure 5.14 Cell and organism analysis based on FOFLs. (a) Lasing from liquid crystal-modified polymeric optical fiber for the detection of lipase. *Source:* Reproduced with permission from Duan et al. [40] / The Royal Society of Chemistry. (b) Illustration of a CdS nanowire laser in the cell. *Source:* Reproduced with permission from Wu et al. [41] / The Royal Society of Chemistry. (c) Schematic diagram of the chlorophyll laser based on the OFRR. *Source:* Reproduced with permission from Chen et al. [42] / The Royal Society of Chemistry.

PMMA fiber decorated with liquid crystals was implemented to monitor the concentration of lipase that showed a detection limit of 0.01 μg/ml with a detection time of 200 seconds (Figure 5.14a) [40]. It shows great potential for *in vivo* monitoring of lipase in the pancreas and adipose tissue. In addition, great efforts have been made to develop biomaterial-based FOFLs. Chen et al. presented the first chlorophyll laser by using chlorophyll, an essential material for the photosynthesis process in plants, as a laser dye, which is an effective way to monitor activity in plants (Figure 5.14c) [42]. The FDA-approved dye indocyanine green (ICG) was used to show a blood laser and investigate the primary serological components of blood [43]. Blood lasers provide an excellent strategy to achieve blood analysis for disease diagnosis. Another interesting phenomenon is that biological tissues can provide optical feedback to the laser through random scattering, and such random biolasers have been applied in studying nanoscale structural changes and morphological changes of tissues [44]. As a novel sensing platform, FOFLs have unique properties that are promising for monitoring the function of organs, including artificial organs or organs-on-a-chip, and for aiding pathogenic studies of diseases, including neurodegenerative diseases and cardiovascular diseases.

5.5 Conclusion

This chapter introduces the laser principle, sensing mechanism, typical optical fiber microresonators, and unique biochemical sensing applications of FOFLs. The FOFL offers high sensitivity, fast, high throughput, and disposable biodetection, thanks to its high *Q*-factor, high repeatability, high integration, and low cost, making it promising for high-performance biochemical sensing technology. FOFL is one of the multidisciplinary frontiers that has a deep rooting in the fields of optics, materials science, biochemistry, and microfluidics. Innovations in optical fiber structures and materials, sensing mechanisms, and the exploitation of new biochemical kits are helpful to promote the development of FOFLs. It has been believed that FOFL has great potential in the fields of biomedicine, environmental monitoring, and food safety.

References

1 Zhi, Y.Y., Yu, X.C., Gong, Q.H. et al. (2017). *Adv. Mater.* 29: 1604920.
2 Shen, B.Q., Yu, X.C., Zhi, Y. et al. (2016). *Phys. Rev. A* 5 (2): 024011.
3 Siegman, A.E. (1986). *Lasers*. Sausalito, CA: University Science Books.
4 Fan, X. and Yun, S.H. (2014). *Nat. Methods* 11: 141.
5 Yang, X., Gong, C., Zhang, C. et al. (2021). *Laser Photonics Rev.* 16: 2100171.
6 Moon, H.J., Chough, Y.T., and An, K. (2000). *Phys. Rev. Lett.* 85: 3161.
7 Gong, C., Yang, X., Tang, S.J. et al. (2022). *Optics* https://doi.org/10.48550/arXiv.2207.02409.
8 Yu, J., Liu, Y., Luo, M. et al. (2017). *IEEE Photonics J.* 9: 7105510.
9 Yan, D., Liu, B., Zhang, H. et al. (2018). *IEEE Photon. Technol. Lett.* 30: 1202.
10 Shi, H., Yu, J., Guo, H. et al. (2020). *Laser Phys. Lett.* 17: 025101.
11 Gong, C., Gong, Y., Chen, Q. et al. (2017). *Lab Chip* 17: 3431.
12 Jiang, X., Yang, Q., Vienne, G. et al. (2006). *Appl. Phys. Lett.* 89: 143513.
13 Jiang, X., Song, Q., Xu, L. et al. (2007). *Appl. Phys. Lett.* 90: 233501.
14 Gu, F., Xie, F., Lin, X. et al. (2017). *Light Sci. Appl.* 6: e17061.
15 Guo, X., Zhen, S., Ouyang, T. et al. (2021). *J. Mater. Chem. C* 9: 4888.
16 Li, H., Hao, X., Li, Y. et al. (2021). *J. Mater. Sci.* 56: 570.
17 Rowland, K.J., Afshar, S.V., and Monro, T.M. (2008). *Opt. Express* 16: 17935.
18 Shapira, O., Kuriki, K., Orf, N.D. et al. (2006). *Opt. Express* 14: 3929.
19 Stolyarov, A.M., Wei, L., Shapira, O. et al. (2012). *Nat. Photonics* 6: 229.
20 Meng, X., Ma, J., Xie, K. et al. (2021). *Opt. Mater.* 115: 111027.
21 Gong, C., Gong, Y., Zhang, W.L. et al. (2018). *IEEE J. Sel. Top. Quantum Electron.* 24: 0900206.
22 Chen, X., Zhao, X., Guo, Z. et al. (2020). *Opt. Express* 28: 15161.
23 Gerosa, R.M., Sudirman, A., de Menezes, L.S. et al. (2015). *Optica* 2: 186.
24 Luan, F., Gu, B., Gomes, A.S.L. et al. (2015). *Nano Today* 10: 168.
25 Ta, V.D., Saxena, D., Caixeiro, S., and Sapienza, R. (2020). *Nanoscale* 12: 12357.

26 Shi, X., Ge, K., Tong, J.H., and Zhai, T. (2020). *Opt. Express* 28: 12233.
27 Luan, J., Seth, A., Gupta, R. et al. (2020). *Nat. Biomed. Eng.* 4: 518.
28 Wu, X., Oo, M.K.K., Reddy, K. et al. (2014). *Nat. Commun.* 5: 3779.
29 Zhang, X., Lee, W., and Fan, X. (2012). *Lab Chip* 12: 3673.
30 Lee, W., Chen, Q., Fan, X., and Yoon, D.K. (2016). *Lab Chip* 16: 4770.
31 Tan, X., Chen, Q., Zhu, H. et al. (2020). *ACS Sens.* 5: 110.
32 Zhang, N., Wang, Y., Sun, W. et al. (2018). *Nanoscale* 10: 2045.
33 Yang, X., Luo, Y., Liu, Y. et al. (2020). *Lab Chip* 20: 923.
34 Ta, V.D., Chen, R., Ma, L. et al. (2013). *Laser Photonics Rev.* 7: 133.
35 Zhang, S., Shi, X., Yan, S. et al. (2021). *ACS Sens.* 6: 3416.
36 Martino, N., Kwok, S.J.J., Liapis, A.C. et al. (2019). *Nat. Photonics* 13: 720.
37 Tan, X., David, A., Day, J. et al. (2018). *ACS Sens.* 3: 2327.
38 Yang, X., Gong, C., Wang, Y. et al. (2021). *Lab Chip* 21: 1686.
39 Humar, M. and Yun, S.H. (2015). *Nat. Photonics* 9: 572.
40 Duan, R., Li, Y., He, Y. et al. (2020). *Analyst* 145: 7595.
41 Wu, X., Chen, Q., Xu, P. et al. (2018). *Nanoscale* 10: 9729.
42 Chen, Y.C., Chen, Q., and Fan, X. (2016). *Lab Chip* 16: 2228.
43 Chen, Y.C., Chen, Q., and Fan, X. (2016). *Optica* 3: 809.
44 Song, Q., Xu, Z., Choi, S.H. et al. (2010). *Biomed. Opt. Express* 1: 1401.

6

Two Micrometer Ultrafast Fiber Laser

Tianshu Wang

Institute of Space Optoelectronics Technology, Changchun University of Science and Technology, Changchun, 130012, China

6.1 Introduction

At present, the 2 μm band gain fibers with relatively mature technology mainly include three types: thulium-doped fiber (TDF), holmium-doped fiber (HDF), and thulium–holmium codoped fiber (THDF). The difference in energy level structure leads to slightly different pumping methods and output characteristics.

The energy level structure of the Tm^{3+} ion is shown in Figure 6.1. The energy level transition of $^3F_4 \rightarrow ^3H_6$ corresponds to the light radiation in the 2 μm band, and its spontaneous emission spectral range can reach more than 300 nm. With the increase of the doping concentration of the gain fiber, the cross-relaxation process between the Tm^{3+} ions can significantly improve the conversion efficiency between the pump light and the signal light. With the maturity of high-power semiconductor laser technology in the 790 nm band, industrial and high-power 2 μm lasers mostly use lasers in this band as the pump source. However, the thermal management of the pump in the 790 nm band is more complicated, so scientific research and some special fields also use the 1500-band laser cascade as the pump source.

The energy level structure of the Ho^{3+} ion is shown in Figure 6.2, which contains two absorption peaks at 1.1 and 1.9 μm, and the optical radiation in the 2.1 μm band is generated during the transition of the $^5I_7 \rightarrow ^5I_8$ energy level. In recent years, in silica-based doped fibers, the 1.9 μm band is mostly used as the pump source for holmium-doped fibers. However, thulium-doped fibers can also emit lasers in this band, and thulium–holmium codoped fibers appear.

By doping Tm^{3+} and Ho^{3+} (two rare earth ions) simultaneously in the silica fiber, a gain fiber in the 2 μm band, namely a thulium–holmium codoped fiber, is formed. Its absorption wavelength is mainly determined by Tm^{3+}, so the same pumping method as thulium-doped fiber can be used. The introduction of Ho^{3+} enables a

Optical and Electronic Fibers: Emerging Applications and Technological Innovations, First Edition.
Edited by Lei Wei.

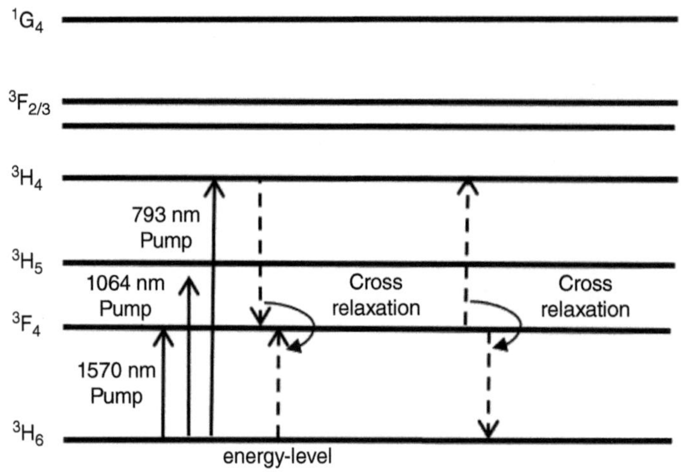

Figure 6.1 Schematic diagram of energy level structure of Tm^{3+} ion [1].

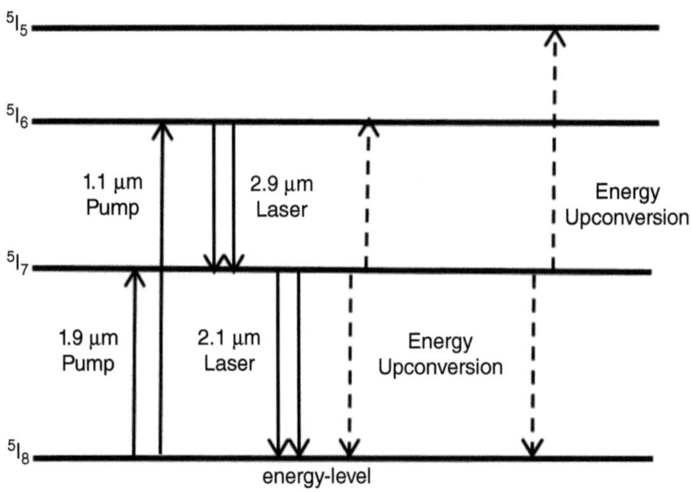

Figure 6.2 Schematic diagram of Ho^{3+} ion energy level structure [1].

certain energy exchange with the corresponding energy level of Tm^{3+} in the process of energy level transition, and the process of energy exchange is also different under the condition of different pump wavelengths. Taking the pumping mode of 1570 nm as an example, as shown in Figure 6.3, when Tm^{3+} absorbs the pump light in the 1570 nm band, it will transition from the ground state $^3H_6$ to the $^3F_4$ energy level. Since the $^3F_4$ energy level of Tm^{3+} is quite similar to the $^5I_7$ energy level of Ho^{3+}, energy exchange occurs between Tm^{3+} ions and Ho^{3+} ions, so that the ground state Ho^{3+} ions are excited to the $^5I_7$ energy level, forming a population inversion and generating radiation in the 2 µm band.

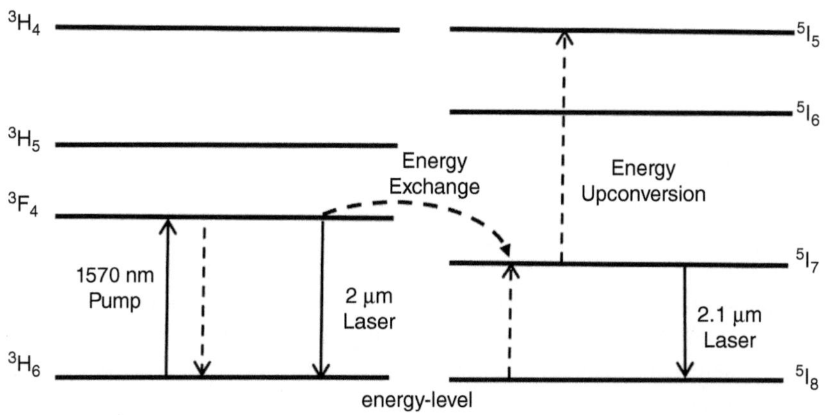

Figure 6.3 Schematic diagram of the energy level structure of thulium and holmium co-doping [1].

The generation of ultrashort laser pulses is based on the mode-locking effect, which refers to the phase locking of multiple longitudinal modes randomly distributed in the resonator. Generally, a fiber laser composed of a pump source, a gain fiber, and an optical resonator contains multiple longitudinal modes, and the phases between the longitudinal modes are randomly distributed. Therefore, the optical field output by the laser is an incoherent superposition of each longitudinal mode, which results in a continuous laser output from the fiber laser. In a cavity, the number of longitudinal modes depends on the gain bandwidth and cavity length, and the frequency separation is $\Delta\nu = c/nL$, where c is the propagation speed of light in vacuum, n is the refractive index of the medium in the cavity, and L is the cavity length.

A stable pulse sequence can be achieved after phase locking of each longitudinal mode in the resonator. This technology of locking the longitudinal mode phase is called mode locking technology. Mode-locking technology can generally be divided into active mode-locking technology and passive mode-locking technology. Active mode-locking technology is a mode-locking method that controls the parameters in the resonant cavity by applying a periodic signal to achieve a fixed phase between longitudinal modes. Active mode locking is easy to achieve high repetition frequency mode locking pulse output. Compared with active mode-locking, passive mode-locking uses passive mode-locking devices in the resonator to generate mode-locking pulses, which can realize an all-fiber structure, easy to integrate, low cost, and can achieve femtosecond-scale pulses. However, femtosecond ultrafast pulse is challenging to realize by active mode-locking structure. In passive mode-locking technology, the mode-locking device is generally a saturable absorber; and its working principle is shown in Figure 6.4. The saturable absorber has high transmittance to the part with large pulse energy and low transmittance to the part with small pulse energy; that is, it allows most of the pulse center energy to pass through, but does not allow other energy to pass through. In this way, when the pulse passes

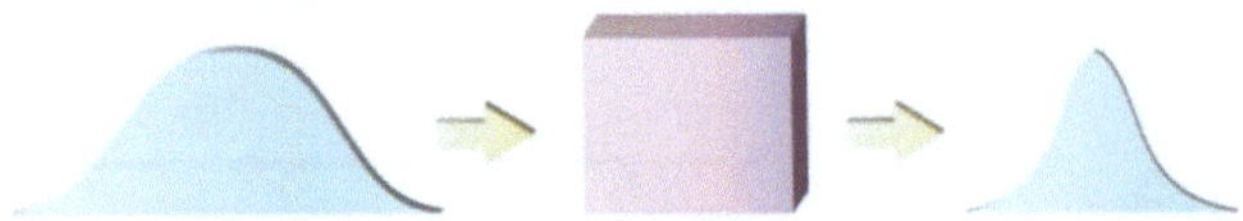

Figure 6.4 Principle of saturable absorber [1].

through the saturable absorber, the pulse width will be narrowed, and the laser will enter a mode-locked state after continuous cycling in the resonator.

Currently, saturable absorbers can be classified into two types. One is natural saturable absorbers, such as carbon nanotubes, SESAM, black scales, graphene, transition metal oxides, and the other is artificial saturable absorbers, such as NPR and NALM structures.

6.2 Mode-locked Fiber Laser

The generation of ultra-short laser pulses is based on the mode-locking effect. This section will introduce active mode-locking and passive mode-locking such as nonlinear polarization rotation (NPR) structure, nonlinear optical loop mirror (NALM) structure and other related content.

6.2.1 Active Mode-locked Ultrafast Fiber Lasers

Active mode locking is an important technical means to realize ultrafast fiber lasers. Ultrafast fiber lasers based on active mode locking have the characteristics of narrow pulse width, high repetition frequency and peak power, strong wavelength tunability, and easy integration.

Active mode-locking is a mode-locking structure achieved by using an active modulation mechanism. In most cases, an intensity modulator or a phase modulation can be used as the mode-locking element of the active mode-locking structure. Figure 6.1 shows the structure of a conventional active mode-locked laser. Active mode-locked fiber lasers usually consist of the following components: pump source, gain fiber, $LiNbO_3$ phase or intensity electro-optic modulator (EOM), signal generator, optical isolator (ISO), optical filter, and polarization controller.

In terms of active mode locking, some representative research results are as follows (Figure 6.5).

In 2019, Wanzhuo Ma reported that a multiwavelength actively mode-locked fiber laser at 2.07 μm [2]. Stable activelymode-locked pulses can be obtained with up to 10.078 GHz repetition rate. The shortest measured pulse width is 16 ps. The wavelength channels can be accurately tuned and with up to 27 channels can be obtained under a stable mode-locked operation. Furthermore, the output power and threshold characteristics were specifically discussed (Figure 6.6).

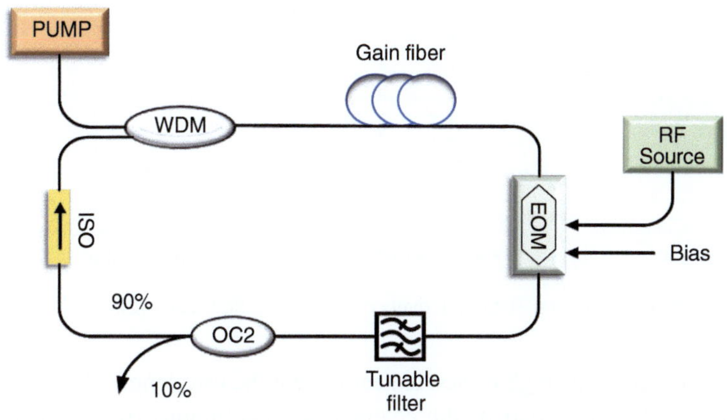

Figure 6.5 Typical active mode-locking structure diagram [1].

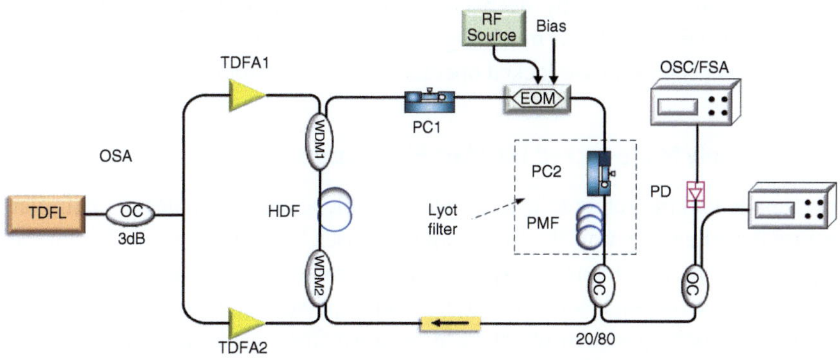

Figure 6.6 Schematic of 2.07 μm multiwavelength AML fiber laser [2].

Keeping the mode-locked laser at single-channel operation, the output power as a function of pump power under different repetition rates is shown in Figure 6.7a. As the repetition increases, there exists a slight increase in both the threshold and the slope efficiency. It can be explained that the average gain of single pulse decreases with the increasing modulation frequency, so that much pump power is needed for stable mode-locked operation. The specific thresholds were 650, 675, 712, and 769 mW under 20.35 MHz, 251.16 MHz, 2.2 GHz, and 10.078 GHz repetition, respectively. Fixing the repetition rate at 20.35 MHz, the output power as a function of pump power with different wavelength channels is shown in Figure 6.7b. The starting pump powers to achieve stable modelocking operations with different wavelength channels are 650, 749, 973, and 1263 mW, respectively. One can see that there exist great differences among the starting pump powers. It can be explained that, with the increase of wavelength channels, a greater pump power is needed to ensure all wavelength channels are mode-locked simultaneously. Noted that same wavelength channels or more can also be obtained if laser operates below the starting pump power. However, the pulse train obtained usually presents strong fluctuation,

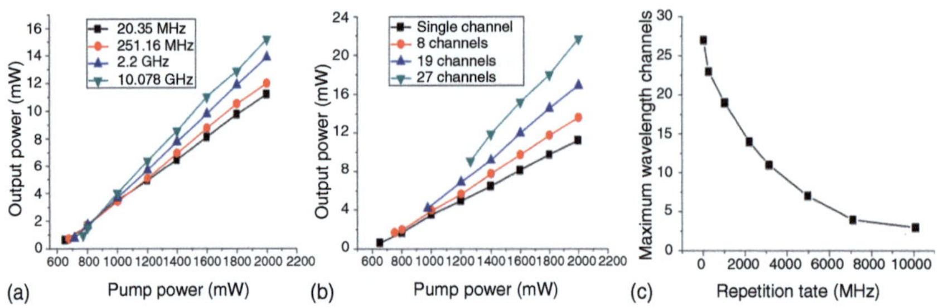

Figure 6.7 Output power characteristics of actively mode-locked fiber laser [2].

and the RF spectrum presents high noises, indicating that the wavelength channels are not all mode-locked, and part of the channels are continuous components. Fixing total pump power at 2 W, the maximum wavelength channels obtained under different repetition rates are shown in Figure 6.7c. It can be seen that the maximum wavelength channel gradually decreases to three when repetition rate reaches 10.078 GHz. It indicates that a much higher pump power is also needed to ensure stable multiwavelength mode-locked operation as the repetition rate increases.

6.2.2 Passively Mode-locked Ultrafast Fiber Lasers

Passive mode locking refers to inserting a saturable absorbing material or similar structure into the resonator of a free oscillation laser, and adjusting the loss in the cavity through its nonlinear absorption characteristics. When the mode locking conditions are satisfied, a series of mode locking pulses can be obtained. Typical passive mode-locking structures of 2 μm ultrafast fiber lasers include nonlinear polarization-rotation mode-locking, nonlinear magnifying ring mirror mode-locking, and hybrid mode-locking.

6.2.2.1 Nonlinear Polarization Rotation Effect

The figure below shows the structure of a conventional nonlinear polarization rotation (NPR) mode-locked fiber laser [3]. During the optical fiber transmission in the cavity, the self-phase modulation (SPM) and cross-phase modulation (XPM) in the cavity will introduce nonlinear phase shifts in the two orthogonal directions of the elliptically polarized light. The magnitude of the nonlinear phase shift is intensity dependent, the middle part of the pulse usually has a higher intensity, the rising and falling parts have lower intensity, and the difference in the magnitude of the polarization rotation introduces an intensity modulation of the net gain in the cavity. By properly adjusting PC2, when the light pulse is reincident to the PD-ISO, the middle part of the pulse has a higher transmittance than the two wings, resulting in a mode-locking effect similar to that of a saturable absorber (Figure 6.8).

The representative research results of author's research team are as follows.

In 2016, Qingsong Jia presents a high-order HML based on passively mode-locked mechanism of NPR in a Tm-doped ring cavity fiber laser, and the 406.8-MHz

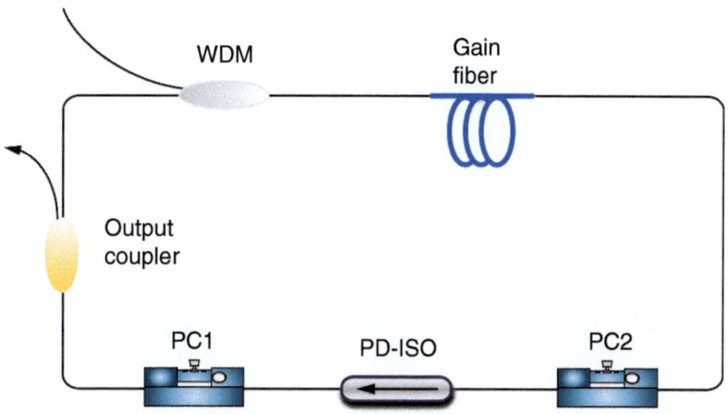

Figure 6.8 NPR mode locking structure diagram [3].

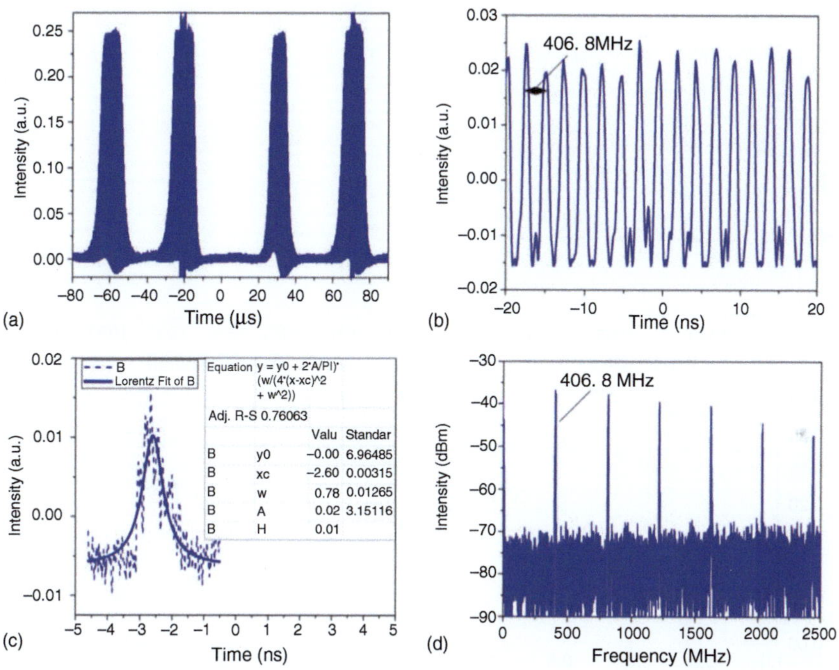

Figure 6.9 (a) QS ML pulse train, (b) pulse train with a repetition rate of 406.8 MHz, (c) pulse width of 128th order HML, and (d) RF spectrum with a repetition rate of 406.8 MHz Jia et al. [4] / SPIE.

repetition rate is obtained[4]. Passively, high-order harmonic mode-locked pulse train is observed in three different conditions, its formation can be explained as the interaction between solitons and the split of solitons under high pump power. With the increase of pump power, tunable multiwavelength output was also obtained in the experiment (Figures 6.9 and 6.10).

In 2017, Wanzhuo Ma demonstrated a widely tunable mode-locked thulium-doped fiber laser at 2 μm[5]. As for the continuous-wave laser, the tuning range can

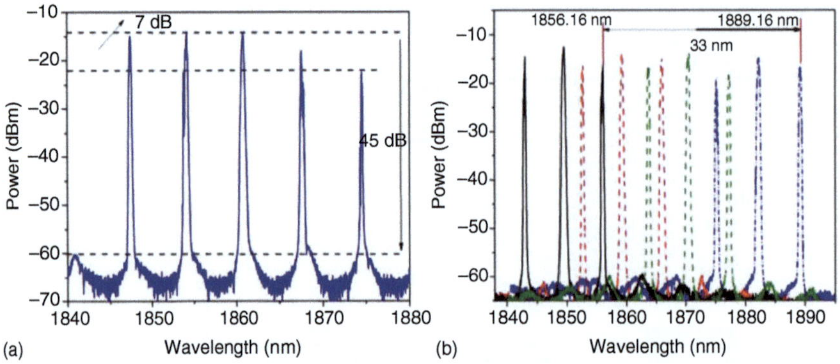

Figure 6.10 (a) Five wavelengths output spectrum and (b) tunable range of three wavelengths output Jia et al. [4] / SPIE.

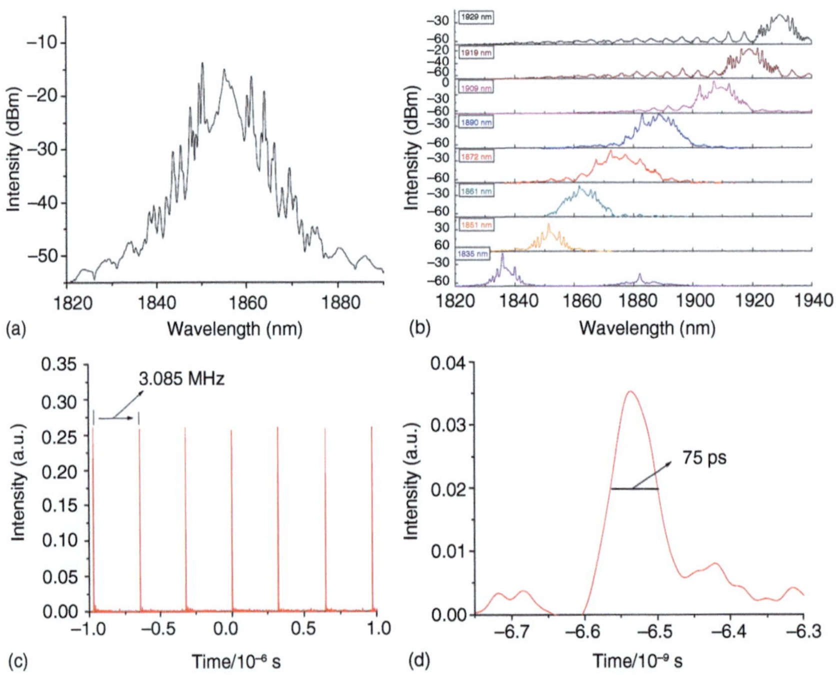

Figure 6.11 (a) mode-locked spectrum, (b) tunable single-wavelength mode-locked spectrum from 1835 to 1929 nm, (c) pulse repetition rate of single wavelength mode-locked output, and (d) FWHM of single pulse Ma et al. [5] / Optical Society of America.

reach 127 nm (1823–1950 nm). As for the mode-locked laser, the tuning ranges of single-wavelength and dual-wavelength mode-locked lasers are 94 nm (1835–1929 nm) and 87 nm (1831–1918 nm), respectively, with a 3.085 MHz repetition rate and 75 ps pulse width (Figures 6.11 and 6.12).

In 2017, Wanzhuo Ma have proposed the switchable generation of DS and NLP in a full-fiberized and low-cost Tm-doped fiber laser [5]. The net dispersion of the ring

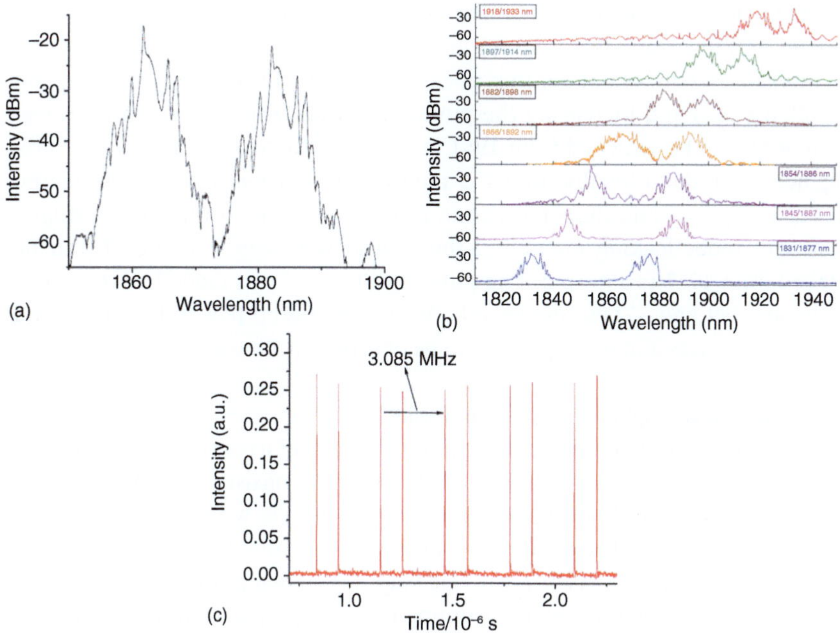

Figure 6.12 (a) Dual-wavelength mode-locked spectrum, (b) tunable spectrum from 1831 to 1918 nm, and (c) pulse train of dual-wavelength mode-locked laser Ma et al. [5] / Optical Society of America.

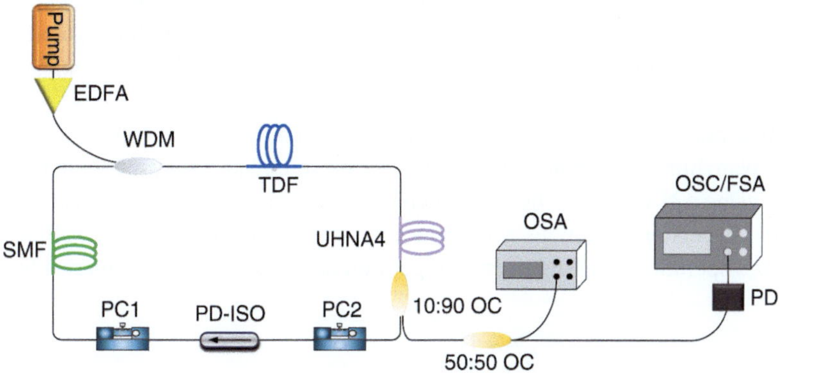

Figure 6.13 Schematic configuration of the normal-dispersion Tm-doped mode-locked fiber laser [6].

cavity is compensated with up to $+0.06\,\text{ps}^2$ using UHNA4 fiber. By properly adjusting the cavity parameters (such as PCs and pump power), the switchable generation of DS and NLP can be achieved. The 3 dB spectrum bandwidth of the DS pulse is greater than 50 nm, and the corresponding pulse duration can be dechirped down to 193 fs. The achieved NLP has a 3 dB spectrum width of 105.6 nm and duration of 121 fs. With higher pump power and proper cavity polarization state, the 3 dB spectrum width of NLP operation extends even up to 150.7 nm (Figure 6.13).

In 2019, Desheng Zhao used similar structure and proposed a SP fiber laser based on nonlinear polarization rotation (NPR) [6]. Stable SP operation around 1961 nm can be achieved at a pump power of 572 mW. By adjusting polarization controllers (PCs) and increasing the pump power to 727 mW, the central wavelength of SP can be switched from 1961 to 1980 nm due to an NPR-induced birefringence filter. Additionally, under a pump power of 990 mW, loosely bound states of two pulses and three pulses can be observed at 1980 nm. For the single-wavelength mode-locking operation, the central wavelength of the SP can be switched from 1961 to 1980 nm. The pulse durations of two mode-locking wavelengths are 389 and 371 fs, respectively. With high pump power, loosely bound states of two pulses and three pulses can be observed at 1980 nm.

6.2.2.2 Nonlinear Amplified Loop Mirror (NALM)

The following figure shows the mode-locking structure of a conventional nonlinear magnifying ring mirror (NALM). The NALM structure consists of an amplification ring and a one-way ring. The amplification ring and the one-way ring are connected by a 2 × 2 3 dB coupler. The gain fiber is located in the amplification ring, and the one-way operation of the laser in the one-way ring is realized by connecting the polarization independent isolator. Assuming the laser propagation in the one way ring clockwise and the coupler 1 port incident to enlarge ring, again by coupler is divided into the same intensity, from the three ports and four ports output, respectively, by two light in clockwise (the CW) and counterclockwise (CCW) direction of two transmission in the zoom ring, due to the asymmetry of two beams of light amplification location. The coherent superposition is formed when the two beams return to the coupler again, thus forming the intensity modulation with net gain in the cavity. By adjusting the PC in the resonator, the mode-locking is realized (Figure 6.14).

The representative research results of author's research team are as follows:

In 2018, Wanzhuo Ma demonstrated a square-wave soliton passively mode-locked fiber laser using a NALM [7]. The net dispersion is specifically designed in a small anomalous dispersion region to achieve soliton operation. Square-wave soliton mode-locked operation at a fundamental repetition rate of 3.2 MHz was obtained by

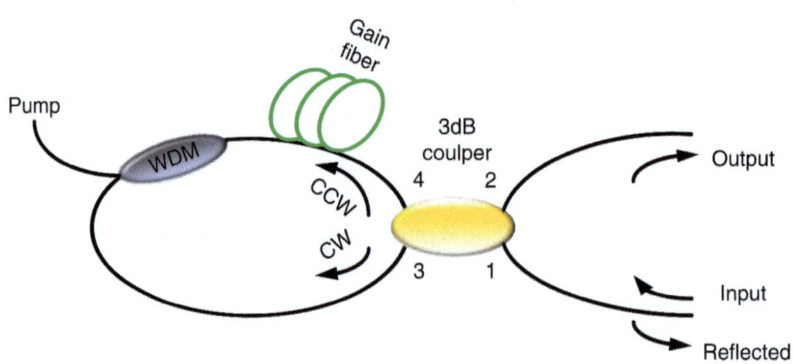

Figure 6.14 Schematic diagram of NALM mode-locking structure [1].

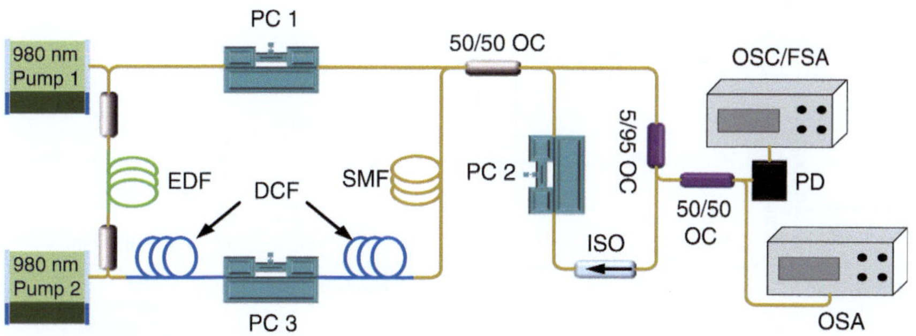

Figure 6.15 Schematic of a passively mode-locked SWP fiber laser [7].

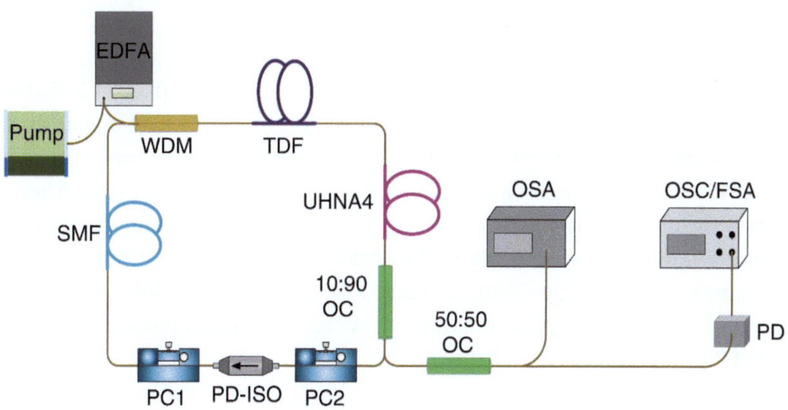

Figure 6.16 Square-wave pulse laser structure diagram Zhao et al. [8] / SPIE.

fixing the pump power in appropriate region. The optical spectrum presents several subsidebands, and it is different from the smooth and broad spectrum of rectangular NLPs. Pulse-breaking and stable dual-SWP with fixed interval can be observed. Furthermore, harmonic mode-locked operation can be achieved at a higher pump power. The harmonic mode-locked pulses can also operate in a dual-SWP state (Figure 6.15).

In 2020, Yiwu Zhao demonstrated the performance of square-wave pulses in a modified figure-eight Tm-doped fiber laser. The cavity was specially designed by employing a polarization-dependent isolator (PD-ISO) in a unidirectional loop (UL) with a cavity length of 17.2 m [8]. The pulse energy can be varied from 2.29 to 3.4 nJ. By reducing the net dispersion to $-1.033\,\text{ps}^2$ using an ultrahigh-numerical-aperture fiber for dispersion management, the 3 dB bandwidth of the spectrum can be broadened to 14.78 nm, and the duration of the AC spike is reduced from 680 to 421 fs. The maximum single-pulse energy can reach 4.97 nJ, and the output power and threshold are also improved (Figure 6.16).

In 2020, Long Han demonstrated the DSR pulses in the 2 μm fiber with the NALM mechanism from an all-PM fiber laser [9]. The cavity contains dual amplifiers and two spools of gain fiber. This configuration increased to two adjustment freedom

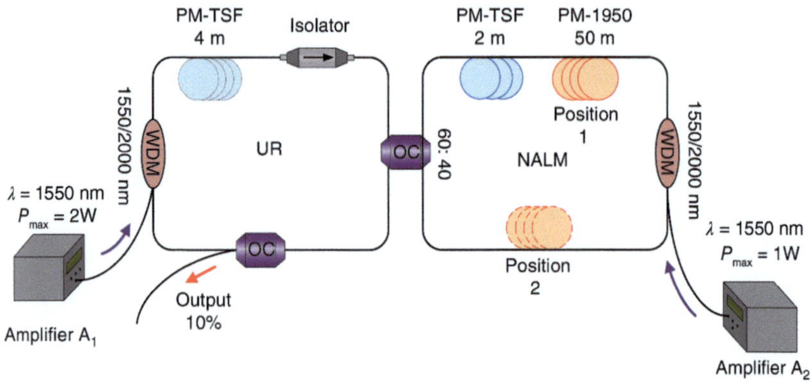

Figure 6.17 DSR pulse laser structure diagram Han et al. [9] / SPIE.

degrees for the all-PM fiber laser. When the amplifier A1 power was fixed at 2 W, the pulse width increases from 2.2 to 3.5 ns with the rise of the amplifier A2 power, and the single-pulse energy rises from 3.7 to 6.8 nJ. The threshold of the mode-locked laser varies with different positions of the PM-1950 fiber in the NALM loop. Then, they observed the characteristics of the output DSR pulse in different lengths of the PM1950 fiber (50, 80, and 110 m) when the position is fixed. The output average power and single-pulse energy—both increased linearly with amplifier power in different situations. The DSR pulse maximum average power and pulse energy are 45.5 mW and 29.1 nJ, respectively (Figure 6.17).

6.2.2.3 2D Material Mode Locking

In recent years, with the development of condensed matter physics and material preparation technology, carbon nanotubes, graphene, topological insulators, and other materials as a saturable absorption materials have been successfully applied in laser lock mode, especially the new developed two-dimensional nanomaterials with narrow band gap, ultrafast electronic relaxation time and high damage threshold, and other characteristics showing excellent saturable absorption properties. The research of mode-locked laser using this material has also become one of the hot research topics. Some representative research results are as follows.

In 2022, Yingkui Li demonstrated an all-fiber Tm-doped laser operating in the anomalous dispersion regime [10], which is capable of generating broadband noise-like pulses. The mode-locking is achieved via a microfiber decorated with single-wall nanotubes. The full width at half maxima (FWHM) is 22 nm, and the pulse train is 58.05 ns; the SNR is up to 56 dB. And the trace of autocorrelation is 6.4 ps and 643 fs corresponding to the pedestal and peak, respectively (Figures 6.18–6.20).

6.2.2.4 Hybrid Mode-locked

We can also analyze some representative research results. In 2017, Qingsong Jia presented a mode-locking fiber laser with high repetition rate by high-order HML pulse trains using NPR and SESAM hybrid mode locking [11]. Stable high-order HML pulse trains can be produced under lower pump power in this configuration.

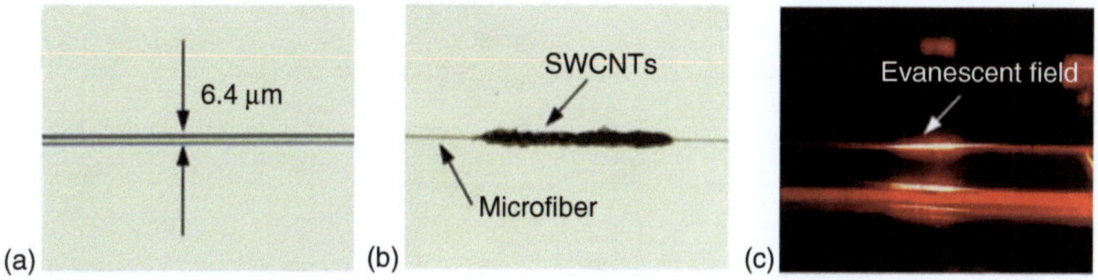

Figure 6.18 Schematic of the micro-fiber based SWCNTs, (a) the waist diameter of micro-fiber, (b) the image of micro-fiber decorated with SWCNTs, and (c) the presentation of the evanescent field [10].

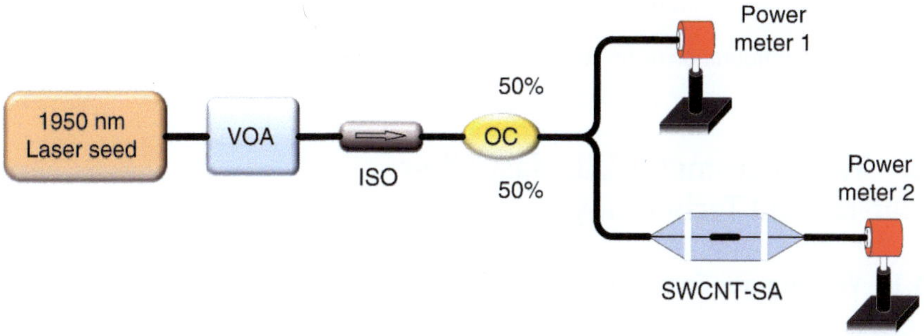

Figure 6.19 Twin detector measurement setup. VOA: variable optical attenuator; and SWCNT-SA: single-walled carbon nanotubes saturable absorber [10].

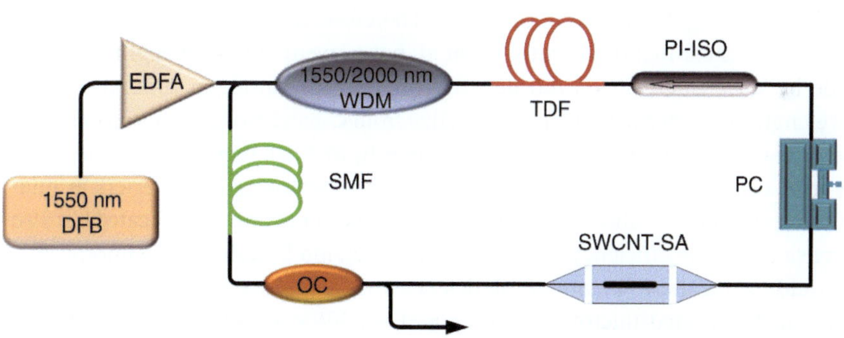

Figure 6.20 Schematic diagram of thulium-doped fiber laser. DFB: distributed feedback laser; EDFA: Erbium-doped fiber amplifier; WDM: wavelength division multiplexer; TDF: thulium-doped fiber; PI-ISO: polarization-independent isolator; PC: polarization controller; SWCNT-SA: single-walled carbon nanotubes saturable absorber; and OC: optical coupler [10].

The repetition rate of HML pulse train is 1.78 GHz. The central wavelength of HML can be varied in a wide range of 1839–1890 nm. And multiwavelength HML operation is also obtained. Furthermore, HML pulse trains with higher repetition rate and pulses with high-quality can be obtained according to the proposed method by using a SESAM with deeper modulation depth and less recovery time (Figure 6.21).

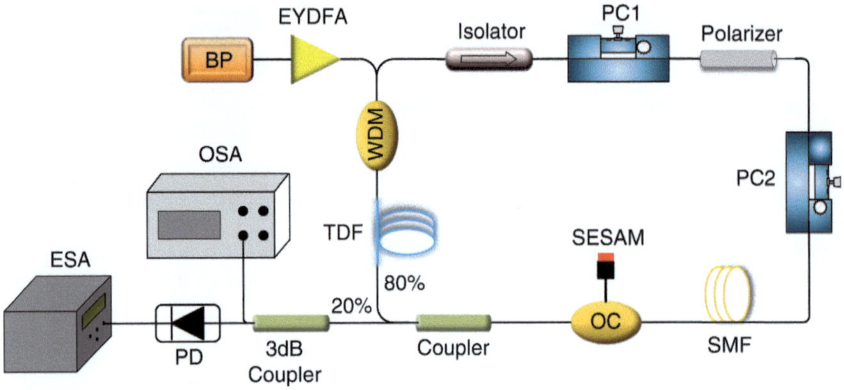

Figure 6.21 Configuration of Tm-doped mode locking fiber laser [11].

6.3 Two Micrometer Ultrafast Fiber Laser-related Technology

6.3.1 Wavelength Conversion

Currently, 1.9 μm lasers have emerged as a promising platform for both science and industry, because of their inherent high raw power and eye-safe spectrum [12]. It has shown greater potential for many application areas, such as optical wireless communications, chemical, and biomolecular sensing to reduce the transmission loss or improving the sensitivities [13–17]. Therefore, the research on mid-infrared band is very meaningful. Recently, Tu et al. have invented a wavelength converter made by silicon-rich silicon nitride/silicon hybrid waveguide in 2021. In 2019, a wavelength converter with the tunable range from C-band to mid-infrared band has been created by Alamgir et al. [18]. Subsequently, in 2020, they have demonstrated the new significant progress on fabrication of the wavelength converters at mid-infrared [19]. But the fabrication of these converters is not only complicated but also expensive, and the efficiency of wavelength conversion is desired to be improved.

To address the above-mentioned deficiencies, an all-optical wavelength converter based on GO-coated microfiber is proposed via FWM effect by Tianshu Wang's research team [20] (Figure 6.22).

The experimental layout of all-optical wavelength converter based on GO is shown in Figure 6.22. It is divided into three parts, including the pump laser, signal laser, and wavelength converter. The structure of pump laser and signal laser are identical; the only one distinction between the pump laser and signal laser is the wavelength of the output spectra controlled by the tunable filter (TF). As portrayed in the first part, the laser from distributed feedback source (DFB) operating at C-band is amplified by an erbium–ytterbium codoped fiber amplifier (EYDFA). The amplified light from EYDFA acts as the pump source and can inject into the 4 m thulium-doped fiber (TDF) via a 1550/2000 nm wavelength division multiplexer (WDM). The polarization insensitive isolator (PI-ISO) ensures that the laser can

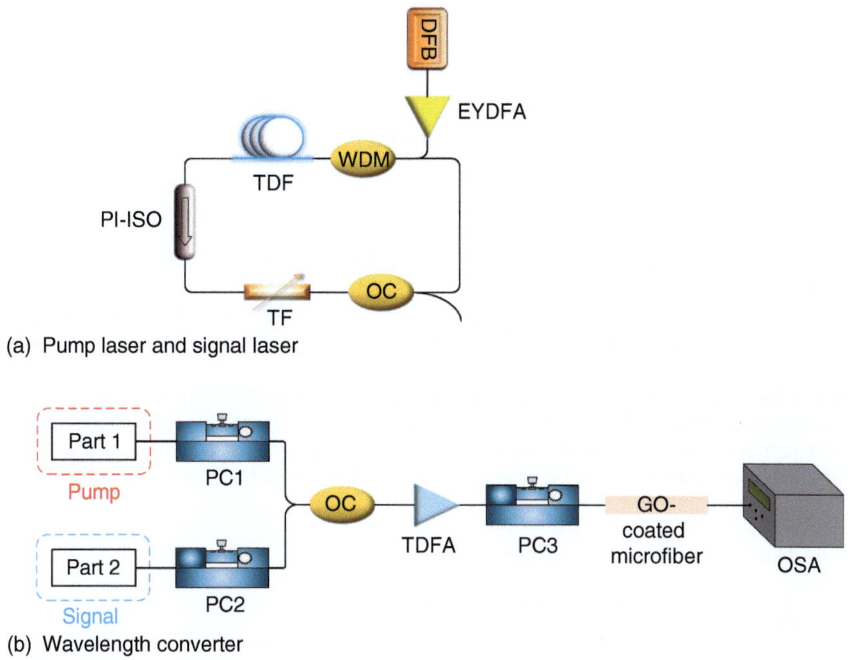

Figure 6.22 Schematic of all-optical wavelength converter based on GO: (a) the configuration of homemade thulium-doped fiber laser regarded as pump laser and signal laser; and (b) the configuration of wavelength converter Du et al. [20] / Optical Society of America / CC BY 4.0.

propagate unidirectionally. The TF can be used to select the desired wavelength. After the wavelength filtering, the 80% port of the 20/80 optical coupler (OC) serves as the feedback of the ring cavity; and the 20% port is considered as the output laser. The second part takes similar configuration and function compared with the first part. The wavelength converter employed in the third part comprises a pair of polarization controllers (PC1 and PC2), a 50/50 OC, a thulium-doped fiber amplifier (TDFA), PC3 spliced with the TDFA output, and a sample of GO-coated microfiber. PC1, 2, and 3 dominate the polarization of the pump and signal light, simultaneously. And the 50/50 OC couples the pump and signal light together. The TDFA is responsible for providing gain amplification of the two lights, and the final all-optical wavelength conversion happens in the location of GO-coated microfiber. Insertion loss of the microfiber sample without materials is estimated to be 1 dB approximately. When the microfiber is decorated with GO, the insertion loss increased to about 4 dB. The modulated spectrum is observed and recorded by an optical spectrum analyzer (Figures 6.23 and 6.24).

A GO-based all-optical wavelength converter operating at 1.9 μm is proposed and demonstrated. The following analysis is obtained through the analysis of the above experimental results. The conversion efficiency of −45.52 dB and the tunable wavelength interval of 6 nm can be realized by this converter. As for the characteristic of steady, the fluctuation of conversion efficiency is ±0.41 dB within two hours. The results indicate that there is a potential for this converter to be an effective device for

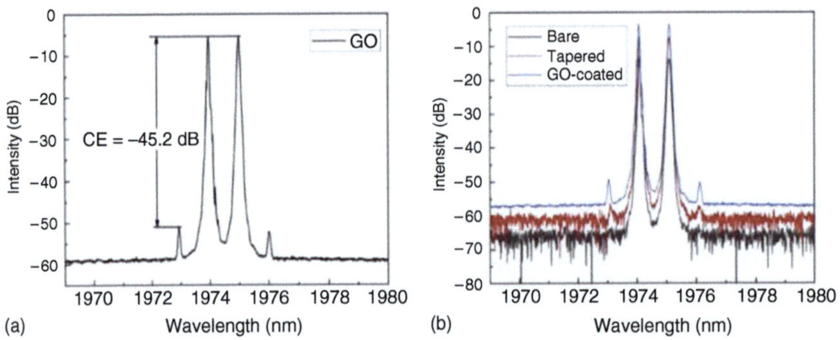

Figure 6.23 Experimental optical spectra of wavelength conversion signal of the GO-coated microfiber: (a)The spectrum of the effect of the GO-coated microfiber on the FWM effect; and (b) The spectrum of wavelength conversion contrast of bare fiber, tapered fiber, and GO-coated microfiber Du et al. [20] / Optical Society of America / CC BY 4.0.

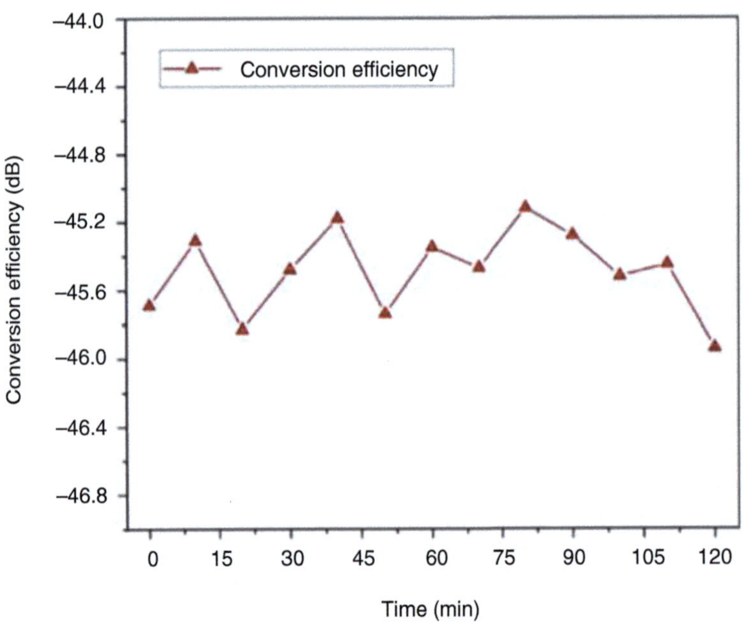

Figure 6.24 Time stability of all-optical wavelength conversion Du et al. [20] / Optical Society of America / CC BY 4.0.

all-optical networks. This work can be beneficial for easing the blocking of wavelength in optical and enriching the kinds of communication device of 1.9 μm.

6.3.2 Pulse Shaping and Evolution

The development of optical communication systems is moving towards ultrafast and ultralong optical communication systems. However, when optical pulses are transmitted in optical communication systems, they are affected by loss, dispersion, and nonlinear effects, which will lead to bit errors and distortions. In order to improve the transmission distance and transmission rate of optical pulses, it is necessary to evolve the pulses through pulse shaping or evolution to make them more suitable for transmission in high-speed and long-distance optical communication systems.

Various research institutions are very concerned about pulse shaping and evolution technology, among which author's research team has been engaged in related fields for many years, and has made in-depth explorations in this aspect. The following is a brief introduction to the relevant research in this field by the team.

In 2018, Wanzhuo Ma proposed a QML square-wave fiber laser at 1.9 μm [21]. The experiment structure is shown below. Continuous DSR pulse with 4.4 MHz repetition rate was obtained by using a nonlinear amplifying loop mirror. Q-switched mode-locked operation was achieved by properly setting the pump power and the polarization controllers. The Q-switched repetition rate can be varied from 41.6 to 74 kHz by increasing pump power. Moreover, the characteristics of output energy were also discussed. The corresponding average single-pulse energy of QML pulse can be varied from 2.67 to 5.2 nJ, which was almost three times higher than continuous SWP. Such improvement was also presented in the output peak power (Figure 6.25).

In 2019, Desheng Zhao demonstrated improvements on pulse energy and peak power of soliton pulse in a thulium-doped fiber laser [6]. The experimental structure diagram is shown in the following figure. By adjusting the polarization controllers (PC1 and PC2), a conventional soliton can be observed with a threshold of 370 mW. Under the pump power of 550 mW, conventional soliton pulse can convert to Gaussian pulse with high pulse energy and peak power by rotating the PC2. As a result of dispersion management, excess accumulation of nonlinear phase shifts can be avoided in the cavity. The pulse energy can be increased from 77 pJ to 1.38 nJ, and the peak power can be increased from 106.9 W to 2.67 kW (Figure 6.26).

In 2019, Runmin Liu reported on a dispersion-managed thulium-doped fiber (TDF) laser operating in the near zero-dispersion regime, utilizing the NPR

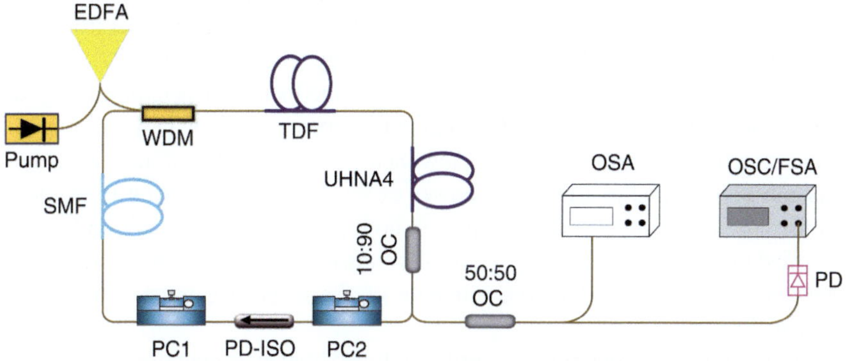

Figure 6.25 Schematic of Q-switched mode-locked DSR thulium-doped fiber laser Ma. [21] / Optical Society of America / CC BY 4.0.

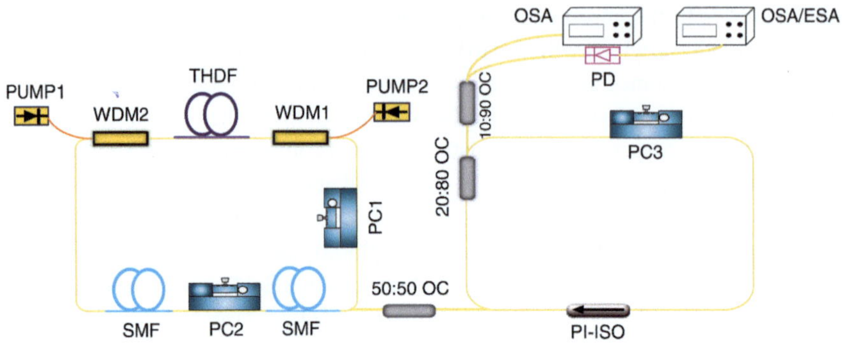

Figure 6.26 Schematic of thulium-doped fiber laser [6].

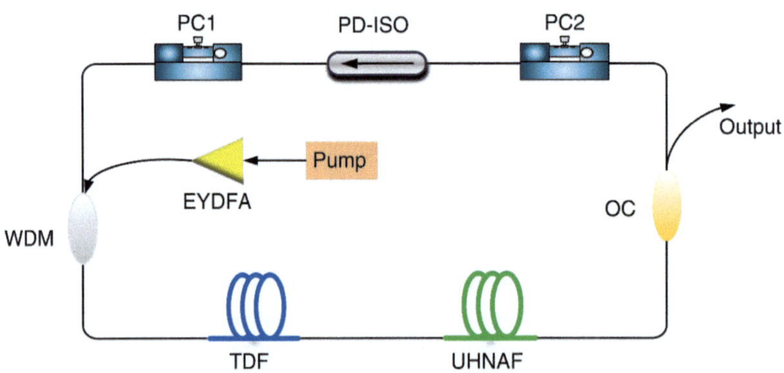

Figure 6.27 Schematic of thulium-doped fiber laser Liu. [22] / SPIE.

technology [22]. The schematic diagram is shown below. Under the pump power of 492 mW, the single pulse can evolve into the dispersion managed soliton (DMS) pair by adjusting polarization controllers (PCs), and the pulse duration of the single pulse is 382 fs. Meanwhile, multipulse-bound DMSs are systematically investigated. The experimental results, especially the bound states composed of four and five pulses, might have potential application of extending the coding alphabet from the traditional binary coding scheme to the higher coding scheme (Figure 6.27).

In 2019, Peng Lin demonstrated a passively mode-locked dispersion-managed thulium-doped fiber (TDF) laser in the near-zero dispersion regime [23]. The experimental structure diagram is shown below. Stable, self-started solitons with sidebands can be obtained as the pump power exceeds the mode locking threshold. With further increasing of the pump power, both multisolitons and soliton bunch are formed in the cavity. Moreover, the results showed that the separation of soliton in soliton bunch is related to the intensity of the dispersive wave component (Figure 6.28).

In 2019, Runmin Liu reported on a thulium-doped fiber laser operating in anomalous dispersion regime by employing NPR technology [24]. The experimental structure diagram is shown below. Stable, single soliton as well as soliton pairs with the

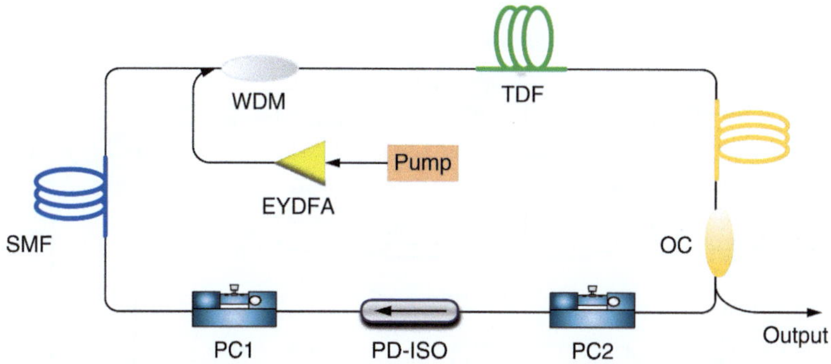

Figure 6.28 Experimental structure diagram of 2-μm dispersion-managed soliton fiber laser Lin. [23] / SPIE.

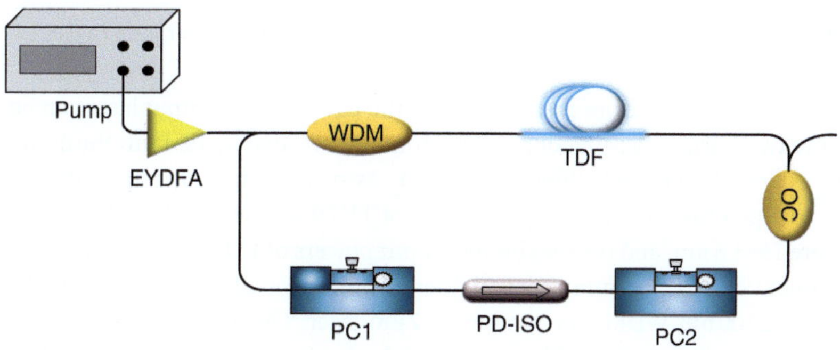

Figure 6.29 Scheme of the NPR-based mode-locked thulium-doped fiber laser [24].

phase difference of 0, and can be achieved by adjusting PCs within the same cavity dispersion and pump power. With the increasing of pump power, soliton triplets, soliton quartets, and soliton quintets can be observed. To the best of our knowledge, it is the first time that soliton molecules composed of five bound solitons were observed around 2 μm (Figure 6.29).

6.4 Two Micrometer Ultrafast Fiber Laser Communication

6.4.1 FSO Communication

The first report on the 2 μm free-space laser communication experiment is the 2 μm free-space data transmission holmium-doped active mode-locked fiber system laser by Wang Tianshu's research team [25].

The experimental structure is illustrated in Figure 6.30. A 1565 nm laser diode (LD1) is amplified by an erbium-doped fiber amplifier (EDFA1) to 2 W and backward pumps a segment of 3 m thulium-doped fiber via a 1550/2000 nm wavelength division multiplexer (WDM). A thulium-doped fiber laser with a central wavelength

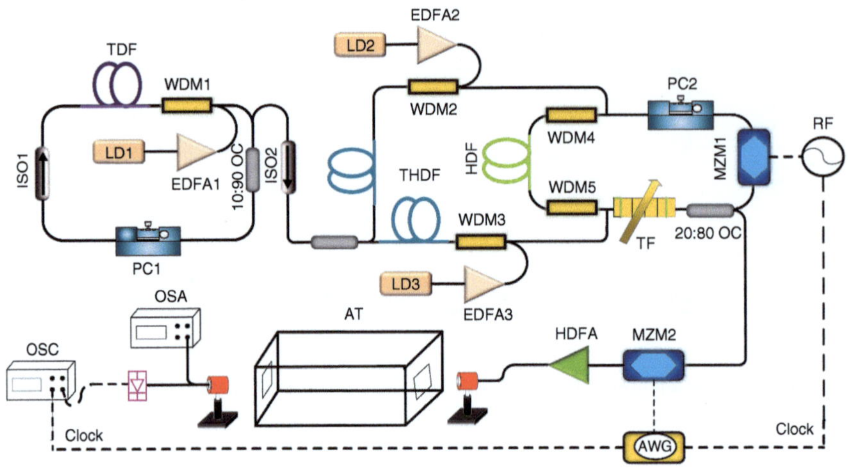

Figure 6.30 Experimental structure of a 2 μm free-space optical communication [25].

of 1.9 μm can be obtained from the 10% port of the coupler. The output laser can be divided into two parts by a 3 dB coupler and amplified by two home-built thulium–holmium codoped fiber amplifiers (THDFA). Two segments of 2 m thulium–holmium codoped fiber are used as gain media of THDFA. The wavelengths of LD2 and LD3 are 1564.6 nm, and the maximum output powers of EDFA2 and EDFA3 are 1 W. A segment of 1.2 m holmium-doped fiber is bidirectionally pumped by TDFLs via two 1900/2100 nm WDMs for producing 2 μm laser. The actively mode-locked pulse sequence can be generated by a Mach–Zehnder intensity modulator (MZM) that is driven by a radio frequency signal generator. A tunable filter is used to suppress super-mode noise and to realize wavelength tunable. The 2 μm pulses can be modulated into pseudorandom binary pulse sequence by MZM2, which is driven by an arbitrary waveform generator, and amplified by a holmium-doped fiber amplifier. The modulation eye pattern can be observed by an optical wide-range oscilloscope. We can use a simulated atmospheric turbulent device (AT) based on the hot air convective Atmospheric turbulence principle to measure the influence of a weak turbulent channel on 2 μm carriers. The pulses can be detected by a 2 μm photodetector (PD) with a 12 GHz bandwidth, and monitored by a 2.5 GHz oscilloscope and a 3 GHz radio frequency signal analyzer. The optical spectrum can be observed by an optical spectrum analyzer, and the pulse duration can be obtained by a second harmonic autocorrelator. The bit error rate (BER) can be obtained by a MATLAB program that calculates the offline data from OSC.

The stable pulse sequence can be modulated by a digital signal at a data rate of 2.02 Gb/s and transmitted in a simulated weak turbulent channel. The received sensitivity of the 2 μm link is −19.52 dBm, and the power penalty at a BER of 3.8×10^{-3} (FEC) is 2.67 dB compared with BTB. According to a comparison of the average receiving power standard deviation and BER curves between a 2 μm laser communication system and a 1.55 μm laser communication system, the 2 μm laser is more suitable for propagation in turbulence than the 1.55 μm laser.

6.4.2 Somke Channel Communication

Free-space optical (FSO) communication has garnered significant attention owing to its high bandwidth, low cost, resistance to electromagnetic interference, and flexible system erection [26–28]. Although FSO communications offer prominent advantages and are feasible for various applications, its commercialization is hindered by transmission link instabilities [29]. Atmospheric turbulence caused by temperature inhomogeneity and pressure fluctuations affect the link reliability and quality [30]. In addition, aerosol particles floating in the atmosphere are typically smaller than 10 μm and have particle size distributions similar to those of optical wavelengths in FSO communication [31]. According to the Mie scattering model, this can potentially result in the scattering and absorption of visible and infrared optical beams [32]. Besides the smoke attenuation caused by scattering, the extinction coefficient of different wavelength lasers were also related to the effective radius of the particle [33, 34].

It is important to study the transmission characteristics of infrared wavelengths in the real outdoor atmosphere channel for FSO links. However, outdoor smoke channels are time consuming to build and offer low reproducibility for dense smoke ($V < 0.5$ km) [35]. Moreover, the aerosol along the link is uneven and random, thereby complicating atmospheric control and characterization [32]. Therefore, an indoor simulated smoke chamber is proposed by Tianshu Wang research team to analyze the transmission characteristics of multiple wavelengths under the stable condition of a uniform aerosol distribution [36].

The schematic diagram is shown in Figure 6.31a. The laser sources comprised two semiconductor lasers operating at 0.85 and 1.06 μm, as well as two custom-designed fiber lasers operating at 1.55 and 2.04 μm. The lengths of the erbium-doped fiber and holmium-doped fiber were both 1.5 m. The output power of the pump operating at 980 and 1920 nm were 0.8 and 1.5 W, respectively. The actively mode-locked pulse sequence can be generated using a Mach–Zehnder intensity modulator (MZM), which is driven by a frequency continuously tunable radio-frequency signal generator. A Fabry–Perot tunable filter was added into the cavity to suppress the supermode noise generated by high-order harmonic mode locking. Further details of the actively mode-locked holmium-doped fiber laser are provided in [48]. The output mode-locked pulses can be modulated into a pseudorandom binary pulse sequence by MZM2, which was driven by an arbitrary waveform generator. An indoor air chamber measuring 4 m × 0.5 m × 0.5 m was used to simulate the smoke channel. Two circular transmission windows of diameter 50.8 mm were installed at both the ends of the smoke chamber. The smoke can spread rapidly and maintain suspension by controlling four chassis fans at the bottom of the chamber and two pumping valves at the top of the chamber. The smoke can be generated using a smoke generator, in which a mixture of glycerol, ethylene glycol, propylene glycol, and water was heated to 300 °C to produce dry smoke particles. The smoke can be drained outdoors through the smoke outlet, and the link visibility can be controlled by the smoke density. Laser transmitting and receiving were realized by two collimators on both sides of the smoke chamber. Furthermore, two additional mirrors were added to

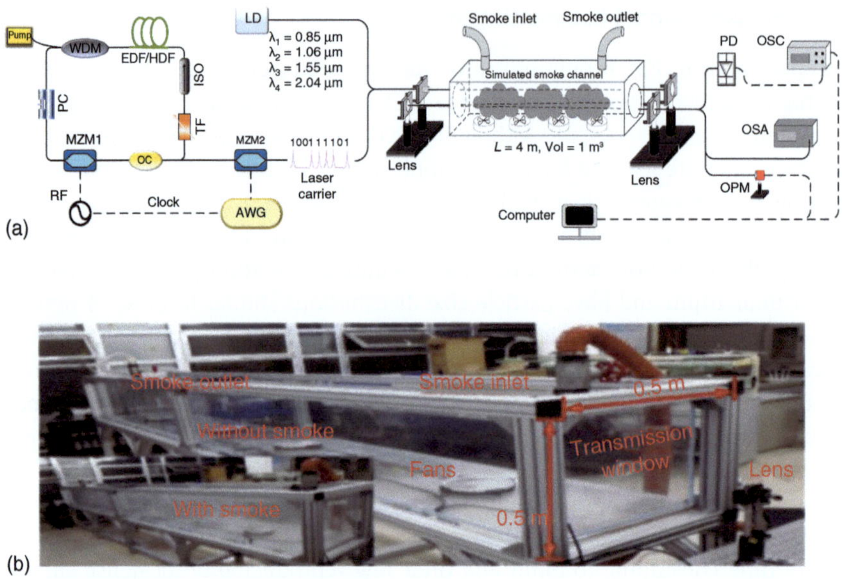

Figure 6.31 (a) Schematic diagram of free-space data transmission system in indoor simulated smoke channel. (b) Photograph of simulated smoke channel. Inset shows the presence of smoke Lin et al. [36] / Optical Society of America.

increase two reflections, which further improved the homogeneity of smoke. The length of the transmission link was 12 m. The received power was detected by optical power meter, and the received power of 0.85 and 1.06 μm laser at dense smoke condition can be obtained by OPM. The received optical spectrum was observed via an optical spectrum analyzer. The mode-locked pulse sequence was detected using 12 GHz photodetector with an operating wavelength range of 1.2–2.2 μm, whereas it was monitored using a 4 GHz oscilloscope and a 44 GHz radio frequency signal analyzer. The eye pattern of the data transmission system was observed using an optical wide-range oscilloscope. Due to the optical eye patterns of 2.04 μm laser carrier is not within the measuring range of the OWO, the eye patterns in this study were all electrical eye patterns observed after the photodetector. The BER was obtained using a MATLAB program that calculated the offline data from the OSC. The minimum BER was 2.63×10^{-7}, which was limited by the hardware conditions of the computer.

In this study, a 4 m long indoor smoke chamber was established to study the transmission characteristics of an FSO link. The smoke can be produced by heating a glycerol-based liquid using a smoke generator. The attenuation of four wavelengths ($\lambda_1 = 0.85$ μm, $\lambda_2 = 1.06$ μm, $\lambda_3 = 1.55$ μm, and $\lambda_4 = 2.04$ μm) in the smoke channels with different visibilities were analyzed and compared with the Ijaz model. The experimental data indicated the strong dependency of the wavelength on the smoke attenuation. In addition, the 1.55 and 2 μm mode-locked fiber lasers were modulated by a 4.04 Gbps digital signal and transmitted in the smoke channels based on $V = 0.5$, 0.05, and 0.005 km. The characteristics of the bit-error rates (BERs) and eye patterns under different smoke conditions were measured and analyzed. As the visibility of the smoke channel decreased from 0.5 to 0.005 km, the optical signal-to-noise ratios (OSNRs) of the 1.55 and 2.04 μm FSO systems decreased by 12.91 and 5.72 dB, respectively, and the corresponding power penalties were 4.08 and 2.73 dB, respectively. The experimental results show that the 2 μm band FSO system can be adopted in the dense smoke channel to improve the reliability of the communication link.

6.5 Conclusion

This chapter introduces the fundamentals of 2 μm ultrafast fiber lasers. Some representative research results of 2 μm ultrafast fiber lasers are introduced, and finally its application in space optical communication is analyzed. It is hoped that it will be helpful to research scholars interested in this field.

References

1 Wang, T. and Ma, W. (2021). *Ultrafast Mode Locked Fiber Laser Technology*. Beijing: Chinese Science Publishing & Media.
2 Ma, W., Wang, T., Wang, F. et al. (2018). *IEEE Photon. Technol. Lett.* 31: 242.
3 Matsas, V., Newson, T., Richardson, D., and Payne, D.N.J.E.L. (1992). *Electron. Lett.* 28: 1391.
4 Jia, Q., Wang, T., Ma, W. et al. (2016). *Opt. Eng.* 55: 106121.
5 Ma, W., Wang, T., Zhang, Y. et al. (2017). *Appl. Opt.* 56: 3342.
6 Zhao, D., Wang, T., Ma, W. et al. (2019). *IEEE Photon. Technol. Lett.* 31: 909.
7 Ma, W., Wang, T., Su, Q. et al. (2018). *Laser Phys. Lett.* 15: 065102.
8 Zhao, Y., Zhao, D., Liu, R. et al. (2020). *Opt. Eng.* 59: 066106.
9 Han, L., Wang, T., Chen, B. et al. (2020). *Opt. Eng.* 59: 116107.
10 Li, Y., Du, L., Li, X. et al. (2022). *IEEE Photonics J.* 14: 1.
11 Qingsong, J., Tianshu, W., Wanzhuo, M. et al. (2017). *IEEE Photonics J.* 9: 1.
12 Eckerle, M., Kieleck, C., Świderski, J. et al. (2012). *Opt. Lett.* 37: 512.
13 Li, Z., Heidt, A., Daniel, J. et al. (2013). *Opt. Express* 21: 9289.
14 Zhong, C., Ma, H., Sun, C. et al. (2021). *Opt. Express* 29: 23508.
15 Zhou, H., Xu, S., Wu, S. et al. (2020). *Opt. Express* 28: 10280.
16 Hui, Z., Chen, L., Pan, D. et al. (2021). *Optik* 241: 166957.
17 Kawano, Y., Ito, Y., Ito, S. et al. (2021). *Macromolecules* 54: 1934.
18 Alamgir, I., St-Hilaire, F., and Rochette, M.J.O.L. (2020). *Opt. Lett.* 45: 857.
19 Alamgir, I., Abdukerim, N., and Rochette, M. (2019). *The European Conference on Lasers and Electro-Optics*, cd_p_37. Optical Society of America.
20 Du, L., Ding, X., Han, D. et al. (2021). *Opt. Express* 29: 40286.
21 Ma, W., Wang, T., Su, Q. et al. (2018). *Opt. Express* 26: 12514.
22 Liu, R., Wang, T., Zhao, D. et al. (2019). *Appl. Opt.* 58: 6464.
23 Lin, P., Wang, T., Ma, W. et al. (2019). *Opt. Eng.* 58: 116106.

24 Liu, R., Wang, T., Ma, W. et al. (2019). *IEEE Photon. Technol. Lett.* 31: 341.

25 Lin, P., Wang, T., Ma, W. et al. (2020). *IEEE Photon. Technol. Lett.* 32: 223.

26 Qu, Z. and Djordjevic, I.B.J.O.L. (2016). *Opt. Lett.* 41: 3285.

27 Zhou, H., Xie, W., Zhang, L. et al. (2019). *Opt. Express* 27: 27062.

28 Tan, J., Zhao, Z., Wang, Y. et al. (2018). *Opt. Express* 26: 2099.

29 Kaushal, H. and Kaddoum, G.J.I. c. s (2016). *IEEE Commun. Surv. Tutor.* 19: 57.

30 Yang, Y., Mandehgar, M., and Grischkowsky, D.R.J.I.P.T.L. (2014). *IEEE Photon. Technol. Lett.* 27: 383.

31 Ijaz, M., Ghassemlooy, Z., Le Minh, H. et al. (2012). *2012 International Workshop on Optical Wireless Communications (IWOW)*, 1. IEEE.

32 Ijaz, M., Ghassemlooy, Z., Pesek, J. et al. (2013). *J. Light. Technol.* 31: 1720.

33 Libich, J., Perez, J., Zvanovec, S. et al. (2017). *Appl. Opt.* 56: 336.

34 Grabner, M. and Kvicera, V.J.O. e. (2011). *Opt. Express* 19: 3379.

35 Strickland, B.R., Lavan, M.J., Woodbridge, E., and Chan, V.J.A. o. (1999). *Appl. Opt.* 38: 424.

36 Lin, P., Wang, T., Ma, W. et al. (2020). *Opt. Express* 28: 39216.

7

Advanced Fibers for Optogenetic Modulation

Minghui Du and Shifeng Zhou

State Key Laboratory of Luminescent Materials and Devices, School of Materials Science and Engineering, South China University of Technology, No. 381, Wushan Road, Tianhe District, Guangzhou, 510640, China

7.1 Introduction

Thousands of neurons communicate with each other by electrophysiological signals in the brain's nervous system [1]. In the early days, the modulation of neural activities was mainly conducted by metal electrodes, including tungsten electrode, platinum electrode, tetrodes, Michigan arrays, and Utah arrays [2–7]. With the emergence of optogenetic technology in 2005, researchers began to use light to manipulate the neural activities and simultaneously record the electrophysiological signals with electrodes [8]. The optogenetic technology is integrated with optics, software control, gene manipulation and electrophysiology technologies [9], and its aim to use light to control neural activities, which opens new routes to the research of function of specific neurons and neural circuits in behaving animals. Owing to its advantages of precise cell targeting, high spatial-temporal resolution, and millisecond-level rapid response, optogenetics has been widely used in the fields of neuroscience and brain science. In the future, it can be potentially applied to treat a variety of neurological and psychiatric diseases, such as Parkinson disease, Alzheimer disease, spinal cord injury, and schizophrenia [8–10].

As a bridge between the nervous system and the outside world, fiber-based devices play an important role in the development and application of optogenetic technology. It can not only deliver the pump light to the specific neurons for excitation or suppression of the target neuron, but also obtain the information of neural activities, such as electrophysiology signals and imaging [8, 11]. The success of optogenetics is considerably contributed by the development of the fiber-based probes. Over the past decades, a variety of fiber-based probes have been developed, including glass fiber probes, nonstretchable polymer fiber probes, and stretchable polymer fiber probes. The functions of these fiber probes evolve from initial optical

Optical and Electronic Fibers: Emerging Applications and Technological Innovations, First Edition.
Edited by Lei Wei.

stimulation to simultaneous optical stimulation and electrical recording, and even to the multifunction of neural recording, optical stimulation, and drug delivery. Here the advances of these fiber-based probes are summarized, and the future prospects are also discussed.

7.2 Basic Principle of the Optogenetic Technology

Generally, the emergence of optogenetic technology was triggered by the discovery of light-gated opsin genes, which can confer to specific neurons with light-response capability and cell specificity [9]. The optogenetic technology usually involves three core characteristics: the first is opsins, which can respond to the outer optical stimulation; the second is cell specificity, which can target the specific neurons by expressing opsin genes; and the third is the readout of neural signals. This technology can activate or inhibit the physiological activity of specific neurons by expressing the opsins in specific nerve cells. The principle of activation or inhibition lies in the selection of opsins [12, 13]. Currently, the commonly used photosensitive channel proteins can be classified into two categories according to their functions. One is the activation type, such as channelrhodopsin 2 (ChR2), which is sensitive to blue light. Another is the inhibitory type, such as halorhodopsin and archaeorhodopsin (NpHR), which is sensitive to yellow light [14].

As shown in Figure 7.1, if the channel proteins expressed in the target neuron cell belong to the activation type, the channel can be opened and cations flow in while the neuron is irradiated by blue light (e.g. 473 nm) [8]. This results in the generation of depolarization potential and induces the action potential to activate the target neurons. If the inhibitory type proteins are transferred into target cells, anion influx will occur when cells are irradiated by yellow light (e.g. 598 nm) [16]. This will result in hyperpolarization potential, which is difficult to release action potential and inhibits the activity of the target neurons. Therefore, the manipulation of neural activities such as activation or inhibitory lies in the selection of light-sensitive channel protein.

Currently, the commonly used methods to introduce photosensitive proteins into target neurons include viral vector, transgene, and electroporation [15]. After the expression of light-sensitive channels, fiber-based probes are implanted into the target brain region, which is used to deliver light to specific areas and stimulate the neurons. Then, the precise manipulation of neuron activities can be achieved by the controlling of light parameters, such as wavelength, intensity, and frequency. In addition, the fiber-based probe must deliver enough energy to opsin-mediated neuron for excitation or inhibition [17]. The area of light irradiation can be tailored by tuning the parameters of the fiber probe, such as the numerical aperture and the incident light power [18].

Except for light delivery ability, more functions are subsequently integrated into a single fiber probe, such as electrical recording, drug delivery, which offers more opportunities for multimodality neural manipulation. Thus, the development of

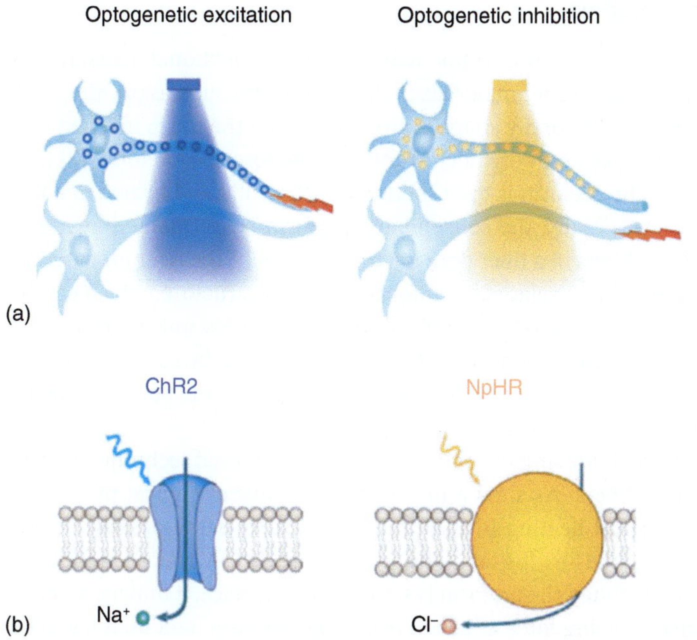

Figure 7.1 (a) The schematic diagram of optogenetic excitation and inhibition with a blue light-activated channel rhodopsin and a yellow light-activated halorhodopsin, respectively. *Source:* Reproduced with permission [8] Copyright 2010, Springer Nature. (b) Schematic diagram of cation and anion flows, respectively. *Source:* Reproduced with permission [15] Copyright 2010, Springer Nature.

fiber-based probes has important influence on the development and application of optogenetic technology. Next, we will introduce the contents about the design and fabrication of various fiber-based probes.

7.3 Fabrication Techniques of Advanced Fibers

Generally, the fiber fabrication process contains two steps. The first step is fabrication of a macroscopic preform, whose composition and structure can be rationally designed according to the preset goals. The shape of the preform can be square, rectangular, elliptic, circular, etc. The constituent materials can be metals, semiconductors, crystals, polymers, and even microdevices (e.g. μLEDs) [19, 20]. The second step is the processing of the macroscopic preform into uniform and thin fibers whose diameter can be precisely controlled, and the final fiber can preserve the original shape and structure of the prefabricated preform. In the past decades, with the development and progress of materials processing technologies, various methods have been developed for fabrication of fibers, as presented in the following contents.

7.3.1 Rod-in-tube Method

The rod-in-tube method is a common approach to fabricate functional fibers. In this approach, various functional core materials in different forms such as powder, bulk materials, thin films, etc. are firstly shaped into a rod. Then the rod is inserted into a cladding tube (e.g. glass and polymer) to form a macroscopic preform [21], as shown in Figure 7.2. The preform is then solidified into a solid and heated in a furnace to softening state. Finally, the consolidated preform is thermally drawn into thin and long uniform fibers in a fiber drawing tower. Usually, the fiber prepared by the rod-in-tube method has a relatively uniform core–clad structure, which can be controlled by the preassigned size and proportion of the macroscopic preform. The advantages of this approach are that the preform structure can be easily processed, and it is suitable for a variety of materials, especially for glasses and polymers owing to their good machining properties.

Currently, the thermal drawing is one of the most widely used technique to process the preform into fibers. As shown in Figure 7.3, a macroscopical preform is usually fed into a vertical heating furnace to be heated to soften or melt in this method [22]. Generally, the heating temperature is about 50 °C higher than the T_g of the cladding materials. When the preform is soften in the furnace, it undergoes neck down by an external drawing force attached below the preform, and then it can be drawn into uniform and long fibers.

During the thermal drawing process, the diameter of the fiber (D_f) is dependent on the drawing speed (V_d) of the fiber, the feeding speed of the preform (V_f), and the initial diameter of the preform (D_p), it can be represented as follows:

$$D_f = D_p\sqrt{V_f / V_d} \tag{7.1}$$

Using this simple relation, various diameters of fibers with sophisticated architecture and complex functionalities such as photonic crystal fibers (PCFs) and photonic bandgap (PBG) fibers, optoelectronic fibers, electronic fibers, and

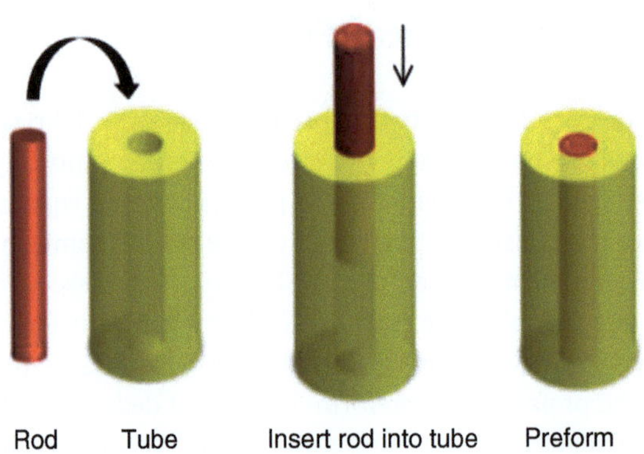

Figure 7.2 The rod-in-tube approach. *Source:* Reproduced with permission [21] Copyright 2015, Springer Nature.

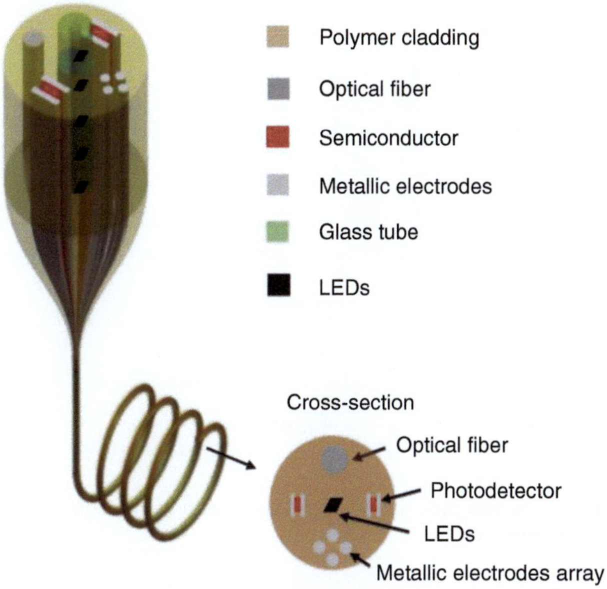

Figure 7.3 The thermal drawing technique. *Source:* Reproduced with permission [22] Copyright 2018, JOHN WILEY & SONS, INC.

elastomer fibers can be successfully obtained. Moreover, the final fiber can maintain the geometry as the preform, enabling the product highly predictable and controllable. Furthermore, recent research shows that functional materials such as metals, crystals, and even μLEDs, which are nonthermoplastic can also be incorporated into the fibers [23, 24], which greatly expand the candidate materials and extend the functionality of the conventional fibers.

7.3.2 Molten-core-in-tube Method

Different from the rod-in-tube approach, the melting temperature of the core materials is controlled to be lower than the fiber drawing temperature in molten-core-in-tube method [25, 26], as shown in Figure 7.4. This method was firstly developed by the group of John Ballato in the year of 1995 [26]. In this method, the core materials are molten during the whole drawing process, while the cladding materials remain in softened state. In order to prevent excessive volatilization of the fiber core material, the boiling point of the core materials should be higher than the drawing temperature. The commonly used core materials are glass, semiconductors, and ceramics, etc. Because the core material is in a molten state during the drawing process, the mismatch in the thermal expansion coefficient of heterogeneous materials can be well solved. For example, the group of John Ballato, Yoel Fink and Zhongmin Yang, et al. fabricated many kinds of composite fiber with semiconductor (e.g. Se, $Se_{0.8}Te_{0.2}$, Sb_2Se_3, As_2Se_3, etc.) as core, and glass or polymer as cladding

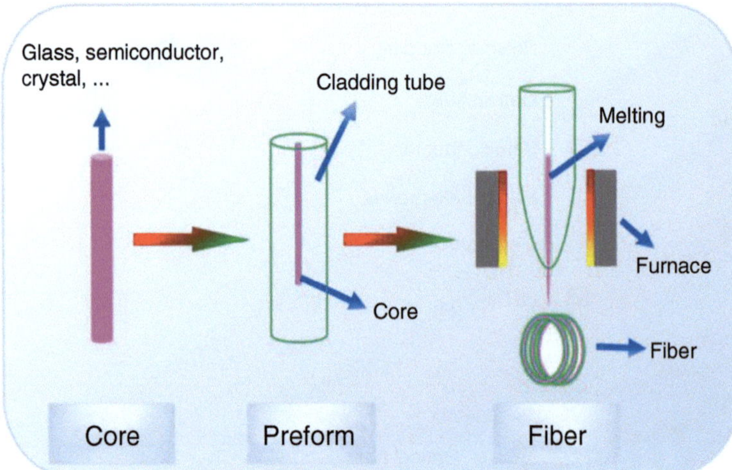

Figure 7.4 The molten-core-in-tube approach. *Source:* Reproduced with permission [25] Copyright 2020, Royal Society of Chemistry.

(e.g. phosphate glass, borosilicate glass, polymethyl methacrylate, polyether sulfone, etc.) [19, 23, 25–30].

In addition, the core materials may interact with cladding materials at high temperatures, making this method having potential to obtain functional cores with extreme components, which may greatly expand the range of materials selection. For example, John Ballato and coworkers fabricated Si core fibers utilizing the reaction between SiC and SiO_2 in quartz cladding in 2011 [31]. Subsequently, this group prepared amorphous and crystalline oxide–glass composite fibers by in-situ chemical reaction using Bi_2O_3 and GeO_2 as fiber cores and sodium borosilicate glass as cladding [32]. In 2013, Yoel Fink et al. fabricated high-melting-point ZnSe-polymer fibers by the reaction of low-melting-point constituent materials $Se_{97}S_3$ and $Sn_{85}Zn_{15}$ [24]. In 2015, this group reported the fabrication of crystal Si core glass cladding fibers utilizing the reaction between Al and SiO_2 in the cladding during the drawing process [33].

7.3.3 Thin-film Rolling Method

The thin-film rolling method is usually used to fabricate polymer fibers. As shown in Figure 7.5, in this method, the raw materials to be pulled into the fibers are firstly processed into thin films, and then rolled into rods and consolidated to make the constituent materials in close contact [34]. This method is be well-suited for a variety of thermoplastic polymer materials due to their excellent flexibility and machinability. In addition, it has unique advantages for preparing PCFs and PBG fibers. For example, in 2002, Burak Temelkuran and coworkers developed an omnidirectional dielectric mirror fiber with the composite of multilayer polymer and chalcogenide semiconductor compound by this method [35, 36]. The structure of alternating composite of multilayer high-index glass and low-index polymer constructs a large photonic band gap. The transmission loss is less than 1 dB/m when used to transmit

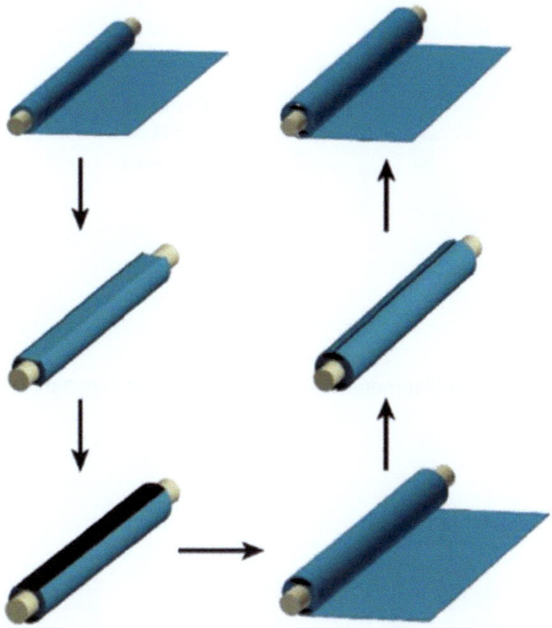

Figure 7.5 Thin-film rolling approach. *Source:* Reproduced with permission [34] Copyright 2019, JOHN WILEY & SONS, INC.

carbon dioxide laser at 10.6 μm wavelength, which is several orders of magnitude lower than the traditional core–shell structure fibers.

Generally, the rheological behavior and elongation of constituent materials at the drawing temperature are the key factors affecting the fiber preparation. The materials need to be sufficiently softened to be drawn to avoid the excessive internal stress. This needs the match of the thermal expansion coefficient and wettability between the different constituent materials.

7.3.4 Extrusion Method

The extrusion technique is generally used to create fibers with fixed cross-sectional structures by pushing soft materials through a die under pressure. As shown in Figure 7.6, in this approach, the billet is heated in a sleeve until it reaches the viscous state [21]. Then, the billet is slowly extruded from the sleeve mold through a screw or piston. Next, a cylindrical bar can be obtained, whose diameter is determined by the diameter of the mold opening. In addition, the cross-section structure of the extrusion bar can be adjusted by designing the outlet structure of the die, which is not limited to the circle. In 2004, Daniel J. Gibson et al. used this method to fabricate a variety of hollow-waveguide fibers with full dielectric layer by extrusion [37]. In this experiment, a multilayer fiber with intermittent refractive index distribution was prepared by As_2Se_3–PSU composite. In 2012, Guangming Tao and coworkers prepared multimaterial chalcogenide fibers using this method [38].

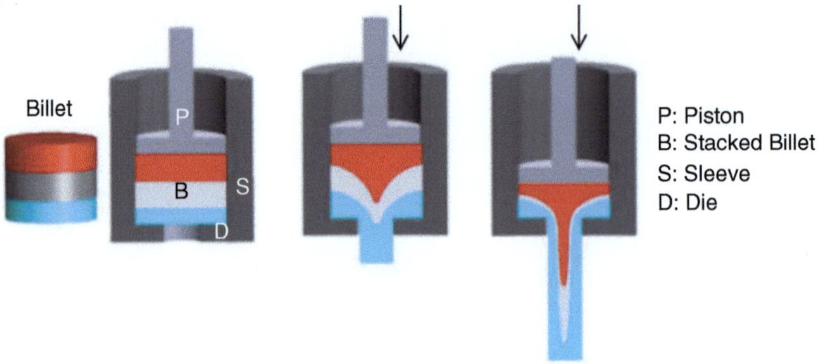

Figure 7.6 The extrusion technique. *Source:* Reproduced with permission [21] Copyright 2015, Springer Nature.

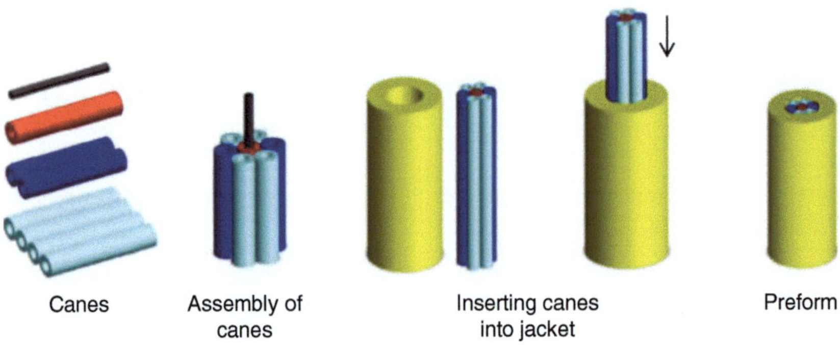

Figure 7.7 Stack-and-draw method. *Source:* Reproduced with permission [21] Copyright 2015, Springer Nature.

7.3.5 Stack-and-draw Method

Stack-and-draw method is usually used to fabricate microstructured fibers, such as PCFs and PBG fibers. As shown in Figure 7.7, this method has virtually unlimited structure of the preform, and the number of stackings can be extended, so the fiber with very thin internal core can be prepared [21]. It has unique advantages to fabricate PCFs and PBG fibers, which are widely used as optical waveguide, and for optical imaging technology. In 2006, E. F. Chillcce and coworkers first fabricated tellurite PCFs by this method [39]. In 2009, Liao Meisong and coworkers fabricated tellurite glass cladding/chalcogenide glass fiber with a low transmission loss of 18.3 dB/m at 1.55 μm using this method [40].

7.3.6 3D Printing Approach

3D printing is the latest type of technology to fabricate functional fibers. This method can assemble a variety of materials into a single macroscopic preform with various structures at the same time [41], as shown in Figure 7.8. Moreover, it can

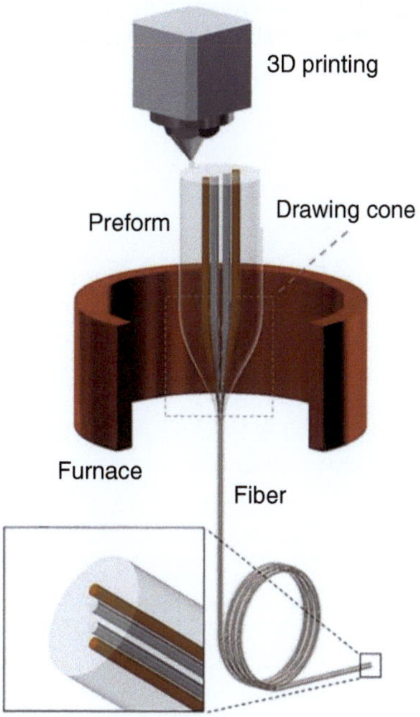

Figure 7.8 The 3D printing approach. Faccini de Lima et al. [41] / Springer Nature / CC BY 4.0.

digitally arrange the constituent materials. The advantages of 3D printing are obvious, which can efficiently, cheaply, and quickly prepare various structured preforms, especially special-shaped structures, offering more opportunities for the devices in various fields such as micronano electronics, smart fabrics, Internet of Things, and biosensors. In 2015, Cook Kevin et al. firstly combined 3D printing and the fiber drawing technique. They used fused deposition modeling technology to print a cylindrical preform with six holes using polystyrene [42]. The attenuation of the obtained fiber was comparable to those fibers drawn from preforms made by conventional methods. Toal Jr P M et al. further printed multimaterial performs with various structures [43]. Inspired by the Fused Deposition Modeling technology, Klein John and coworkers firstly developed a 3D extrusion system for molten glass material, and they successfully printed complex-structure glass with varying colors [44]. In 2016, Kaufman Joshua J et al. fabricated preforms with complex geometry by 3D printing of the fluorescent dye-mixed core material (cyclin olefin polymer) and cladding material (polystyrene), and assembling them into preforms. Subsequently, the preform was thermally drawn into fibers [45].

7.3.7 Double-crucible Technique

Double-crucible technique is a simple method to fabricate fibers with low-cost. In this method, the fiber core and cladding were heated by two different crucibles [46], as shown in Figure 7.9. When the core and cladding materials are heated to the melting state, the fibers can be obtained by flowing down the melts from the middle coaxial leakage nozzle at the same time. The diameter of the fiber is directly determined

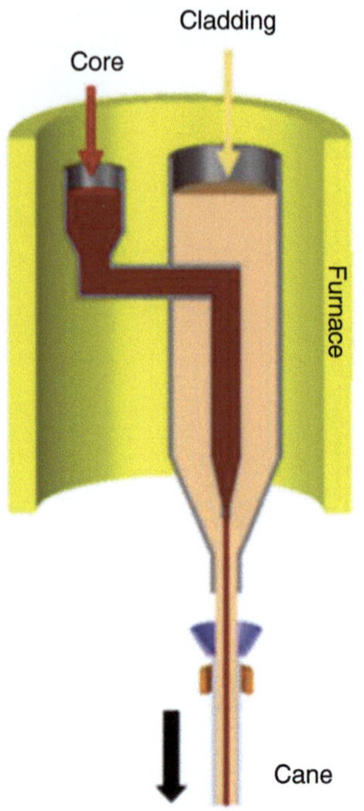

Figure 7.9 Double-crucible approach. *Source:* Shabahang et al. [46] / Springer Nature / CC BY 4.0.

by the size of the leakage nozzle, the drawing speed, and the drawing temperature. Compared with the conventional rod-in-tube method, the double-crucible method exhibits several prominent advantages: (1) wide selection of raw materials; (2) suitable for batch continuous production, which can reduce costs; (3) the bonding of the core–shell interface is smoother and better. Mossadegh R., et al. prepared a single-mode chalcogenide glass fiber by this process in 1998, and the loss of the fiber is as lower as 1 dB/m at 2.7 μm [47]. In 2017, Shabahang S and coworkers fabricated a low-loss chalcogenide glass preform by the double-crucible method, and then the preform was drawn into fibers by thermal drawing, achieving a transmittance of more than 97% at 1.5 μm [46].

However, due to the high processing temperature, the fiber core material is easy to volatilize, and the element diffusion between the core and the cladding layer is difficult to avoid. At the same time, the gas is easy to be introduced by the external feeding process, which may lead to the existence of bubbles in the final fiber. Therefore, the optical performance of the final fiber is still required to be improved.

7.3.8 High-pressure Chemical Vapor Deposition Technique

High-pressure chemical vapor deposition is a method to fabricate fibers based on phase transformation. As shown in Figure 7.10, the capillary glass tube with ultra-high length to diameter ratio is used as a container for chemical reaction, and the precursor of gas-phase fiber core materials is transmitted along the axial direction of optical fiber under high pressure [48]. Then, the precursor is deposited on the

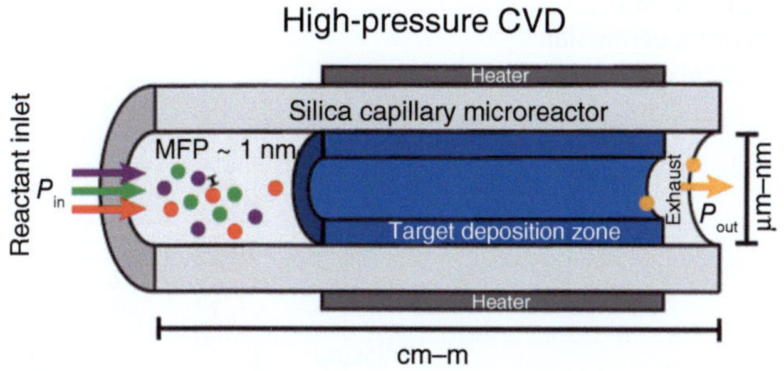

Figure 7.10 High-pressure chemical vapor deposition approach. *Source:* Reproduced with permission [48] Copyright 2013, JOHN WILEY & SONS, INC.

inner wall of the tube through heat treatment or photocatalysis. In 2006, Sazio Pier J A et al. firstly used silica capillaries and microstructured optical fibers as templates to deposit semiconductors into air holes by chemical vapor deposition, and obtained semiconductor–glass composite optical fibers [49]. This method can produce extremely pure semiconductor wires with excellent optical and electrical properties, and the diameters ranging from hundreds of nanometers to several micrometers [49–53].

7.3.9 Pressure-assisted Melt Filling Technique

Pressure-assisted melt filling is a method of pumping high-temperature molten core materials into hollow channel of cladding to fabricate fibers [54], as shown in Figure 7.11. This method needs to meet three conditions: First, the core material is in a molten state, and its viscosity is low than 10 Pa s; second, the filling temperature is lower than the transition temperature of the glass cladding; and, finally, the filling material does not chemically react with the cladding glass. In 2008, Russell P. S. J. and coworkers firstly reported the use of this method to fill the semiconductor Ge into a single hole of a quartz PCFs [55]. In 2011, Nicolai Granzow prepared all-solid chalcogenide wide-band gap PCFs with chalcogenide glass as the fiber core and capillary quartz tube as the cladding by this technique [54]. The advantage of this technique is that few glass is needed. However, the production efficiency is low for this method.

7.3.10 Laser-heated Pedestal Growth Technique

Laser-heated pedestal growth is usually used to grow single-crystal fibers. As shown in Figure 7.12, the CO_2 laser is commonly used to form the single-crystal core materials in this method, and it is possible to grow pure single-crystal fibers with few defects [56]. However, the preparation efficiency of fiber is relatively low due to the slow growth rate. In 2002, Lo Chia Yao prepared Nd: $Y_3Al_5O_{12}$ crystal fiber by this method [57]. The laser experiment demonstrates that the laser output power can

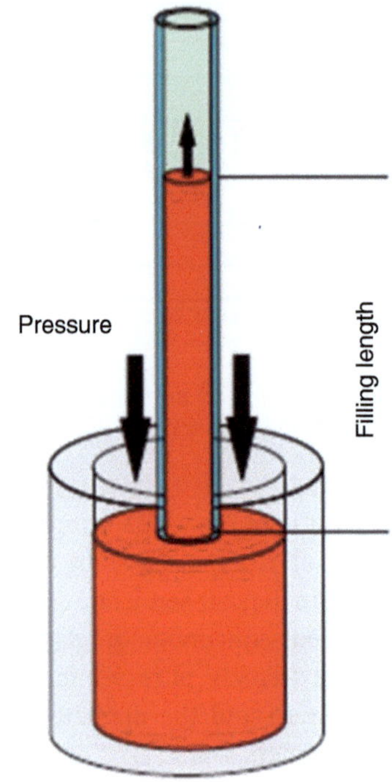

Figure 7.11 Pressure-assisted melt filling technique. *Source:* Reproduced with permission [54] Copyright 2011, Optical Society of America.

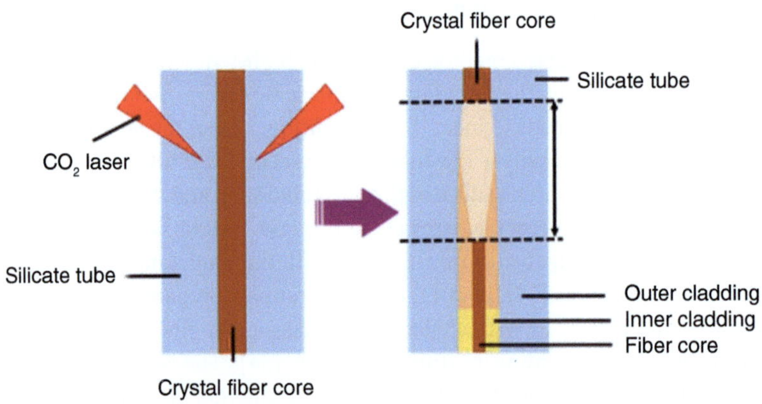

Figure 7.12 Laser-heated pedestal growth technique. *Source:* Reproduced with permission [56] Copyright 2011, Optical Society of America.

reach to 145 mW, and the slope efficiency is 28.9%. In 2008, Lai Chien Chih of the same research group prepared Cr: $Y_3Al_5O_{12}$ crystal fiber by this method [58]. The fabricated fiber has a smaller core diameter, a lower laser threshold, and a transmission loss as low as 0.08 dB/cm. In 2009, Lai Chien Chih prepared Yb: $Y_3Al_5O_{12}$ crystal fiber. The slope efficiency of 7 mm long Yb: $Y_3Al_5O_{12}$ fiber reaches to 76.3%, and the corresponding extraction power was close to 1 W/cm [59].

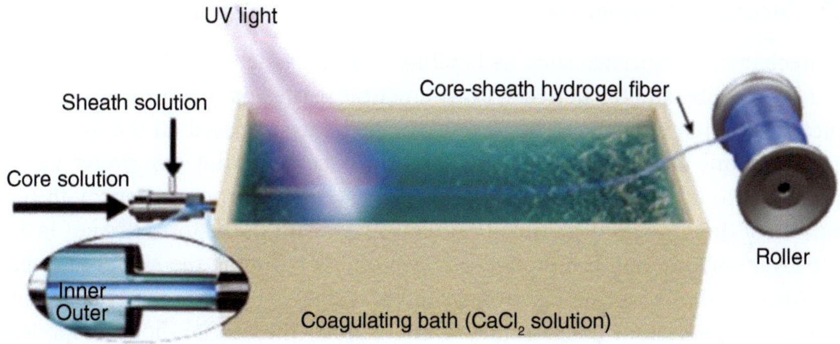

Figure 7.13 The diagram illustration of the integrated dynamic wet spinning technique. *Source:* Chen et al. [60] / Oxford University Press / CC BY 4.0.

7.3.11 Integrated Dynamic Wet Spinning Technique

Integrated dynamic wet spinning is a new technology developed recently by the group of Meifang Zhu [60], as shown in Figure 7.13. Based on the dynamic crosslinking wet spinning process and microchannel spinning process, a variety of core-coated hydrogel fibers were developed. In this process, the diameter, mechanical properties, and optical properties of the fiber can be accurately controlled. The hydrogel fiber prepared by this method has high light transmission, mechanical flexibility, and biocompatibility. The fibers exhibit great promise for a wide range of applications such as optical therapy, sensing, optogenetics, flexible electronics, etc.

7.4 Design Rules of Fiber-based Neural Probes

7.4.1 Biocompatibility

The biocompatibility of the fiber-based probes is one of the most important factors that govern their long-term practical applications. For achievement of the excellent biocompatibility, various types of flexible materials such as hydrogel, elastomer, carbon nanotube (CNT), etc. have been employed as the materials candidates. For the probes with poor biocompatibility, surface coating techniques are usually applied to modify the interface between the neural tissue and the probe. For example, polyethylene glycol (PEG) hydrogel was used to improve the interfacial biocompatibility of the probe [61]. This approach can reduce gliosis near the probe, but the long-term implantation still remains an issue due to the gradual degradation of the coatings. Moreover, anti-inflammatory drugs and bioactive molecules, such as dexamethasone, were also coated on the surface of the optogenetic probe to reduce the neuroglia [62–64]. The integration of neuro-attractive biomolecule and stem cells on the probe surface was demonstrated to be able for preventing gliosis and at the same time favorable for the neurite growth [65–68].

7.4.2 Mechanical Properties

The mechanical properties such as bending stress, elasticity modulus, and stiffness of the probe should match with the neural tissues as much as possible to reduce the foreign body response. These can be discussed based on the following factors. On one hand, glial cell proliferation occurs around the probe when implanted into target region in the brain, and it becomes more obvious with the extension of the implantation time [69]. This may result in the formation of insulating layer surrounding the probe, thus potentially reducing the signal-to-noise ratio (SNR) of the recording electrophysiological signals [70]. On the other hand, the probe implantation process may easily induce the foreign-body response, including blood–brain barrier, cell death, and inflammatory cell proliferation [71–73]. Moreover, the relative displacement between neural tissue and probes during respiration or heartbeat would also cause foreign body response, and it will become prominent with the increase of the mechanical stiffness of the probe [74, 75]. From these points of view, improving the flexibility of the probe can reduce the foreign-body response.

Figure 7.14 shows the Young's modulus and the bending stiffness of materials commonly used as recording electrodes [76]. Generally, the bending stiffness can be

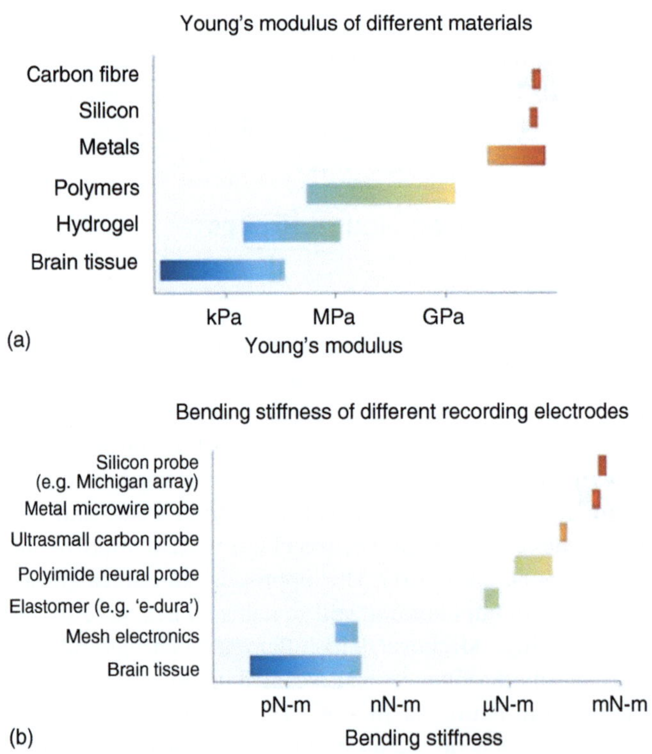

Figure 7.14 Mechanical properties of neural probes. (a) Young's modulus. (b) Bending stiffness. *Source:* Reproduced with permission [76] Copyright 2019, Springer Nature.

used to represent the flexibility of the fiber-based probe, and it is defined as the force (F) required for reaching a certain deflection (d):

$$F = \frac{48EI}{L^3} \bullet d \tag{7.2}$$

where E is the Young's modulus of the probe, I is the moment of inertia, and L is the length of the probe. It can be seen that the bending stiffness of the neuron probe depends on the Young's modulus of the materials and the moment of inertia, and this gave us a guide to design the materials for neural probes.

7.4.3 Optical Properties

The optical properties of the fiber-based probes, including optical loss, transmittance, numerical aperture, are very important parameters for the optogenetic modulation. Generally, the fiber probe must deliver enough energy to stimulate the opsin for excitation or inhibition of specific neuron (e.g. 1 mW mm^2 for ChR2, 7 mW mm^2 for NpHR) [17]. In addition, the area of the affected tissue can be tuned by the rational design of the numerical aperture of the fiber probe. Considering the optical scattering and absorption within the neural tissue, the light with the minimum power of ~10 mW mm^2 should be delivered at the fiber tip [77].

In the early research of optogenetics, the light was mainly delivered by conventional silica fibers, while the neural signals were recorded by various electrodes, such as tungsten, Utah arrays and Michigan probes. Recently, the combinatorial approaches have been proposed by combining the recording arrays with the integrated μLEDs. Fully wireless probes integrated with μLEDs and laser diode chips have emerged to enable optogenetics manipulation in behaving animals. Most recently, tapered silica fibers were introduced for tuning the irradiation depth and area, enabling for delivering light with multiwavelength to the multiple regions at the same time.

7.4.4 Electrical Properties

The electrochemical performance of the fiber probes, including the impedance, charge injection capacity, and pulses stability, are very important for its long-term efficiency neural recording. To improve the electrical properties of the probe, highly conductive metals are widely used as electrodes by depositing on flexible polymer matrix (e.g. polyimide, parylene, PDMS, etc.) via photolithographic or printing techniques [78–83]. Organic conductors such as poly(3,4-ethylene-dioxythiophene): poly (styrene sulfonate) (PEDOT: PSS) are also used as recording electrodes due to their lower moduli and comparable electrical properties to those of metals [84]. In addition, carbon-based materials such as CNT, graphene with high conductivities comparable to those of metals, are increasingly used as recording electrodes. For example, it has been demonstrated that a single carbon fiber enables long-term recordings of isolated action potentials and local field potentials (LFPs) with minimal tissue damage [85].

The electrical properties are also crucial for flexible probe. For improving it, carbon, metal, or organic conductor materials are integrated into polymer matrix to reduce the stiffness of the electrode while maintaining conductivity [86, 87]. For example, silver nanowires were integrated into the styrene–butadiene–strylene elastomer as neuron interface [88]. Polymers doped with CNTs and graphene were used for neural recording [89, 90]. Being compatible with established microfabrication techniques, these carbon-based composites could also be processed into electrodes of low dimensions. These materials are mechanically and chemically stable while simultaneously offering high conductivity, which is suitable for electrophysiological recording.

7.5 Fiber-based Neural Probes for Optogenetics

Fiber-based devices are widely used in optogenetics, from silica fibers to flexible polymer fibers. The development of fiber probes plays an important role in promoting the development and application of optogenetic technology. In the past decades, various fiber-based probes such as glass fiber-based probes, polymer fiber-based probes have been developed. Here, the representative advances of these fiber-based probes are summarized.

7.5.1 Glass Fiber-based Neural Probes

Glass fibers, especially silica fiber, are widely used in optogenetic technology due to its advantages of small size, lightweight, low optical loss, and wide transmission range from visible to infrared band. In 2007, Karl Deisseroth and coworkers firstly applied the multimode silica fiber for *in vivo* optogenetics [91]. The diameter of the used fiber is ~200 μm, and the optical intensity of the fiber end is ~380 mW/mm^2, which is sufficient to stimulate the ChR2 neurons in behaving mice. To simultaneously record the electrophysiological signal upon optical stimulation, Viviana Gradinaru et al. adhered tungsten electrodes (with a diameter of ~125 μm) on the silica fiber. The resulting probe allows for simultaneous optical stimulation and electrical recording in behaving mouse [92]. Moreover, the probe can drive or inhibit an array of distinct circuits in freely moving Parkinsonian rodents. To synchronously record more neural signals from a single probe, Polina Anikeeva et al. integrated 16 nickel–chromium alloy (with a diameter of 12 μm) into a silica fiber, as shown in Figure 7.15a. The composite probe allows for multichannel electrophysiological recording and optical stimulation in freely moving mice for more than six weeks [93]. Moreover, the probe is highly compact and lightweight (~2 g), which makes it suitable for application in behaving mice.

To reduce the size of the probe and obtain the neural stimulation with high spatial and temporal precision, tapered silica fiber was subsequently developed and applied for optogenetics. As shown in Figure 7.15b, a typical tapered fiber probe developed by Jiayi Zhang et al. allows for optical stimulation and electrical recording in living mouse. Specifically, epileptiform events can be successfully triggered by the probe,

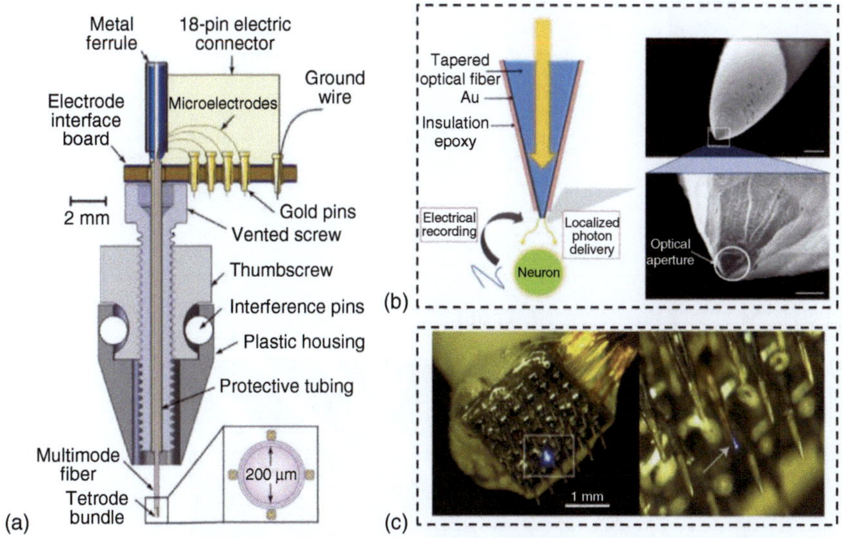

Figure 7.15 (a) The structure of the composite neural probe. *Source:* Reproduced with permission [93] Copyright 2011, Springer Nature. (b) The schematic and SEM images of the tapered fiber probe. *Source:* Reproduced with permission [94] Copyright 2009, IOP Publishing. (c) Image of the array devices. *Source:* Reproduced with permission [95] Copyright 2012, IOP Publishing.

and the electrophysiological signals can be simultaneously recorded [94]. They further fabricated a hybrid device using the tapered fiber, as shown in Figure 7.15c. It was fabricated by assembling tapered optical waveguide and microelectrode array. The hybrid device allows for optical stimulation and electrical recording for more than eight months [95]. Similarly, Se′ bastien Royer et al. integrated silica fibers onto the shanks of commercial Michigan probes to achieve the functions of both recording and optical stimulation [96]. As shown in Figure 7.16a,b, the used Michigan probe have four or eight shanks, which are 250 μm apart and have eight recording sites. The probe can record from 140 well-clustered neurons in the hippocampus and neocortex.

To further obtain more stable and accurate light stimulation, Fan Wu and coworkers integrated microscopic light-emitting diodes (μLED) and electrodes onto silicon probes [97]. As shown in Figure 7.16c,d, each μLED and recording site has dimensions similar to a neuron cell, providing confined emission and electrophysiological recording of action potentials. They fabricated and implanted four-shank probes, each integrated with 12 μLEDs and 32 recording electrodes, into the freely moving mice. Spikes can be clearly observed under excitation with a 60 nW light power, and fast population oscillations can be induced at the microwatt range. They showed the independent control of distinct cells ~50 μm apart, demonstrating the spatiotemporal precision of parallel stimulation and recording. The scalability and spatiotemporal resolution of this probe provides versatility and precision for cellular-level circuit analysis in freely moving animals.

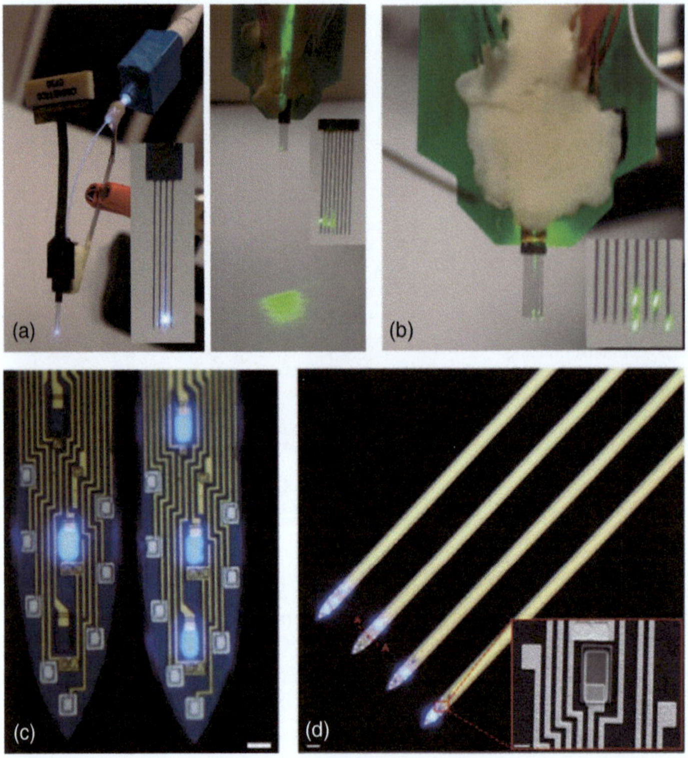

Figure 7.16 (a, b) Four- or eight-shank silicon probes equipped with optical fibers. *Source:* Reproduced with permission [96] Copyright 2010, John Wiley and Sons. (c, d) 3D schematic of the μLED probe, and the probe consists of four shanks, and each shank is integrated with eight recording sites and three interspersed μLED. *Source:* Reproduced with permission [97] Copyright 2015, Elsevier.

The rational control of illumination area and intensity of the probe is important for optogenetic manipulation. Recently, Ferruccio Pisanello and coworkers develop a fiber probe whose illumination area and intensity can be easily tuned [98], as shown in Figure 7.17a. More importantly, the light-emitting position can be tuned by changing the light input angle, thus gifting the probe with the ability to achieve broader and more efficient optogenetic activation of neurons compared to the standard flat-faced fiber stimulation. Except for illumination area and intensity, the number of recording sites in the probe is also very important. James J. Jun et al. developed a multichannel recording probe named Neuropixels, which is composed of 384 recording channels [99], as shown in Figure 7.17b. The probe allows for neuron recording with high spatiotemporal resolution and large coverage volume. The fully integrated and small-sized neuropixels probe allows large populations of neurons to be recorded in freely moving animals, which opens a path towards recording of brain-wide neural activity in behaving animals.

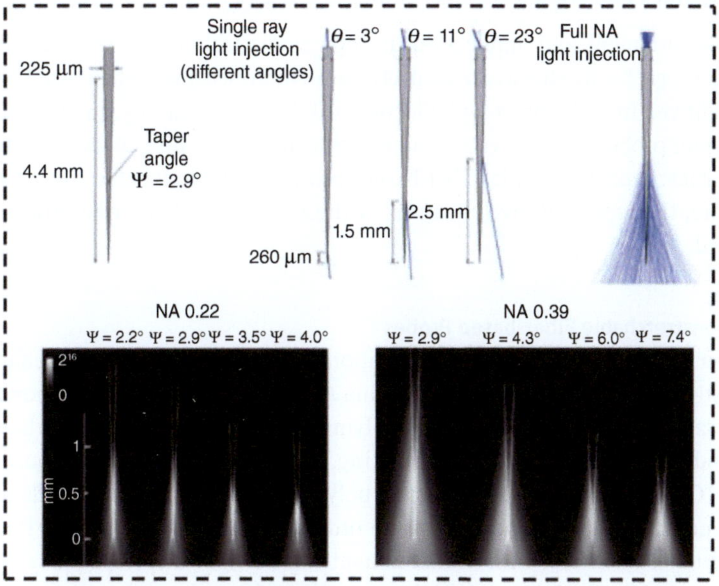

(a)

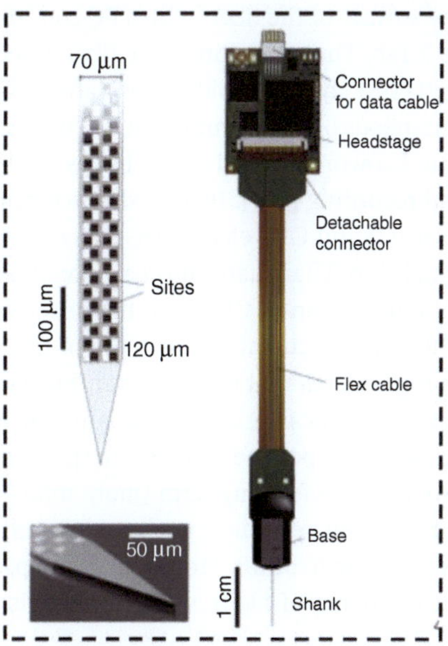

(b)

Figure 7.17 (a) Schematic of the fiber probe and its ray-tracing simulations of emissions from the tip with a single ray into the fiber at different angles. *Source:* Reproduced with permission [98] Copyright 2017, Springer Nature. (b) The configuration of the Neuropixels probe. *Source:* Reproduced with permission [99] Copyright 2017, Springer Nature.

7.5.2 Polymer Fiber-based Neural Probes

With the development and application of optogenetic technology, polymer fiber-based probes have been extensively used as neuron interface due to their advantages of being lightweight and having high biocompatibility, stretchability, and flexibility. These polymer probes can be divided into two categories according to their mechanical properties: nonstretchable fiber-based probes and stretchable fiber-based probes. Here, the representative advances of these polymer fiber-based probes are summarized.

7.5.2.1 Nonstretchable Fiber-based Probes

All-polymer fiber probe for optical stimulation and neuron recoding in behaving mice is firstly developed by the group of Polina Anikeeva in 2014 [100]. As shown in Figure 7.18a, the probe is composed of a polymer waveguide (with a polycarbonate core, PC and a cyclic olefin copolymer cladding, COC), and two conductive polymer electrodes (Carbon black doped conductive polyethylene, CPE). It is highly flexible, and the bending stiffness of the probe is an order of magnitude lower than that of the conventional silica fiber with similar diameter and length. The impedance of the probe is up to 6 MΩ, which is much higher than that of the conventional metal electrodes. To improve the conductivity of the probe, this group further coated conductive silver nanowires (AgNWs) on the surface of the fiber (with a PC core and a COC cladding) [101], as shown in Figure 7.18b. Then, the composite fiber was encapsulated with a polydimethylsiloxane (PDMS) layer to avoid the oxidation of AgNWs. The final fiber probe allows for simultaneous optical stimulation and neuron recording in behaving mice. Flavia Vitale and coworkers used CNT fiber as electrodes for bidirectional neuron stimulation and recording [107]. Its impedance is remarkably lower than that of the conventional metal and CPE electrodes, thus enabling it for recording single-neuron activity. In addition, Yuanyuan Guo et al. integrated carbon nanofiber (CNF) into conductive polyethylene (CPE) as the recording electrode [102], as shown in Figure 7.18c. The electrical conductivity of the resulting fiber probe is 2 orders of magnitude higher than that of the conventional CPE electrode, while maintaining its excellent mechanical flexibility. Minghui Du and coworkers integrated the metal electrodes into polymer fibers [103], as shown in Figure 7.18d. The resulting fiber probe allows for long-term (more than 10 weeks) simultaneous optical stimulation and neural recording at the single-cell level in behaving mice with a signal-to-noise ratio up to 30 dB. Moreover, the impedance of the probe is only 22.71 MΩ cm^2 at the frequency of 1 kHz, which is much lower than conventional polymer electrodes.

To record more neural signals simultaneously by a single probe, it is necessary to increase the recording sites of the probe. Christopher J. Davey et al. developed a multifunctional, tapered polymer fiber probe with high-density electrodes arrays and optical waveguides [104], as shown in Figure 7.18e. The tapered shape of the probe is beneficial to improve the compatibility with chronic implantation due to the taper's overall minimum length. In addition, the optical losses can be reduced to the estimated 25 dB. Similarly, Antje Kilias and coworkers developed a flexible probe with

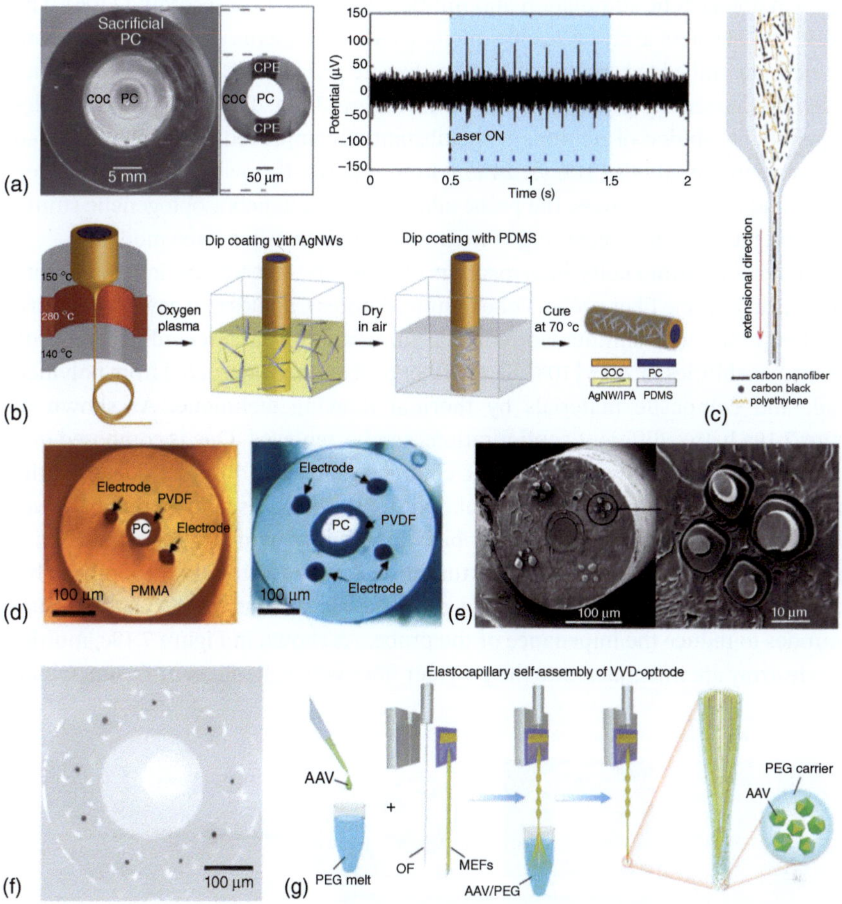

Figure 7.18 (a) The structure of the all-polymer fiber probe. *Source:* Reproduced with permission [100] Copyright 2014, JOHN WILEY & SONS, INC. (b) Fabrication process of the flexible fiber probe. *Source:* Reproduced with permission [101] Copyright 2017, AAAS. (c) A diagram of the in-situ alignment of the CNF during thermal drawing. *Source:* Reproduced with permission [102] Copyright 2017, American Chemical Society. (d) The structure of the metal–polymer composite fiber probe, including two and four metal electrodes. *Source:* Reproduced with permission [103] Copyright 2020, JOHN WILEY & SONS, INC. / CC BY 4.0. (e) SEM images of the multimodal fiber probe. *Source:* Reproduced with permission [104] Copyright 2015, Optical Society of America. (f) Cross-section of the hybrid probe. Spontaneous LFPs measured in a freely behaving mouse using the probe. *Source:* Reproduced with permission [105] Copyright 2018, IOP Publishing. (g) Schematics of the self-assembly process of the multifunctional system. *Source:* Zou [106] / Springer Nature / CC BY 4.0.

multielectrode [105]. As shown in Figure 7.18f, the probe is composed of 9 tin (Sn) electrodes, a polyetherimide (PEI) cladding, and a flexible optical waveguide with a polycarbonate (PC) core and cyclic olefin copolymer (COC) cladding in the center. Sinusoidal stimulation by the fiber probe results in the reliable and coherent neural signals. Recently, Liang Zou and coworkers present a multifunctional system for long-term optogenetic stimulation and multichannel electrical recording [106]. The system consists of an array of flexible microelectrode filaments and fiber optics, both of which are embedded in a flexible polyethylene glycol (PEG) carrier through an efficient self-assembly process, as shown in Figure 7.18g. The final probe enables optogenetic inhibition of neurons with substantially improved precision compared to conventional methods (91% vs. 24%). More importantly, owing to their one-step surgery and low invasiveness, the probe allows for simultaneous optogenetic stimulation and multichannel recording of neuronal populations for three months.

To achieve multimodality interrogation of neurons, researchers integrate more functions in a single fiber probe. For example, Andres Canales and coworkers fabricate fiber probes with simultaneous optical stimulation, neural recording, and drug delivery capabilities in 2015 [108]. These fiber probes are constructed from polymer, metal, and composite materials by thermal drawing technique. As shown in Figure 7.19a,b, two different configurations can be achieved. One is composed of a cylindrical waveguide, two microfluidic channels, and two CPE electrodes; and the other consists of a microfluidic channel, four CPE electrodes, and an optical waveguide. These fiber probes have good biocompatibility, which allows for *in vivo* recordings, stimulation, and drug perturbation for more than two months. This group also attempts to use low-melting-point metals such as tin (Sn) as recording electrodes to reduce the impedance of the probe. As shown in Figure 7.19c, multielectrode array are integrated into the polymer fiber with a diameter of ~5 μm, which

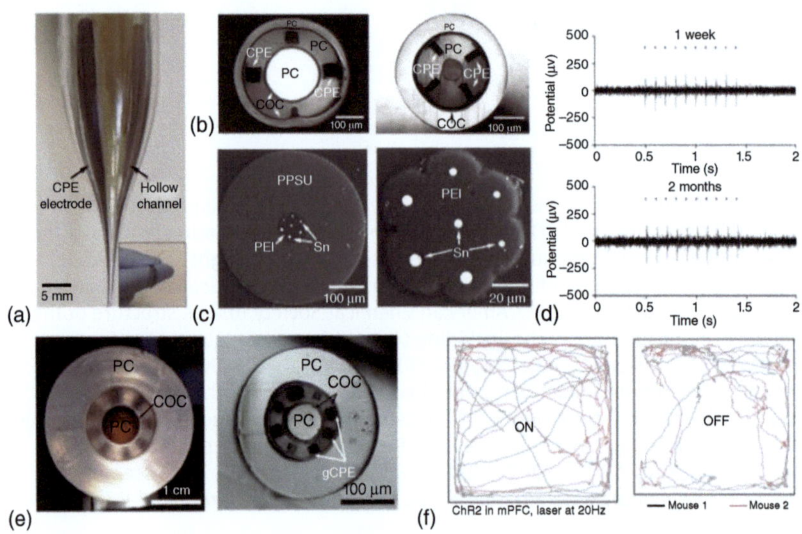

Figure 7.19 (a) A photograph of the drawn multimodal fiber probe. (b) Cross-section of two different fiber structures. (c) SEM images of a multielectrode fiber and the selective etching of the sacrificial layer. (d) The recording electrophysiological signals at one week and two months after implantation. *Source:* (a–d) are reproduced with permission [108] Copyright 2015, Springer Nature. (e) Cross-section of the preform and the drawn fiber. (f) Trajectories of a ChR2-transfected mouse during optical stimulation. *Source:* (e) and (f) are reproduced with permission [109] Copyright 2017, Springer Nature.

allows for single-neuron recordings with high SNR (Figure 7.19d). Subsequently, Seongjun Park et al. develop a probe composed of an optical waveguide, six recording electrodes, and two microfluidic channels by thermal drawing [109], as shown in Figure 7.19e. To improve the conductivity of the probe, they integrate graphite into the CPE electrode, and the resistance of the fiber can be reduced similar to fourfold. Though the composite fiber probe is lightweight (<0.5 g) and with miniature size (<200 μm), it allows for long-term optical, electrical, and chemical interrogation of brain circuits (Figure 7.19f). More importantly, these fiber probes can help to achieve one-step optogenetics by directly delivering viral vectors carrying opsin genes while simultaneously providing neural recording and optical stimulation.

The realization of deep-tissue optogenetics inquiry is essential for understanding the structure and function of the brain. Recently, Shan Jiang and coworkers developed a multifunctional fiber probe, which can provide three-dimensional coverage of brain tissue [110]. In addition, these probes can achieve long-term optical stimulation, electrical recording, and drug delivery with a single-unit resolution. The fabrication process of the probe is shown in Figure 7.20a; a preform is first fabricated and then thermally drawn into fibers. Various fiber structures can be obtained by rational design of the preform. Then, the drawn fiber is machined by a femtosecond laser to gift the function of multichannel optical stimulation, electrical recording, and microfluidic drug delivery, as shown in Figure 7.20b. The light in the optical waveguides is scattered out at the selected locations by partially ablating the cladding and core with the femtosecond micromachining process (Figure 7.20c). Expect for micromachining to create multifunctional interfaces, they further fabricate spatially expandable fiber probes using the helical scaffolding fiber (Figure 7.20d). The resultant fiber probe can record neural singles by different channels (Figure 7.20e). Moreover, these fiber probes can be used for studying the spatiotemporal structure of seizure (Figure 7.20f).

7.5.2.2 Stretchable Fiber-based Probes

Stretchable fibers such as hydrogel fibers and elastomer fibers have attracted increasing attention for applications as neuron interfaces in the recent years, due to their advantages of excellent mechanical properties such as low Young's modulus and compatible with the neural tissues. In 2013, Myunghwan Choi and coworkers firstly report the stretchable hydrogel fibers for *in vivo* optical sensing and therapy [111]. The implanted hydrogels show long-term transparency, biocompatibility, cell viability and light guiding properties (loss of <1 dB/cm) in freely moving mice for more than one week (Figure 7.21a,b). They further demonstrate that the hydrogels can be used for light-controlled therapy in a diabetic mouse with the help of optogenetics. Subsequently, Lulu Wang et al. fabricate a soft and high-stretchable alginate–polyacrylamide (PAAm) hydrogel fiber for chronic optogenetic neuron modulation in free-moving animals [112], as shown in Figure 7.21c–e. *In vivo* experiments show that the hydrogel fiber can modulate animal's behavior by optical stimulation, and it can significantly reduce the inflammatory response compared with the commercial silica fiber. Soon afterwards, this group develop a core–clad structured hydrogel fiber

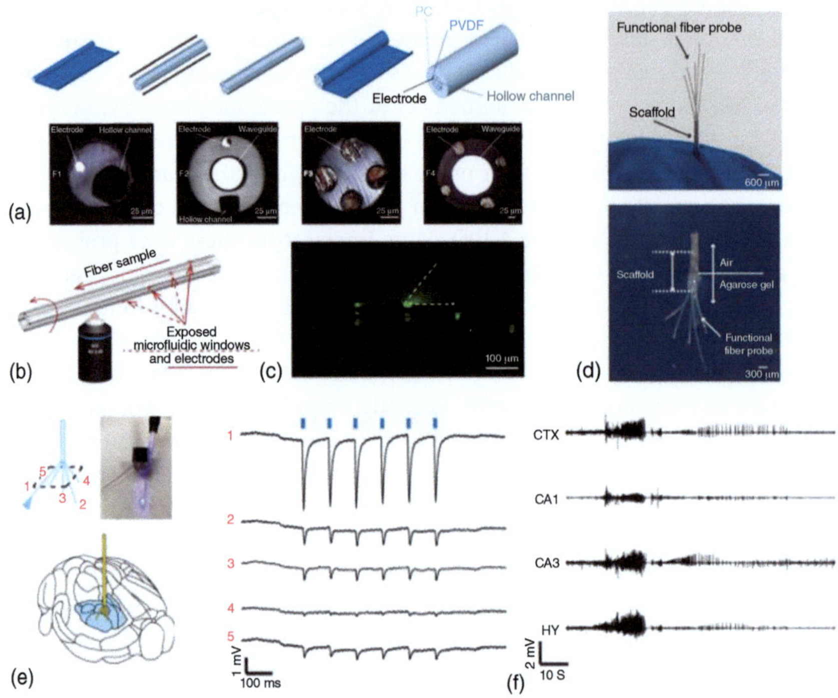

Figure 7.20 (a) The fabrication process and structure of the multifunctional fiber. (b) Micromachining of the fiber with femtosecond laser. (c) Optical image of the fiber immersed in a drop of fluorescein excited by a 473 nm laser. (d) The functional fiber probes before implantation. (e) An illustration of the multimodal fiber probe targeting the thalamus region in a mouse brain and the assembled device coupled to the 473 nm laser. The right is electrophysiological recording during optogenetic stimulation three months after implantation. (f) The seizure signal recorded by the probe. *Source:* Jiang [110] / Springer Nature / CC BY 4.0.

for chronic optogenetic modulations in central and peripheral nervous systems (Figure 7.21f) [113]. While L. Frey et al. used the hydrogel as coating material to improve the biocompatibility of the neural probe [115]. In addition, Sonja Johannsmeier et al. modified the properties of poly- (ethylene glycol) derivative hydrogels by tuning the monomer length and concentration, and achieve compatibility with muscle tissue as well as optimal light guidance and distribution for optogenetic applications (Figure 7.21g,h) [114].

Although hydrogel fibers are widely used in neuron modulation, the large-scale fabrication of core-sheath hydrogel fibers with the desired optical and mechanical properties remains a challenge. Recently, Meifang Zhu et al. reported a versatile concept of integrated light-triggered dynamic wet spinning (ILDWS), which is capable of continuously producing core-sheath hydrogel fibers with tunable fiber diameters, mechanical and optical-propagation properties [60], as shown in Figure 7.22. The low optical loss (0.18 dB/cm at 650 nm), excellent biocompatibility, and tissue-like Young's modulus (<2.6 MPa) of the obtained hydrogel fiber makes it suitable for

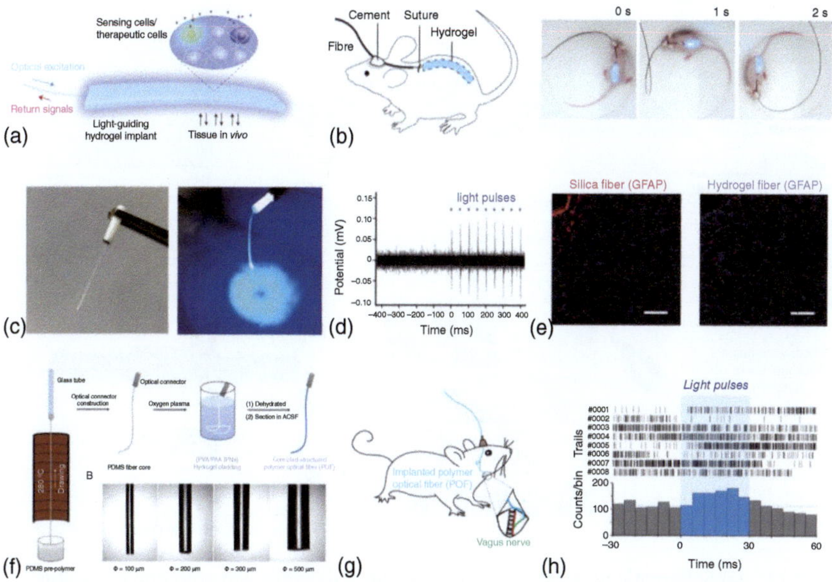

Figure 7.21 (a) Schematic of the light guiding hydrogel for *in vivo* sensing and therapy. (b) Schematic of a hydrogel fiber implanted in a mouse and the probe coupled with a blue light. *Source:* (a) and (b) are reproduced with permission [111] Copyright 2013, Springer Nature. (c) Images of a dehydrated and swollen hydrogel fiber. (d) Electrophysiological recordings using hydrogel fiber *in vivo*. (e) Inflammatory response of the conventional silica and hydrogel fiber. Source: (c–e) are reproduced with permission [112] Copyright 2018, JOHN WILEY & SONS, INC. (f) Fabrication of the hydrogel fiber with core–clad structure. *Source:* Cao [113] / Springer Nature / CC BY 4.0. (g) The implantation of the fiber for optogenetic stimulation. (h) The recording neural signals. *Source:* (g) and (h) are reproduced with permission [114] Copyright 2019, Elsevier.

deep-tissue cancer therapy and brain optogenetic stimulation. Yuwei Zhao and coworkers reported a composite probe, which allows for simultaneously electrical recording and optical stimulation [116]. The probe is composed of a tungsten wire, a composite electrode (integrated with alcohol [PVA], multiwalled carbon nanotubes, poly(3,4-ethylenedioxythiophene) [PEDOT], and polystyrene sulfonate [PSS]) and a polydimethylsiloxane (PDMS) waveguide. The loose surface structure of the hydrogel makes it easy to stick to target neurons. Moreover, the composite hydrogel electrode has a lower impedance and higher charge capacity, which make it available for long-term *in vivo* electrical recording.

To integrate more functions into the stretchable fiber probe for multimodal interrogation of neuron activities, Seongjun Park and coworkers developed a multifunctional probe by integrating various polymer fibers into a soft hydrogel matrix [117], as shown in Figure 7.23a–d. They first fabricate individual functional fibers via thermal drawing, such as optical fibers consisting of PC core and COC cladding, microelectrode arrays containing seven Sn microwires encapsulated within poly(etherimide) (PEI) insulating cladding, and PEI tubes for microfluidic channels. Then, these separated fibers are integrated into the soft hydrogel matrix. The obtained probe is capable of long-term

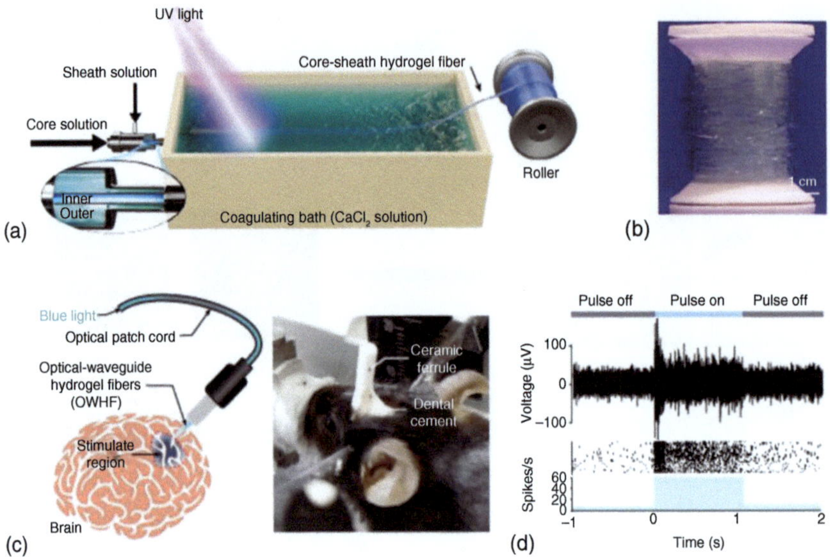

Figure 7.22 (a) Schematic illustration of the integrated dynamic wet spinning apparatus. (b) Photograph of a rolled core-sheath hydrogel fiber. (c) left part is schematic illustration of optogenetic stimulation, and right part is the photo of a mouse implanted with probe. (d) Electrophysiological test before and during optogenetic stimulation. *Source:* Chen et al. [60] / Oxford University Press / CC BY 4.0.

recording of neural activity, delivery of light, and infusion of chemicals into the mouse brain with minimal foreign body responses. In addition, Anthony Tabet et al. proposed a multifunctional hydrogel-neural interface using a solvent evaporation- or entrapment-driven (SEED) integration process [118]. The fabricated probe is integrated with optical, electrical, and microfluidic functions, and all are encased within a copolymer comprised of water-soluble poly (ethylene glycol) and water-insoluble poly(urethane) (PU-PEG), as shown in Figure 7.23e–h. The composite probe can be used for long-term (more than four weeks) simultaneous optogenetics and electrophysiology *in vitro* and *in vivo*.

7.6 Conclusion

In this chapter, an overview and review of the fiber-based devices for neural manipulation are discussed, including the principle of optogenetic technology, the fabrication techniques, and the design principles of the fiber-based devices and recent advances of the representative fiber-based probes. With the development and progress of materials' processing technologies, various techniques are developed for fabrication of fiber-based devices. The function of these fiber devices evolves from initial optical delivery to electro-optical transmission, and even to the integrated functions of optical, electrical, and chemical drug delivery. At the same time, the biocompatibility, mechanical properties, optical properties,

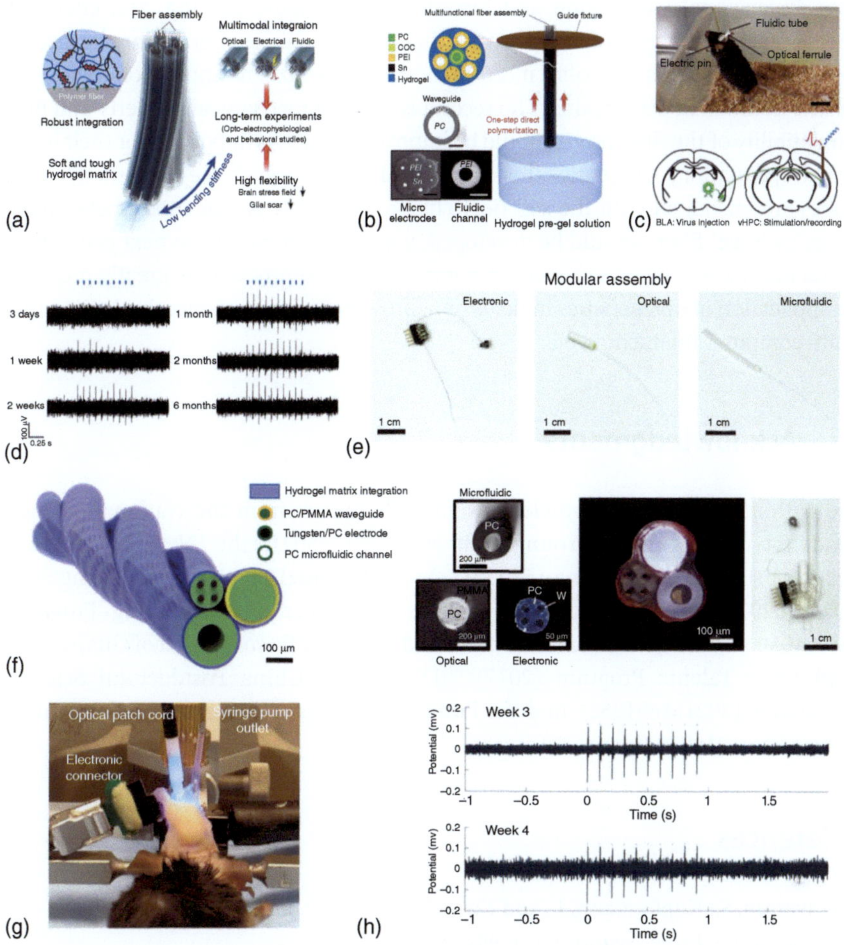

Figure 7.23 (a) Illustration of the hydrogel hybrid probe design. (b) The fabrication process of the hybrid probe. (c) A mouse implanted with the hydrogel hybrid probe, and an illustration of optogenetic modulation and electrophysiological recording. (d) Electrophysiological recordings using the hydrogel hybrid probes. *Source:* Park et al. [117] / Springer Nature / CC BY 4.0. (e) Three separated and fully connectorized modules. (f) Various components of the assembled hydrogel fibers. (g) Image of the experimental mouse. (h) The recording invoked action potentials in a Thy1-ChR2 mouse. *Source:* (e–h) are reproduced with permission [118] Copyright 2021, American Chemical Society.

and electrical properties of the probes are also improved for long-term *in vivo* optogenetics. The success of optogenetics is considerably contributed by the development of these fiber-based probes.

However, there are still some challenges for fiber-based probes. One is that the functionality of the fiber probes are still less integrated. For example, the current fiber probe cannot simultaneously detect the chemicals secreted by neurons, such as amino acids, dopamine, acetylcholine, etc. upon optical stimulation. More functions such as three-dimensional stimulation and recording, temperature sensing, and chemical detection can be integrated into a single fiber on the premise of ensuring the size of the fiber probe, which may offer more chances to recognize the complex functions of specific neurons in deep brains. The second is that the mechanical property and biocompatibility of the fiber probe should be improved, which is important for their long-term implantation. Currently, the mechanical properties of the commonly used probes are mismatch with the soft nerve tissue. More soft probes such as elastomer fibers, hydrogel fibers should be developed, which is a trend for the next generation neural interfaces. The progress of development of novel probes may greatly promote the application of optogenetics in fields of neuroscience, brain science, brain diseases, brain–computer interaction, etc.

7.7 Acknowledgments

The authors gratefully acknowledge financial support from the National Science Fund for Distinguished Young Scholars (62125502), the National Natural Science Foundation of China (51972113), the Key R&D Program of Guangzhou (202007020003), the Science and Technology Project of Guangdong Province (2021A0505030004), the Local Innovative and Research Teams Project of Guangdong Pearl River Talents Program (2017BT01X137), the China Postdoctoral Science Foundation (2021M691052), and the Fundamental Research Funds for the Central Universities (21623326).

References

1 Iversen, S., Kupfermann, I., and Kandel, E.R. (2000). Emotional states and feelings[J]. *Princ. Neural. Sci.* 4: 982–997.

2 Strumwasser, F. (1958). Long-term recording from single neurons in brain of unrestrained mammals[J]. *Science* 127 (3296): 469–470.

3 Campbell, P.K., Jones, K.E., Huber, R.J. et al. (1991). A silicon-based, three-dimensional neural interface: manufacturing processes for an intracortical electrode array[J]. *IEEE Trans. Biomed. Eng.* 38 (8): 758–768.

4 Sakmann, B. and Neher, E. (1984). Patch clamp techniques for studying ionic channels in excitable membranes[J]. *Annu. Rev. Physiol.* 46 (1): 455–472.

5 McNaughton, B.L., O'Keefe, J., and Barnes, C.A. (1983). The stereotrode: a new technique for simultaneous isolation of several single units in the central nervous system from multiple unit records[J]. *J. Neurosci. Methods* 8 (4): 391–397.

6 Drake, K.L., Wise, K.D., Farraye, J. et al. (1988). Performance of planar multisite microprobes in recording extracellular single-unit intracortical activity[J]. *IEEE Trans. Biomed. Eng.* 35 (9): 719–732.

7 Gray, C.M., Maldonado, P.E., Wilson, M. et al. (1995). Tetrodes markedly improve the reliability and yield of multiple single-unit isolation from multi-unit recordings in cat striate cortex[J]. *J. Neurosci. Methods* 63 (1-2): 43–54.

8 Deisseroth, K. (2011). Optogenetics[J]. *Nat. Methods* 8 (1): 26–29.

9 Boyden, E.S. (2015). Optogenetics and the future of neuroscience[J]. *Nat. Neurosci.* 18 (9): 1200–1201.

10 Fenno, L., Yizhar, O., and Deisseroth, K. (2011). The development and application of optogenetics[J]. *Annu. Rev. Neurosci.* 34: 389–412.

11 Lee, J.H., Durand, R., Gradinaru, V. et al. (2010). Global and local fMRI signals driven by neurons defined optogenetically by type and wiring[J]. *Nature* 465 (7299): 788–792.

12 Nagel, G., Szellas, T., Huhn, W. et al. (2003). Channelrhodopsin-2, a directly light-gated cation-selective membrane channel[J]. *Proc. Natl. Acad. Sci. U.S.A.* 100 (24): 13940–13945.

13 Zhang, F., Vierock, J., Yizhar, O. et al. (2011). The microbial opsin family of optogenetic tools[J]. *Cell* 147 (7): 1446–1457.

14 Yizhar, O., Fenno, L.E., Davidson, T.J. et al. (2011). Optogenetics in neural systems[J]. *Neuron* 71 (1): 9–34.

15 Sizemore, R.J., Seeger-Armbruster, S., Hughes, S.M. et al. (2016). Viral vector-based tools advance knowledge of basal ganglia anatomy and physiology[J]. *J. Neurophysiol.* 115 (4): 2124–2146.

16 Tian, H., Xu, K., Zou, L. et al. (2021). Multimodal neural probes for combined optogenetics and electrophysiology[J]. *iScience* 103612.

17 Stark, E., Koos, T., and Buzsáki, G. (2012). Diode probes for spatiotemporal optical control of multiple neurons in freely moving animals[J]. *J. Neurophysiol.* 108 (1): 349–363.

18 Fan, B. and Li, W. (2015). Miniaturized optogenetic neural implants: a review[J]. *Lab Chip* 15 (19): 3838–3855.

19 Loke, G., Yan, W., Khudiyev, T. et al. (2020). Recent progress and perspectives of thermally drawn multimaterial fiber electronics[J]. *Adv. Mater.* 32 (1): 1904911.

20 Yan, W., Dong, C., Xiang, Y. et al. (2020). Thermally drawn advanced functional fibers: new frontier of flexible electronics[J]. *Mater. Today* 35: 168–194.

21 Tao, G., Abouraddy, A.F., Stolyarov, A.M. et al. (2015). *Multimaterial Fibers[M]. Lab-on-fiber Technology*, 1–26. Cham: Springer.

22 Yan, W., Page, A., Nguyen-Dang, T. et al. (2019). Advanced multimaterial electronic and optoelectronic fibers and textiles[J]. *Adv. Mater.* 31 (1): 1802348.

23 Rein, M., Favrod, V.D., Hou, C. et al. (2018). Diode fibres for fabric-based optical communications[J]. *Nature* 560 (7717): 214–218.

24 Hou, C., Jia, X., Wei, L. et al. (2013). Direct atomic-level observation and chemical analysis of ZnSe synthesized by in situ high-throughput reactive fiber drawing[J]. *Nano Lett.* 13 (3): 975–979.

25 Kang, S., Fang, Z., Huang, X. et al. (2017). Precisely controllable fabrication of Er^{3+}-doped glass ceramic fibers: novel mid-infrared fiber laser materials[J]. *J. Mater. Chem. C* 5 (18): 4549–4556.

26 Ballato, J. and Snitzer, E. (1995). Fabrication of fibers with high rare-earth concentrations for faraday isolator applications[J]. *Appl. Opt.* 34 (30): 6848–6854.

27 Gibson, U.J., Wei, L., and Ballato, J. (2021). Semiconductor core fibres: materials science in a bottle[J]. *Nat. Commun.* 12 (1): 1–15.

28 Han, B., Luo, Q., Zhang, P. et al. (2022). Multifunctional single-crystal tellurium core multimaterial fiber via thermal drawing and laser recrystallization[J]. *J. Am. Ceram. Soc.* 105 (3): 1640–1647.

29 Luo, Q., Tang, G., Sun, M. et al. (2020). Single crystal tellurium semiconductor core optical fibers[J]. *Opt. Mater. Exp.* 10 (4): 1072–1082.

30 Ordu, M. and Basu, S.N. (2020). Recent progress in germanium-core optical fibers for mid-infrared optics[J]. *Infrared Phys. Technol.* 111: 103507.

31 Liu, Y., Sun, M., Tang, G. et al. (2019). Multifunctional GeSe core fibers[J]. *Mater. Lett.* 247: 193–196.

32 Morris, S., Hawkins, T., Foy, P. et al. (2011). Reactive molten core fabrication of silicon optical fiber[J]. *Opt. Mater. Exp.* 1 (6): 1141–1149.

33 Ballato, J., McMillen, C., Hawkins, T. et al. (2012). Reactive molten core fabrication of glass-clad amorphous and crystalline oxide optical fibers[J]. *Opt. Mater. Exp.* 2 (2): 153–160.

34 Dong, C., Page, A.G., Yan, W. et al. (2019). Microstructured multimaterial fibers for microfluidic sensing[J]. *Adv. Mater. Technol.ogies* 4 (10): 1900417.

35 Temelkuran, B., Hart, S.D., Benoit, G. et al. (2002). Wavelength-scalable hollow optical fibres with large photonic bandgaps for CO_2 laser transmission[J]. *Nature* 420 (6916): 650–653.

36 Hart, S.D., Maskaly, G.R., Temelkuran, B. et al. (2002). External reflection from omnidirectional dielectric mirror fibers[J]. *Science* 296 (5567): 510–513.

37 Gibson, D.J. and Harrington, J.A. (2004). Extrusion of hollow waveguide preforms with a one-dimensional photonic bandgap structure[J]. *J. Appl. Phys.* 95 (8): 3895–3900.

38 Tao, G., Shabahang, S., Banaei, E.H. et al. (2012). Multimaterial preform coextrusion for robust chalcogenide optical fibers and tapers[J]. *Opt. Lett.* 37 (13): 2751–2753.

39 Chillcce, E.F., Cordeiro, C.M.B., Barbosa, L.C. et al. (2006). Tellurite photonic crystal fiber made by a stack-and-draw technique[J]. *J. Non-Cryst. Solids* 352 (32-35): 3423–3428.

40 Liao, M., Chaudhari, C., Qin, G. et al. (2009). Fabrication and characterization of a chalcogenide-tellurite composite microstructure fiber with high nonlinearity[J]. *Opt. Express* 17 (24): 21608–21614.

41 Faccini de Lima, C., van der Elst, L.A., Koraganji, V.N. et al. (2019). Towards digital manufacturing of smart multimaterial fibers[J]. *Nanoscale Res. Lett.* 14 (1): 1–16.

42 Cook, K., Canning, J., Leon-Saval, S. et al. (2015). Air-structured optical fiber drawn from a 3D-printed preform[J]. *Opt. Lett.* 40 (17): 3966–3969.

43 Toal, P.M. Jr., Holmes, L.J., Rodriguez, R.X. et al. (2017). Microstructured monofilament via thermal drawing of additively manufactured preforms[J]. *Addit. Manufac.* 16: 12–23.

44 Klein, J., Stern, M., Franchin, G. et al. (2015). Additive manufacturing of optically transparent glass[J]. *3D Print. Addit. Manufac.* 2 (3): 92–105.

45 Kaufman J J, Bow C, Tan F A, et al. 3D printing preforms for fiber drawing and structured functional particle production[C]. Australian Conference on Optical Fibre Technology. Optical Society of America, 2016: AW4C. 1.

46 Shabahang, S., Tan, F.A., Perlstein, J.D. et al. (2017). Robust multimaterial chalcogenide fibers produced by a hybrid fiber-fabrication process[J]. *Opt. Mater. Exp.* 7 (7): 2336–2345.

47 Mossadegh, R., Sanghera, J.S., Schaafsma, D. et al. (1998). Fabrication of single-mode chalcogenide optical fiber[J]. *J. Lightwave Technol.* 16 (2): 214–217.

48 Sparks, J.R., He, R., Healy, N. et al. (2013). Conformal coating by high pressure chemical deposition for patterned microwires of II–VI semiconductors[J]. *Adv. Funct. Mater.* 23 (13): 1647–1654.

49 Sazio, P.J.A., Amezcua-Correa, A., Finlayson, C.E. et al. (2006). Microstructured optical fibers as high-pressure microfluidic reactors[J]. *Science* 311 (5767): 1583–1586.

50 Ballato J, Morris S, Hawkins T, et al. Advances in semiconductor core optical fibers[C]//European Conference and Exposition on Optical Communications. Optical Society of America, 2011: Mo. 2. LeCervin. 6.

51 He, R., Sazio, P.J.A., Peacock, A.C. et al. (2012). Integration of gigahertz-bandwidth semiconductor devices inside microstructured optical fibres[J]. *Nat. Photonics* 6 (3): 174–179.

52 He, R., Day, T.D., Krishnamurthi, M. et al. (2013). Silicon p-i-n junction fibers [J]. *Adv. Mater.* 25 (10): 1460–1460.

53 Sparks, J.R., Sazio, P.J.A., Gopalan, V. et al. (2013). Templated chemically deposited semiconductor optical fiber materials[J]. *Annu. Rev. Mater. Res.earch* 43: 527–557.

54 Granzow, N., Uebel, P., Schmidt, M.A. et al. (2011). Bandgap guidance in hybrid chalcogenide–silica photonic crystal fibers[J]. *Opt. Lett.* 36 (13): 2432–2434.

55 Tyagi, H.K., Schmidt, M.A., Sempere, L.P. et al. (2008). Optical properties of photonic crystal fiber with integral micron-sized Ge wire[J]. *Opt. Express* 16 (22): 17227–17236.

56 Huang, S.L. (2019). Crystalline fibers for fiber lasers and amplifiers. In: *Handbook of optical fibers. Singapore*. Springer.

57 Lo, C.Y., Huang, P.L., Chou, T.S. et al. (2002). Efficient Nd: $Y_3Al_5O_{12}$ crystal fiber laser[J]. *Jpn J. Appl.Phys.* 41 (11A): L1228.

58 Huang, K.Y., Hsu, K.Y., Jheng, D.Y. et al. (2008). Low-loss propagation in Cr 4+: YAG double-clad crystal fiber fabricated by sapphire tube assisted CDLHPG technique[J]. *Opt. Express* 16 (16): 12264–12271.

59 Lai, C.C., Huang, K.Y., Tsai, H.J. et al. (2009). Yb^{3+}: YAG silica fiber laser[J]. *Opt. Lett.* 34 (15): 2357–2359.

60 Chen, G., Wang, G., Tan, X. et al. (2021). Integrated dynamic wet spinning of core-sheath hydrogel fibers for optical-to-brain/tissue communications[J]. *Nat. Sci. Rev.* 8 (9): nwaa209.

61 Zhang, H. and Chiao, M. (2015). Anti-fouling coatings of poly (dimethylsiloxane) devices for biological and biomedical applications[J]. *J. Med. Biol. Eng.* 35 (2): 143–155.

62 Green, R.A., Hassarati, R.T., Bouchinet, L. et al. (2012). Substrate dependent stability of conducting polymer coatings on medical electrodes[J]. *Biomaterials* 33 (25): 5875–5886.

63 Kozai, T.D.Y., Jaquins-Gerstl, A.S., Vazquez, A.L. et al. (2016). Dexamethasone retrodialysis attenuates microglial response to implanted probes *in vivo*[J]. *Biomaterials* 87: 157–169.

64 Mercanzini, A., Reddy, S.T., Velluto, D. et al. (2010). Controlled release nanoparticle-embedded coatings reduce the tissue reaction to neuroprostheses[J]. *J. Controlled Release* 145 (3): 196–202.

65 Eles, J.R., Vazquez, A.L., Snyder, N.R. et al. (2017). Neuroadhesive L1 coating attenuates acute microglial attachment to neural electrodes as revealed by live two-photon microscopy[J]. *Biomaterials* 113: 279–292.

66 Azemi, E., Lagenaur, C.F., and Cui, X.T. (2011). The surface immobilization of the neural adhesion molecule L1 on neural probes and its effect on neuronal density and gliosis at the probe/tissue interface[J]. *Biomaterials* 32 (3): 681–692.

67 Rao, V.R., Leonard, M.K., Kleen, J.K. et al. (2017). Chronic ambulatory electrocorticography from human speech cortex[J]. *NeuroImage* 153: 273–282.

68 Azemi, E., Gobbel, G.T., and Cui, X.T. (2010). Seeding neural progenitor cells on silicon-based neural probes[J]. *J. Neurosurg.* 113 (3): 673–681.

69 Di, J., Yao, S., Ye, Y. et al. (2015). Stretch-triggered drug delivery from wearable elastomer films containing therapeutic depots[J]. *ACS Nano* 9 (9): 9407–9415.

70 Lu, Y.B., Iandiev, I., Hollborn, M. et al. (2011). Reactive glial cells: increased stiffness correlates with increased intermediate filament expression[J]. *FASEB J.* 25 (2): 624–631.

71 Sawyer, A.J. and Kyriakides, T.R. (2013). Nanoparticle-based evaluation of blood–brain barrier leakage during the foreign body response[J]. *J. Neural Eng.* 10 (1): 016013.

72 Christensen, M.B., Pearce, S.M., Ledbetter, N.M. et al. (2014). The foreign body response to the utah slant electrode array in the cat sciatic nerve[J]. *Acta Biomater.* 10 (11): 4650–4660.

73 Hayn, L. and Koch, M. (2015). Suppression of excitotoxicity and foreign body response by memantine in chronic cannula implantation into the rat brain[J]. *Brain Res. Bull.* 117: 54–68.

74 Grill, W.M., Norman, S.E., and Bellamkonda, R.V. (2009). Implanted neural interfaces: biochallenges and engineered solutions[J]. *Annu. Rev. Biomed. Eng.* 11: 1–24.

75 Polikov, V.S., Tresco, P.A., and Reichert, W.M. (2005). Response of brain tissue to chronically implanted neural electrodes[J]. *J. Neurosci. Methods* 148 (1): 1–18.

76 Hong, G. and Lieber, C.M. (2019). Novel electrode technologies for neural recordings[J]. *Nat. Rev. Neurosci.* 20 (6): 330–345.

77 Aravanis, A.M., Wang, L.P., Zhang, F. et al. (2007). An optical neural interface: *in vivo* control of rodent motor cortex with integrated fiberoptic and optogenetic technology[J]. *J. Neural Eng.* 4 (3): S143.

78 Kim, B.J., Kuo, J.T.W., Hara, S.A. et al. (2013). 3D Parylene sheath neural probe for chronic recordings[J]. *J. Neural Eng.* 10 (4): 045002.

79 Kuo, J.T.W., Kim, B.J., Hara, S.A. et al. (2013). Novel flexible parylene neural probe with 3D sheath structure for enhancing tissue integration[J]. *Lab Chip* 13 (4): 554–561.

80 Delivopoulos, E., Chew, D.J., Minev, I.R. et al. (2012). Concurrent recordings of bladder afferents from multiple nerves using a microfabricated PDMS microchannel electrode array[J]. *Lab Chip* 12 (14): 2540–2551.

81 Srinivasan, A., Tahilramani, M., Bentley, J.T. et al. (2015). Microchannel-based regenerative scaffold for chronic peripheral nerve interfacing in amputees[J]. *Biomaterials* 41: 151–165.

82 FitzGerald, J.J., Lago, N., Benmerah, S. et al. (2012). A regenerative microchannel neural interface for recording from and stimulating peripheral axons *in vivo*[J]. *J. Neural Eng.* 9 (1): 016010.

83 Minev, I.R., Musienko, P., Hirsch, A. et al. (2015). Electronic dura mater for long-term multimodal neural interfaces[J]. *Science* 347 (6218): 159–163.

84 Ludwig, K.A., Langhals, N.B., Joseph, M.D. et al. (2011). Poly (3, 4-ethylenedioxythiophene) (PEDOT) polymer coatings facilitate smaller neural recording electrodes[J]. *J. Neural Eng.* 8 (1): 014001.

85 Kozai, T.D.Y., Langhals, N.B., Patel, P.R. et al. (2012). Ultrasmall implantable composite microelectrodes with bioactive surfaces for chronic neural interfaces[J]. *Nat. Mater.* 11 (12): 1065–1073.

86 Li, B., Cao, X., Ong, H.G. et al. (2010). All-carbon electronic devices fabricated by directly grown single-walled carbon nanotubes on reduced graphene oxide electrodes[J]. *Adv. Mater.* 22 (28): 3058–3061.

87 Rakhilin, N., Barth, B., Choi, J. et al. (2016). Simultaneous optical and electrical *in vivo* analysis of the enteric nervous system[J]. *Nat. Commun.* 7 (1): 1–7.

88 Choi, S., Han, S.I., Kim, D. et al. (2019). High-performance stretchable conductive nanocomposites: materials, processes, and device applications[J]. *Chem. Soc. Rev.* 48 (6): 1566–1595.

89 Keefer, E.W., Botterman, B.R., Romero, M.I. et al. (2008). Carbon nanotube coating improves neuronal recordings[J]. *Nat. Nanotechnol.* 3 (7): 434–439.

90 Wang, Y., Zhu, H., Yang, H. et al. (2018). Nano functional neural interfaces[J]. *Nano Res.* 11 (10): 5065–5106.

91 Gradinaru, V., Thompson, K.R., Zhang, F. et al. (2007). Targeting and readout strategies for fast optical neural control in vitro and *in vivo*[J]. *J. Neurosci.* 27 (52): 14231–14238.

92 Gradinaru, V., Mogri, M., Thompson, K.R. et al. (2009). Optical deconstruction of parkinsonian neural circuitry[J]. *Science* 324 (5925): 354–359.

93 Anikeeva, P., Andalman, A.S., Witten, I. et al. (2012). Optetrode: a multichannel readout for optogenetic control in freely moving mice[J]. *Nat. Neurosci.* 15 (1): 163–170.

94 Zhang, J., Laiwalla, F., Kim, J.A. et al. (2009). Integrated device for optical stimulation and spatiotemporal electrical recording of neural activity in light-sensitized brain tissue[J]. *J. Neural Eng.* 6 (5): 055007.

95 Wang, J., Wagner, F., Borton, D.A. et al. (2011). Integrated device for combined optical neuromodulation and electrical recording for chronic *in vivo* applications[J]. *J. Neural Eng.* 9 (1): 016001.

96 Royer, S., Zemelman, B.V., Barbic, M. et al. (2010). Multi-array silicon probes with integrated optical fibers: light-assisted perturbation and recording of local neural circuits in the behaving animal[J]. *Eur. J. Neurosci.* 31 (12): 2279–2291.

97 Wu, F., Stark, E., Ku, P.C. et al. (2015). Monolithically integrated μLEDs on silicon neural probes for high-resolution optogenetic studies in behaving animals[J]. *Neuron* 88 (6): 1136–1148.

98 Pisanello, F., Mandelbaum, G., Pisanello, M. et al. (2017). Dynamic illumination of spatially restricted or large brain volumes via a single tapered optical fiber[J]. *Nat. Neurosci.* 20 (8): 1180–1188.

99 Jun, J.J., Steinmetz, N.A., Siegle, J.H. et al. (2017). Fully integrated silicon probes for high-density recording of neural activity[J]. *Nature* 551 (7679): 232–236.

100 Lu, C., Froriep, U.P., Koppes, R.A. et al. (2014). Polymer fiber probes enable optical control of spinal cord and muscle function *in vivo*[J]. *Adv. Funct. Mater.* 24 (42): 6594–6600.

101 Lu, C., Park, S., Richner, T.J. et al. (2017). Flexible and stretchable nanowire-coated fibers for optoelectronic probing of spinal cord circuits[J]. *Sci. Adv.* 3 (3): e1600955.

102 Guo, Y., Jiang, S., Grena, B.J.B. et al. (2017). Polymer composite with carbon nanofibers aligned during thermal drawing as a microelectrode for chronic neural interfaces[J]. *ACS Nano* 11 (7): 6574–6585.

103 Du, M., Huang, L., Zheng, J. et al. (2020). Flexible fiber probe for efficient neural stimulation and detection[J]. *Adv. Sci.* 7 (15): 2001410.

104 Davey, C.J., Argyros, A., Fleming, S.C. et al. (2015). Multimodal optogenetic neural interfacing device fabricated by scalable optical fiber drawing technique[J]. *Appl. Opt.* 54 (34): 10068–10072.

105 Kilias, A., Canales, A., Froriep, U.P. et al. (2018). Optogenetic entrainment of neural oscillations with hybrid fiber probes[J]. *J. Neural Eng.* 15 (5): 056006.

106 Zou, L., Tian, H., Guan, S. et al. (2021). Self-assembled multifunctional neural probes for precise integration of optogenetics and electrophysiology[J]. *Nat. Commun.* 12 (1): 1–9.

107 Vitale, F., Summerson, S.R., Aazhang, B. et al. (2015). Neural stimulation and recording with bidirectional, soft carbon nanotube fiber microelectrodes[J]. *ACS Nano* 9 (4): 4465–4474.

108 Canales, A., Jia, X., Froriep, U.P. et al. (2015). Multifunctional fibers for simultaneous optical, electrical and chemical interrogation of neural circuits *in vivo*[J]. *Nat. Biotechnol.* 33 (3): 277–284.

109 Park, S., Guo, Y., Jia, X. et al. (2017). One-step optogenetics with multifunctional flexible polymer fibers[J]. *Nat. Neurosci.* 20 (4): 612–619.

110 Jiang, S., Patel, D.C., Kim, J. et al. (2020). Spatially expandable fiber-based probes as a multifunctional deep brain interface[J]. *Nat. Commun.* 11 (1): 1–14.

111 Choi, M., Choi, J.W., Kim, S. et al. (2013). Light-guiding hydrogels for cell-based sensing and optogenetic synthesis *in vivo*[J]. *Nat. Photonics* 7 (12): 987–994.

112 Wang, L., Zhong, C., Ke, D. et al. (2018). Ultrasoft and highly stretchable hydrogel optical fibers for *in vivo* optogenetic modulations[J]. *Adv. Opt. Mater.* 6 (16): 1800427.

113 Cao, Y., Pan, S., Yan, M. et al. (2021). Flexible and stretchable polymer optical fibers for chronic brain and vagus nerve optogenetic stimulations in free-behaving animals[J]. *BMC Biol.* 19 (1): 1–13.

114 Johannsmeier, S., Torres, M.L., Ripken, T. et al. (2018). Hydrogels for efficient light delivery in optogenetic applications[C]. Optogenetics and Optical Manipulation. *Int. Soc. Opt. Photonics* 2018: 10482, 104820Q.

115 Frey, L., Bandaru, P., Zhang, Y.S. et al. (2018). A dual-layered microfluidic system for long-term controlled in situ delivery of multiple anti-inflammatory factors for chronic neural applications[J]. *Adv. Funct. Mater.* 28 (12): 1702009.

116 Zhao, Y., Wang, K., Li, S. et al. (2018). Polydimethylsiloxane (PDMS)-based flexible optical electrodes with conductive composite hydrogels integrated probe for optogenetics[J]. *J. Biomed. Nanotechnol.* 14 (6): 1099–1106.

117 Park, S., Yuk, H., Zhao, R. et al. (2021). Adaptive and multifunctional hydrogel hybrid probes for long-term sensing and modulation of neural activity[J]. *Nat. Commun.* 12 (1): 1–12.

118 Tabet, A., Antonini, M.J., Sahasrabudhe, A. et al. (2021). Modular integration of hydrogel neural interfaces[J]. *ACS Central Sci.* 7 (9): 1516–1523.

8

Novel Functional Fibers for Neural Interfacing

Shan Jiang[1,2,3] *and Xiaoting Jia*[3,4,5]

[1] *Stanford University, Department of Materials Science and Engineering, Stanford, CA 94305, USA*
[2] *Stanford University, Wu Tsai Neurosciences Institute, 290 Campus Drive, Stanford, CA 94305, USA*
[3] *Virginia Tech, Bradley Department of Electrical and Computer Engineering, 1185 Perry St, Blacksburg, Blacksburg, VA 24061, USA*
[4] *Virginia Tech, Department of Materials Science and Engineering, 445 Old Turner St, Blacksburg, VA 24061, USA*
[5] *Virginia Tech, School of Neuroscience, 210 Drillfield Dr, Blacksburg, VA 24061, USA*

8.1 Introduction

The scientific study of fibers has traditionally belonged to the domain of materials science, but it is hard to ignore the similarities between optical fibers and the field of neuroscience. The nervous system contains specialized cells called neurons that communicate by transmitting electrochemical signals along their axons. Modern telecommunication systems also use high aspect ratio structures to transmit information; here synthetic fibers are the conduits of information in the form of optical signals. The prospect of creating a connection between the neuronal and the electronic communication networks could lead to a new class of neuron-integrated electronic devices, and holds great promise for both research and clinic.

Together with the development of genetic tools for modulating neural activities, the advancement in optical fibers has revitalized the research in neuroscience. To date, much progress has been made toward the techniques that can interface with animals' nervous systems. The three major functional components embedded in fibers have provided a powerful platform for scientists to sense and modulate neural activities optically, electrically, and chemically. Optical interfaces mainly involve fiber photometry and the application in optogenetics. For electrical interfaces, integration of microwires and conductive coating to conventional optical fiber have been widely expanded, while thermally drawn multifunctional fiber-based neural probe enables scalable fabrication. Microfluidic channels embedded in fibers allow for chemical modulation of neural activities. Other than these multimodalities, polymer fiber can also serve as a scaffold in tissue engineering.

Optical and Electronic Fibers: Emerging Applications and Technological Innovations, First Edition.
Edited by Lei Wei.

Over the past decades, significant progress has been made in developing fiber-based devices for neural interfacing [1, 2]. Based on the method of interfacing, the fiber devices can be broadly categorized into optical, electrical, chemical, scaffolding fibers, as well as the ones combining multiple modalities. In this chapter, we will discuss the current novel fiber-based technology and its applications in neuroscience. Specifically, we start with the introduction of genetic manipulation tools for optical interfaces. We then present advances in conventional silica fibers for optical recording and optogenetic manipulation. Thermally drawn multimaterial fibers, including the design principle, fabrication methods, and its versatile applications, are overviewed next. We end this chapter with our perspective of fiber-based neural probes in neuroscience studies and related biomedical fields.

8.2 Genetic Manipulation-enabled Optical Approaches

Scientists have long sought to observe and modulate neural activities with light. Throughout the years, the invention of genetically encoded calcium indicators (GECIs) has allowed the visualization of neuronal signaling via monitoring the fluorescence change in response to the cellular calcium-level fluctuation [3]. With this technique, researchers can correlate the fluorescence intensity with the neural activities in a behavior assay over a prolonged timescale [4]. The optical interfaces for measuring the calcium levels will be briefly introduced in Section 8.3.1.

Other than the endogenous neural signal detection, the investigation of brain structures and functions requires causal manipulation of neural activities underlying complex circuits. Electrical stimulation and neurospecific drug delivery have been exploited in many studies for developing the fundamental neuroscience and finding the right treatment to brain disorders. However, these intervention approaches have limited spatial or temporal resolution, as a large group of cells near the interfacing site will be disturbed. Later, a technique called optogenetics was developed that offered a unique pathway to interact with genetically opsin-modified neurons with cell-type specificity and millisecond control using light at specific wavelengths [5].

When illuminated by light at a certain wavelength, those opsin-transduced neurons will be either stimulated to exhibit action potential firing as the on-state (channelrhodopsin-2, ChR2) with blue light or suppressed as the off-state (halorhodopsin, NpHR) with yellow light. And this process desires optical pathways for delivering light to the targeted brain region [6]. This neural-manipulative mechanism involved in optogenetics has allowed for both precise neuron-type-specific targeting and high temporal resolution with millisecond light pulses. Furthermore, bidirectional communication between the targeted individual cells and the implanted device is essential for neuroscientists to gain a deeper understanding of the neuronal function and dynamics. Recent novel optical platforms are overviewed in the following sections.

8.3 Conventional Silica Fiber

Commercial silica fibers have offered a broad range of opportunities for biomedical studies due to its wide availability and superior light transmitting properties [7, 8]. Particularly, the geometry makes it suitable for delivering light over a long distance for deep-brain fluorescence imaging and optogenetic manipulation. In this section, we will briefly introduce application of conventional silica fiber for direct optical recording and modulation with the integration with electronics.

8.3.1 Direct Optical Readout

In conjunction with the well-developed bioengineering tools described above that permit the visualization of neural activities based on the calcium dynamics, various imaging technologies have been developed such as two-photon microscopy, miniature microscopes, and fiber-based fluorescence imaging [9, 10]. The former two imaging techniques hold both great advantages and limitations. In particular, two-photon microscopy can achieve calcium imaging of superficial layers of the brain with superior resolution on head-fixed animals, whereas miniature microscopes can provide deep-brain dynamics with compromised resolution on freely behaving animals.

As for the fiber-based optical recording technique, it mainly involves fiber photometry [11–13] (Figure 8.1a) and fibered fluorescent microscopy (FFM) [9, 18]. Fiber photometry exploits optical fibers to collect calcium-mediated fluorescent signals from a population of neurons via photon counting, which precludes the single-cell resolution. The excitation pulse delivery and emitted photon collection are the key steps involved within this technique. Many efforts, such as designing optical path and photosensor, have been made to improve this technique concerning the weight of the implant and optical recording scalability. On the other hand, FFM is able to capture actual fluorescence images by the employment of fiber bundles, thus enabling a better spatial resolution from individual neurons.

8.3.2 Integration with Electronics

Apart from the light collection through the fibers discussed above, direct *in vivo* light delivery is the most widely used application of fiber-based neural interface, capitalizing their light transmission with high efficiency to a localized region [8]. Typically, a short segment of commercially available optical fiber was inserted through a guided cannula, which can also serve as the drug channel for the delivery of opsin-containing viral vectors. Then a longer segment of optical fiber was coupled to the implanted fiber for direct light delivery. In the *in vivo* optogenetic experiments, behavioral analysis of the modulated subject can serve as an output measurement. However, simultaneous electrical recording of the optically stimulated neurons would be the most straightforward and effective method to evaluate the gene transduction. Hence, various types of engineering tools of integrated

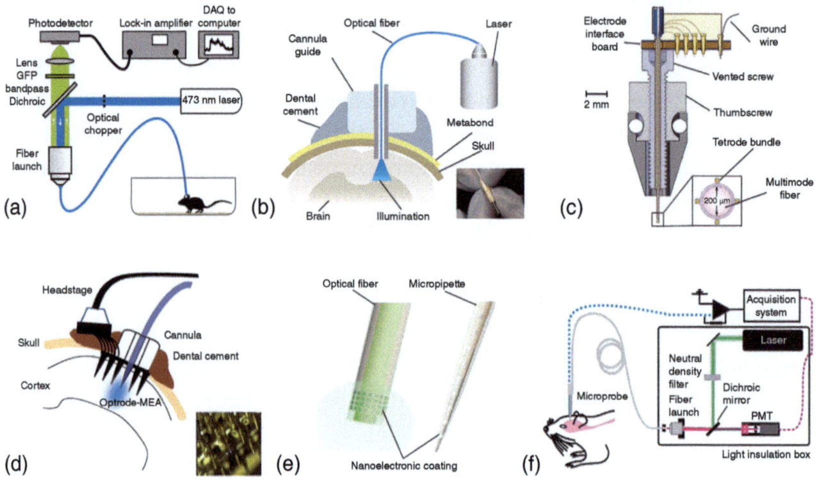

Figure 8.1 Neural interfaces of conventional silica fibers. (a) An illustration of fiber photometry setup for studying neural dynamics in freely behaving animals Gunaydin et al. [11] / with permission of Elsevier. (b) A schematic of the conventional *in vivo* optogenetic stimulation system [8]. (c) A diagram of the optetrode with one center optical fiber and four surrounding tetrodes Anikeeva et al. [14] / Springer Nature. (d) Optical fiber outfitted into a Utah array Wang et al. [15] / IOP Publishing. (e) Schematics of nanoelectronic coatings on an optical fiber forming a multifunctional probe Zhao et al. [16] / American Chemical Society. (f) A tapered dual-core fiber with electrolyte as electrical signal readout interface LeChasseur et al. [17] / Springer Nature.

optical, electrical, and fluidic components have been invented, demonstrating their utilities in modulating neural activities with simultaneous recording capability.

Since the first *in vivo* electrophysiology recording using insulated metal microelectrodes in the 1950s, the past decade has witnessed remarkable development of neural interfaces. The microfabrication techniques have permitted the revolutionary miniaturization and dense electrode packaging of the silicon-based neural probes [19, 20]. In the meantime, many groups have conceived novel electrode techniques utilizing new materials or fabrication methods for better biocompatibility or more functionality. Many of these device platforms were tested via *in vivo* neural recordings [21, 22].

As for the integration of electrical recording with optical stimulation, a first generation of optrode-integrating silica fibers with metal wires was presented by Zhang et al. with one optical fiber and one tungsten electrode (Figure 8.1b) [8]. A notable increase of the electrical interfacing channels was demonstrated in an optetrode device platform consisting of 16 metal wires (4 tetrode bundles) and 1 silica fiber, featured with a customized drive design (Figure 8.1c) [14]. Since the aligned tetrode bundles were separated on the edges of the center optical waveguide, different neural activities were observed during activation/inhibition optical modulation in transgenic mice. The authors further demonstrated the utility of the optetrode in behavioral analysis of freely moving animals under optical stimulation.

Besides the optetrode technique, several methods have been used to embed the optical fiber in a Utah array or a Michigan probe. For instance, laser ablation was utilized to create a vacant space inside a 6×6 Utah array (Figure 8.1d) [15]. Then a Teflon cannula was inserted to the hole, guiding the viral injection as well as the insertion of a tapered silica fiber coated with conductive metal layers (titanium/chromium with additional gold). Using a similar strategy, tapered optical fibers was also glued to one or multiple shanks in a multishank silicon probe, forming a neural interface with optical modulation at a bigger spatial dimension [23].

Nonetheless, the approach of combining relatively rigid metal wires and optical fibers would inevitably increase the device dimension and mechanical stiffness, resulting in irreversible tissue damage and a decrease in the quality of the electrophysiology recording. To this end, Xie group developed a multifunctional neural probe with a 30 μm diameter by attaching nanoelectronic device (nanoelectronic coating, NEC) to the surface of a conventional optical fiber (Figure 8.1e) [16]. The conformal attachment of the NEC to the dielectric probe was accomplished by the surface tension when existing from water. In addition to simultaneous electrical recording and optical stimulation, this versatile design also allowed for the integration of the NEC to a micropipette, realizing a miniaturized neural probe with drug delivery capability.

Similar to the patch–clamp approach, a hollow core has been incorporated within a tapered dual-core optical fiber with an end tip diameter of 6–20 μm enabling the study of the correlations between electrophysiological and optical signals with a single neuron resolution (Figure 8.1f) [17]. Inside the hollow core, a 1–3 M NaCl electrolyte solution has been selected to serve as the contacting media for acquiring the electrical activity where tungsten or stainless steel wire has been introduced for electrical connection with the recording system.

8.4 Thermally Drawn Multifunctional Fiber

The compromised approach of integrating insulated recording electrodes to the traditional optical fibers imposes several challenges to researchers, including complex assembly steps to achieve simultaneous optical stimulation and electrical recording. Besides, the relatively large mechanical stiffness of silica fiber leads to immune responses such as neuron death, glial scaring, and astrocyte reactivity [24]. Thus, developing a flexible optoelectronic device with built-in waveguide and recording electrodes has great significance for chronic studies.

Utilizing a thermal drawing process (TDP) similar to the fabrication of conventional silica fibers used in the fiber optics industry, multimaterial fiber-based neural probes have been developed by Anikeeva group, opening a new era for fiber-based neural interfaces [25]. Generally, a macroscopic preform is created first with prearranged materials and geometry. Then the preform can be drawn into hundreds of meters long fibers with an applied temperature and stress (Figure 8.2a). With this unique fabrication method, optical waveguides and electrodes can be incorporated into a fiber in a controllable and scalable manner.

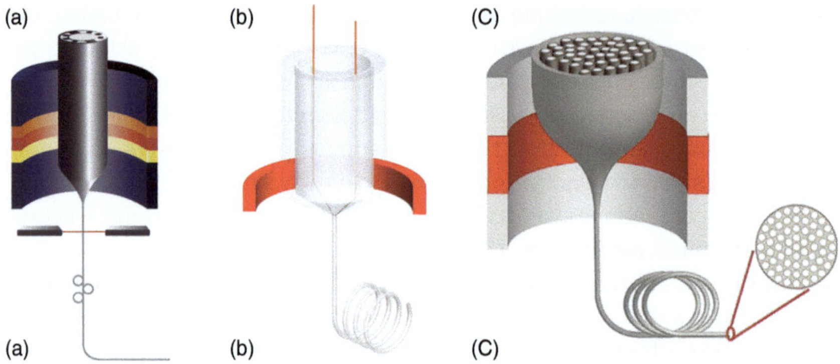

Figure 8.2 Thermal drawing methods for multifunctional fiber production. (a) Typical thermal drawing process [26]. (b) Convergence thermal drawing method for incorporating materials with a high-melting temperature [27]. (c) Two-step thermal drawing method for producing a large number of functional elements with finer features Guo et al. [28] / PLOS ONE / CC BY 4.0.

In this section, we will first discuss the design principles of multimaterial fibers in the context of neural interfacing applications. We will then overview recent fiber-based neural probes with respect to their unique fabrication method, capabilities, and applications.

8.4.1 Design Considerations for Neural Interfacing Applications

8.4.1.1 Material Selection

To accommodate the special requirements of the materials that can be thermally co-drawn, the material choices of the recording electrodes and protective/functional polymer matrix are very limited. The selected materials to constitute polymer waveguides are relatively more straightforward. By adopting polymers that have different refractive indices (RIs), a step-index polymer waveguide can be incorporated inside the multifunctional fibers. Within these polymer waveguides, polycarbonate (PC, $n = 1.58$) is usually chosen as the core material due to its broad transmission across the visible range, whereas its pairing cladding material can be cyclic olefin copolymer (COC, $n = 1.53$), poly(methyl methacrylate) (PMMA, $n = 1.48$), or polyvinylidene difluoride (PVDF, $n = 1.426$).

Incorporating electrodes into the polymer fiber, on the other hand, is usually a more challenging process mainly for two reasons: first, the chosen conductive material's thermomechanical properties need to match that of the codrawn materials to ensure a smooth fabrication process; secondly, a proper conductivity of the electrode material is demanded such that the as-drawn electrodes can enable high-quality electrophysiology recording while maintaining its spatial resolution.

Metals are usually the first choice to serve as recording electrode material due to their high conductivity, and the reported metal-based materials for fiber-based electrodes include Sn [25] and BiSn (metal alloys, 58Bi 42Sn) considering their low melting temperature (T_m) [29]. Due to the multimaterial fiber drawing criteria mentioned above, the melting temperature of the metallic material should be close to the glass transition temperature (T_g) of the main thermoplastics that constitute the majority of the preform. For instance, metal alloy BiSn (T_m = 138 °C) that typically serves as solder pastes can pair with PMMA (T_g = 105 °C) and PC ($T_g \approx 147$ °C), while metal electrode of tin (Sn, T_m = 232 °C) can pair with polymers with higher glass transition temperatures, such as polyethylenimine (PEI, T_g = 217 °C). Furthermore, during the process of drawing metal- or alloy-containing fibers, metals flow inside the predefined channels in a liquid form, leading to potential electrode discontinuity and fiber structure deformation. Therefore, in order to maintain the preform's geometry and the continuity of the electrodes, a sufficient tension must be applied via tuning the capstan pulling speed or the drawing temperatures. With applied high tension and stress on the fiber during the drawing process, the metal/alloys can be restrained tightly by the surrounding polymers, preventing the electrode breakup during drawing.

As an alternative, conductive polymer composites such as carbon black-loaded polyethylene (CPE) can render a more stable electrodes production inside a polymer matrix [25]. Similar to the temperature matching strategy described above, it is necessary to design the preform with respect to different conductive polymer composites and the hosting polymers. The common material combinations of polymer electrodes and hosting polymer matrix are CPE-PC and CPC (conductive polycarbonate)-PEI. Albeit the relatively easier and more stable electrode production, the low conductivity of CPE would inevitably lead to a larger electrode size to compensate the sufficient electrical performance for neural recording, yielding a poor spatial resolution. To this end, many attempts have been made to increase the conductivity of CPE and the most typical approach is doping CPE with carbon-based fillers such as graphite or carbon nanofiber (CNF) [26, 30]. Specifically, those conductive materials were first introduced into CPE by compression and heat, then following an extrusion process of the premixed material, a composite with uniform distribution of the added dopants can be achieved. For instance, 5 wt% graphite was introduced in CPE resulting in a 4.1-fold reduced sheet resistance [26]. Similarly, CNFs were evenly mixed into the CPE electrode material [30]. Owing to the high unidirectional tension from polymer matrix during the TDP, CNFs aligned within the conductive polyethylene in a longitudinal direction, contributing to improved electrical performance after thermal drawing [30]. Attributed to the *in situ* CNF alignment, the resistivity of the CNF composite presented a ~40-fold reduction compared to that before thermal drawing and two orders of magnitude reduction when compared with CPE.

8.4.1.2 Diverse Thermal Drawing Methods

The materials that can be thermally codrawn to constitute a polymer waveguide are very limited owing to the intrinsic thermal properties of those materials. To overcome this challenge, convergence fiber drawing approach was developed, which has opened a new possibility to produce fiber-based neural interfaces with materials that remain in the solid phase when the surrounding polymer matrix are undergoing the transformation from hard state to a viscous glassy state (Figure 8.2b).

With this approach, metal wires that possess a high melting temperature or silica waveguides with a high glass transition temperature will maintain their electrical/optical performance after thermal drawing. For example, researchers have successfully encapsulated copper, tungsten, and gold-coated tungsten wires in an insulating polymer cast for various biomedical applications, such as tumor impedance monitoring and neural recording [27, 31].

Beyond the single-step fabrication of multifunctional fibers discussed above, there remains a challenge for incorporating a large number of functional elements inside one preform due to the fabrication constraints, especially for multiple waveguides and electrodes. To this end, a two-step drawing method has been developed, providing a new strategy for incorporating functional components inside a single fiber (Figure 8.2c) [25, 28]. Typically, segments of thermally drawn functional fibers are bundled and consolidated to construct a new preform. After a second TDP, the resultant fiber consists of functional features (electrodes or optical waveguides) with a much reduced size compared to the first preform (up to ~1000-fold size reduction). The first-reported neural probe using this two-step fiber drawing method was successfully fabricated, containing 7–36 Sn electrodes ($d \approx 5.1\,\mu m$) with an overall size of 85 μm in diameter inside a hosting dielectric polymer of PEI.

In later studies, more attempts have been made for creating fiber-based neural probes with multiple polymer waveguides. For example, Guo et al. created a multifunctional fiber with an optical waveguide bundle of 60 predrawn PC/PMMA fibers in the center for imaging [28]. Other than a bundle of circular waveguides in the center, eight waveguides (PC/COC) in a rectangular shape have also been integrated on the side of a multifunctional fiber probe, tailoring a possible solution for optogenetic manipulation at different locations with individual addressability [29].

8.4.2 Application of As-drawn Fiber-based Probes

All of the above-mentioned fiber probes with embedded electrodes have demonstrated the capability of detecting endogenous neural activities with single-unit resolution. Most of the reported results showed chronic recording capabilities over the course of weeks or months without compromised recording quality. For instance, graphite-enhanced CPE electrode (gCPE, ~20 μm^2) illustrated the ability to track the same neuron over three months [26], while CNF-enhanced CPE electrode (18 μm × 11.3 μm) achieved 15 weeks chronic recording.

When applying multifunctional fibers in opsin-expressing animal brains, the miniaturized fiber probes exhibited unique advantages. First, the channel count and the cross-sectional arrangement of electrodes and waveguides can be easily designed and controlled during preform fabrication. Second, the cleaved fiber probe will ensure that the neighboring electrodes and waveguides are in the same plane, minimizing the distance between the optical stimulation and electrical recording sites in the Z axis. A long-term *in vivo* assessment of the multimodality fiber-based neural probe has confirmed stable optical stimulation and real-time electrical recording for at least two months in mouse brains [25].

Beyond the integration of optical waveguides with recording electrodes inside the multifunctional fiber, microfluidic channels can also be readily created in the fibers in a scalable manner. The first multimodality fiber-based neural probe for simultaneous electrical recording, optical stimulation, and chemical intervention was demonstrated by Canales et al. The neural activities were monitored under optical stimulation while CNQX (a synaptic blocker) was delivered to the targeted brain region directly using the fiber probe. During this process, neurons responded to the optical stimulation regularly till the local infusion of CNQX, which suppressed the optically triggered activities. After a few minutes, the neurons recovered from the drug effect and their activities corresponding to laser pulses returned to the level before drug intervention.

Besides the neural modulation and recording demonstrated using CNQX, the multifunctional fibers can also be utilized for one-step optogenetics. In conventional optogenetics study using wild-type animals, two surgeries are typically required for viral injection and subsequent optrode implantation [26]. By implanting the multifunctional fiber probe into the targeted brain region of a wild-type mouse, viral vectors carrying opsin genes were first delivered to transduce neurons via the hollow channels of the fiber. Electrical recording from the neighboring gCPE electrodes revealed the temporal dynamics of the opsin expression in the mouse brain. Behavioral analysis of the examined subjects was also conducted in conjunction with CNQX infusion under optical stimulation, which facilitated the mapping of projection from basolateral amygdala (BLA) to ventral hippocampus (vHPC).

Despite the physical and functional considerations needed when designing neural probes for brain interfacing, the natural brain–skull interface provides a favorable platform for researchers to fix the implanted devices. However, to interface nerves beyond the brain using conventional rigid probes presents challenges due to the movement of soft tissue and the lack of a stable support for fixation [32]. To that end, the flexible and biocompatible multifunctional fibers serve as an ideal candidate for interfacing soft tissues beyond brain. For instance, stretchable multifunctional fibers have been created for optogenetic stimulation and electrical recording in the spinal cord, accommodating the deformation of the implanted device in the spinal cord during daily motion [33]. This fiber's waveguide was composed of stretchable COC elastomer (COCE; $n = 1.51$) and polydimethylsiloxane (PDMS; $n = 1.41–1.47$), while the electrical component had a ring structure of silver nanowires conformally applied on the surface of the fiber via dip coating. In a more recent work, a polymer fiber (PC/PMMA) has also been applied in the gut lumen for optogenetic manipulation [34].

8.4.3 Advanced Multifunctional Fibers with Post Processing

Although the as-drawn fiber-based probe demonstrates advantages including scalable fabrication and multifunctional integration capability, the limited number of functional components and the restricted interfacing location on the fiber tip make it less ideal for investigating neural activities over a large scale, when compared with commercially available neural probes. To address this challenge, researchers have developed various approaches including postassembly, postmaterial removal, and postdeposition to broaden the applicability of fiber-based neural probes for neuroscience studies [29, 31, 35–37].

To overcome the limitation of the single interfacing area at the fiber tip, Jiang et al. have utilized a femtosecond laser micromachining technology in which ultrashort laser pulses are delivered to the side of the multifunctional fibers to expose functional interfacing sites along the fiber length (Figure 8.3a) [29]. The created depth-dependent fiber probes have significantly broadened the application of functional fiber-based neural probes. The depth-dependent fiber probe had not only multiple exposed electrodes on the side of the thin, flexible fiber probe, but also optical excitation windows and microfluidic channel openings. The drug delivery capability at multiple depths was demonstrated by injecting four different food colors through four hollow channels in a fiber, respectively, with channel openings exposed at different locations along the fiber length. In addition, eight side-illuminating ports were created on a fiber containing eight polymer waveguides, four electrodes, and one center fluidic channel. The resulted fiber probe shares a similar linear architecture with microfabricated silicon-based probes such as Neuropixels or Neuronexus probes with additional integrated multiple waveguides and microfluidic channels for optogenetic and pharmacological studies. *In vivo* study in transgenic mice using the depth-dependent fiber probes further validated the versatility and scalability of this technology.

The exposure of the functional elements in the longitudinal direction has also been realized by in-fiber photolithography [36]. A photopolymeric composite of

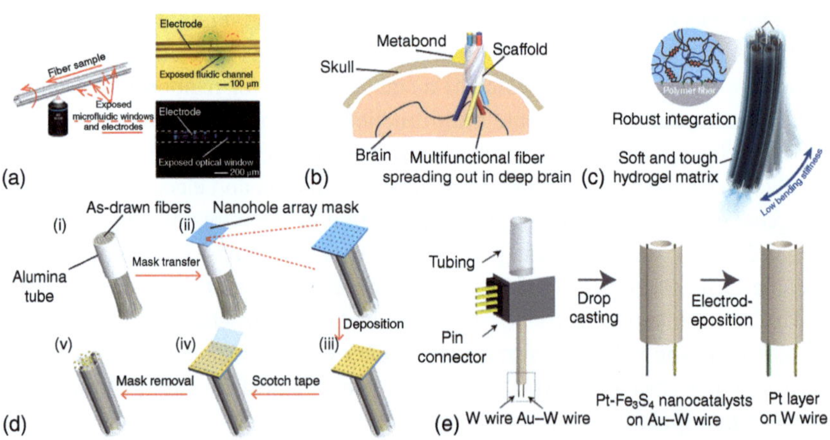

Figure 8.3 Post-processing enabled advanced multifunctional fiber probes. (a) Depth-dependent fiber probes produced by femtosecond laser micromachining Jiang et al., 2020 / Springer Nature / CC BY 4.0. (b) A schematic of spatially expandable fiber probes inside a mouse brain Jiang et al. [29] / Springer Nature / CC BY 4.0. (c) Adaptive hydrogel hybrid fiber probe fabricated via a one-step polymerization process Park et al. [37] / Springer Nature / CC BY 4.0. (d) A diagram showing the post-deposition of nano-optoelectrodes to the fiber tips via soft nanolithography Jiang et al. [35] / American Chemical Society. (e) An electrocatalytic fiber probe enabled by surface deposition via drop casting and electroplating Park et al. [31] / Springer Nature.

thiol–epoxy/thiol–ene was thermally cured during the fiber drawing process and was further photolithographically patterned by subsequent UV light curing and development, enabling access to the functional component within the fiber from the side surface.

One challenge that remains unsolved in the neural interfacing field lies within the limited spatial coverage in the brain in three dimensions without causing significant surgical damage. Many technologies have afforded comparative three-dimensional interfaces, but the lateral distribution of electrodes is restricted by either the multi-module stacking or the diameter of the shuttle device. Utilizing a unique helical scaffolding structure, Jiang et al. were able to deliver functional fiber probes to multiple distant regions of the mouse brain via a single surgical burr hole, offering a substantially larger spatial coverage in the brain (Figure 8.3b) [29]. The scaffold was fabricated from closely packed and consolidated polycarbonate tubes that underwent translational and rotational TDP, forming five or seven intertwined helical channels. During the implant surgery, the scaffold was first inserted and fixed to a predetermined brain region, and then individual functional fiber probes were inserted through the helical channels where the extruding angle was controlled by the helical scaffold. Discrete brain activities from cortical and subcortical brain regions were captured by the spatially expandable fiber probes during spike and local field potential recording in anesthetized mice. Furthermore, in a virus-induced temporal lobe epilepsy mouse model, varying electrographic seizure activities have been detected across distinct brain regions.

Another strategy for assembling functional fibers has been developed more recently (Figure 8.3c) [37]. Although the flexible thin fibers exhibit comparable biocompatibility to standard stainless steel wires, the stiffness of fiber-based probe and the tissue response would inevitably increase when the fiber probe becomes larger to accommodate more functionalities. In order to further reduce the foreign body response from the fiber devices, a new fiber-based neural interface platform has been invented by the employment of a coated hydrogel layer to bond a bundle of multifunctional fibers tightly together through a one-step polymerization process. This method allowed for an adaptive integration of multiple functional fibers in a tissue-favorable hydrogel housing, leading to a reduction of the mechanical stiffness and an improved recording performance. The comparison of the recording results and the tissue immune response between a bare polymer fiber and a hydrogel-integrated fiber proved the better biocompatibility of the fiber-based hydrogel probe.

Besides the postprocessing methods discussed above, integrating other functionality with the multifunctional fiber probe has also been achieved by postdeposition (Figure 8.3d) [35]. In this work, a silica fiber was first incorporated into a multifunctional fiber in the center via convergence fiber drawing method. Then periodic nano-optoelectrodes were deposited onto the tips of the multifunctional fiber via soft nanolithography. The nano-optoelectrodes-integrated probes can not only enable potential plasmonic sensing of biochemical molecules but also improve the electrical performance of the embedded electrodes. As a proof-of-concept demonstration, the authors detected resonant frequency shifts from the background environment via refractive index sensing as well as distinctive benzenethiol (BZT) vibrational peaks via surface-enhanced Raman spectroscopy (SERS) measurement from the integrated nano-optoelectrodes. Furthermore, nano-optoelectrodes deposited on the tips of CNF-CPE fiber resulted in decreased electrical impedance, leading to improved LFP recording performance of the surface-modified electrode compared to the unmodified one [35].

Another strategy for depositing additional materials to the drawn fibers is to directly coat functional materials to selected components in fibers via electroplating and drop casting. In recent work by Park et al., a multifunctional fiber with high-melting-point metals (tungsten and gold-coated tungsten microwires) was first drawn via convergence fiber drawing process (Figure 8.3e) [31]. Then nanocatalysts were drop casted onto the gold-coated tungsten wire, while platinum was deposited to the tungsten wire via electroplating. After the surface deposition of the additional material, the modified wires can serve as cathodes and anodes in an electrocatalytic reaction, respectively. Besides the electrodes, the functional fiber also contains a hollow channel in the center. With this special design, the authors successfully demonstrated the local release of nitric oxide (NO) reduced from nitrite (NO^{2-}) via an electrocatalytic reaction, providing a unique approach for studying the role of NO in neuronal signaling.

8.4.4 Tissue Engineering

The majority of the applications reviewed above take advantages of the recording and stimulation capabilities of fiber-based probes, and the hollow channels are mainly used for drug delivery. On the other hand, the functional fibers share a similar morphology with hollow conduits, which have been widely used for peripheral nerve regeneration both in the lab and clinic. Thus, exploiting the versatile design and fabrication process of multimaterial fibers, researchers have developed a number of strategies for tissue engineering applications using hollow-core fibers. For example, the TDP has allowed the scalable fabrication of hollow fibers with rectangular cores with microgrooves, whereas most synthetic nerve scaffolds consist of cylindrical conduits. The results have shown that hollow rectangular fiber scaffolds with micron-sized grooved channels can facilitate nerve repair with axonal guiding capabilities (Figure 8.4a) [38]. Another approach to improve axonal outgrowth and cell survival is to integrate porous structures into the scaffold, which allows for nutrient and oxygen diffusion, and provides an interlocking substrate for the scaffold and the surrounding tissue [41]. To date, several methods have been conveyed to produce porosity in thermally drawn fibers, such as phase separation, *in situ* water evaporation, and salt leaching [39, 42, 43]. Specifically, the porous fiber produced by the salt-leaching method with an additional fuse printing step has exhibited better performance for repairing injured spinal cord with complex geometry. Due to the scalability of fiber drawing and 3D printing, the porous fiber scaffold enables a high throughput technique for constructing hollow nerve conduits with tunable microporous structures at centimeter scales (Figure 8.4b) [39]. Similar strategy has recently been applied to a volumetric muscle loss mouse model with engineered muscle extracellular matrix and induced skeletal myogenic

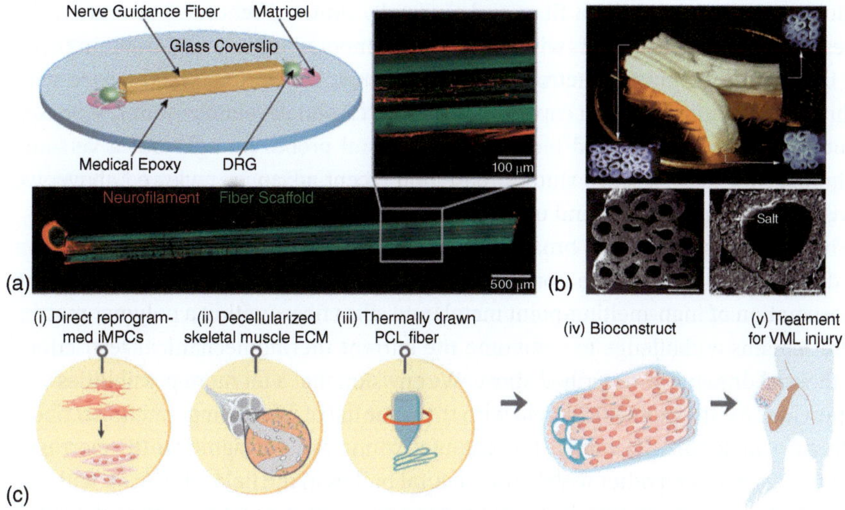

Figure 8.4 Thermally drawn fibers for tissue engineering. (a) Microgrooved fiber scaffolds for nerve guidance Koppes et al. [38] / with permission of Elsevier. (b) Porous fibers created via salt-leaching process which can be fused into 3D printing for a larger scale production Shahriari et al. [39] / JOHN WILEY & SONS, INC. (c) Engineered fiber-based nerve scaffold with muscle favorable materials Jin et al. [40] / John Wiley & Sons.

progenitor cells, showing a promoted muscle regeneration and repair of injured muscle (Figure 8.4c) [40].

Although the facile fabrication makes the fiber-based hollow conduits advantageous for tissue engineering purposes, the utility of other functional elements such as electrodes and optical waveguides in fiber scaffolds has not been investigated. Besides, many reports have shown the effectiveness of electrical stimulation, optogenetic stimulation, and the delivery of growth factor for nerve repair and regeneration [44–46]. The potential challenge of integrating those components for tissue engineering lies within the control over the exposure of the embedded electrodes and the precise light delivery to the desired location. A possible strategy for solving this problem is the femtosecond micromachining technique that has been mainly employed for creating depth-dependent fiber probes for neural recording. However, a detailed optical system needs to be developed for machining structures inside the fiber probes. Nonetheless, simultaneous electrical recording would benefit the multimaterial scaffolding fibers with straightforward real-time monitoring capabilities for the progress of nerve growth [47].

8.5 Conclusions

With the rapidly evolving advancement of molecular biology, photonics, and material science, optical neural interfaces have been revolutionized by generations. Both optical recording and optogenetic manipulation offer unique opportunities for neuroscientists, where optical fibers play significant roles in. The fiber-based neural probes including conventional silica fiber and thermally drawn fibers have permitted the investigation of neural circuits with high spatial temporal resolution. Combining optical fibers with integrated electronics further strengthen the implanted device with bidirectional communication capabilities. The successful development of fiber-based multifunctional, flexible, and biocompatible neural probes brought out a versatile technology for neuroscience studies. Based on recent advances reviewed above, we envision new opportunities and unmet challenges for fiber-based neural probes.

Firstly, new approaches to broaden the limited material selection for the TDP can be developed. Although the convergence fiber drawing method has allowed the incorporation of high-melting-point metals and silica fibers within a polymer matrix, there remains a challenge to overcome the current thermomechanical restrictions for thermal drawing as described above. We envision that a lot more possibilities will be provided for the biomedical field with the potential of fabricating functional fibers with new materials. Secondly, the channel count can be significantly increased (>100) to match with other widely used neural probes in the field. However, with the boost in channel count, the electrical backend connection will be another challenge to be addressed. Traditional manual connection requires precise physical control, which hinders the scalable fabrication. Besides, albeit the successful demonstration of the integration of eight waveguides within a single neural probe for opotogenetic stimulation, the individual addressability has not been established. Potential solutions for the electrodes and waveguides' backend connection could be the involvement of electrical and photonic wire bonding via the introduction of designed silicon chips. Specifically, the electrodes of the fiber can be electrically connected to the chip circuit via wire bonding, while photonic wire bond waveguide can be defined using two-photon lithography to be coupled to cleanroom-fabricated light sources [48, 49]. A low-index cladding material can be adopted as the embedding matrix for optical confinement and electrical insulation. Thirdly, new functionalities such as biological process monitoring can be introduced to the fibers. For example, chemical sensing modality can be conveyed via electrical and optical pathways, including electrochemical sensing and plasmonic sensing. Moreover, applications of the fluidic channel are also largely unexplored. Other than the administration of drugs or other substances, it can also be used for withdrawing fluidic or tissue sample directly from the brain tissue for studying disease mechanisms or therapeutic effects. With the integrated electrical, optical, and fluidic detection capabilities, the multimodality fiber can provide a full spectrum of quantitative biomolecular and electrophysiological information with a high spatial temporal resolution. In addition, robotic fibers and fiber robots are emerging as a powerful tool for both fundamental studies in biomedical fields as well as clinical applications. Last but not the least, the biocompatibility of the fiber-based neural probes can be further improved with adaptive coating or the utilization of biodegradable materials. This would expedite the potential application of multimodality implantable fiber device in clinics, including neurological decease detection and treatment, as well as nerve repair and regeneration. Looking into the future, further innovations are expected with continuing advances in biology, materials, optics, and electronics.

References

1 Jiang, S., Wu, X., Rommelfanger, N.J. et al. (2022). Shedding light on neurons: optical approaches for neuromodulation. *Natl. Sci. Rev.* nwac007. https://doi.org/10.1093/nsr/nwac007.

2 Yan, W., Dong, C.Q., Xiang, Y.Z. et al. (2020). Thermally drawn advanced functional fibers: new frontier of flexible electronics. *Mater. Today* 35: 168–194. https://doi.org/10.1016/j.mattod.2019.11.006.

3 Mank, M. and Griesbeck, O. (2008). Genetically encoded calcium indicators. *Chem. Rev.* 108 (5): 1550–1564. https://doi.org/10.1021/cr078213v.

4 Girven, K.S. and Sparta, D.R. (2017). Probing deep brain circuitry: new advances in in vivo calcium measurement strategies. *ACS Chem. Neurosci.* 8 (2): 243–251. https://doi.org/10.1021/acschemneuro.6b00307.

5 Fenno, L., Yizhar, O., and Deisseroth, K. (2011). The development and application of optogenetics. *Annu. Rev. Neurosci.* 34: 389–412.

6 Deisseroth, K. (2011). Optogenetics. *Nat. Methods* 8 (1): 26–29. https://doi.org/10.1038/nmeth.f.324.

7 Boyden, E.S., Zhang, F., Bamberg, E. et al. (2005). Millisecond-timescale, genetically targeted optical control of neural activity. *Nat. Neurosci.* 8 (9): 1263–1268. https://doi.org/10.1038/nn1525.

8 Zhang, F., Gradinaru, V., Adamantidis, A.R. et al. (2010). Optogenetic interrogation of neural circuits: technology for probing mammalian brain structures. *Nat. Protoc.* 5 (3): 439–456. https://doi.org/10.1038/nprot.2009.226.

9 Flusberg, B.A., Cocker, E.D., Piyawattanametha, W. et al. (2005). Fiber-optic fluorescence imaging. *Nat. Methods* 2 (12): 941–950. https://doi.org/10.1038/nmeth820.

10 Wang, Y., DeMarco, E.M., Witzel, L.S., and Keighron, J.D. (2021). A selected review of recent advances in the study of neuronal circuits using fiber photometry. *Pharmacol. Biochem. Behav.* 201: 173113. https://doi.org/10.1016/j.pbb.2021.173113.

11 Gunaydin, L.A., Grosenick, L., Finkelstein, J.C. et al. (2014). Natural neural projection dynamics underlying social behavior. *Cell* 157: 1535–1551.

12 Kim, C.K., Yang, S.J., Pichamoorthy, N. et al. (2016). Simultaneous fast measurement of circuit dynamics at multiple sites across the mammalian brain. *Nat. Methods* 13 (4): 325–328. https://doi.org/10.1038/Nmeth.3770.

13 Schulz, K., Sydekum, E., Krueppel, R. et al. (2012). Simultaneous BOLD fMRI and fiber-optic calcium recording in rat neocortex. *Nat. Methods* 9 (6): 597–602. https://doi.org/10.1038/nmeth.2013.

14 Anikeeva, P., Andalman, A.S., Witten, I. et al. (2011). Optetrode: a multichannel readout for optogenetic control in freely moving mice. *Nat. Neurosci.* 15 (1): 163–170. https://doi.org/10.1038/nn.2992.

15 Wang, J., Wagner, F., Borton, D.A. et al. (2012). Integrated device for combined optical neuromodulation and electrical recording for chronic in vivo applications. *J. Neural Eng.* 9 (1): 016001. https://doi.org/10.1088/1741-2560/9/1/016001.

16 Zhao, Z., Luan, L., Wei, X. et al. (2017). Nanoelectronic coating enabled versatile multifunctional neural probes. *Nano Lett.* 17 (8): 4588–4595. https://doi.org/10.1021/acs.nanolett.7b00956.

17 LeChasseur, Y., Dufour, S., Lavertu, G. et al. (2011). A microprobe for parallel optical and electrical recordings from single neurons in vivo. *Nat. Methods* 8 (4): 319–325. https://doi.org/10.1038/nmeth.1572.

18 Flusberg, B.A., Nimmerjahn, A., Cocker, E.D. et al. (2008). High-speed, miniaturized fluorescence microscopy in freely moving mice. *Nat. Methods* 5: 935. https://doi.org/10.1038/nmeth.1256. https://www.nature.com/articles/nmeth.1256#supplementary-information.

19 Jun, J.J., Steinmetz, N.A., Siegle, J.H. et al. (2017). Fully integrated silicon probes for high-density recording of neural activity. *Nature* 551 (7679): 232–236. https://doi.org/10.1038/nature24636.

20 Steinmetz, N.A., Koch, C., Harris, K.D., and Carandini, M. (2018). Challenges and opportunities for large-scale electrophysiology with Neuropixels probes. *Curr. Opin. Neurobiol.* 50: 92–100. https://doi.org/10.1016/j.conb.2018.01.009.

21 Kozai, T.D., Langhals, N.B., Patel, P.R. et al. (2012). Ultrasmall implantable composite microelectrodes with bioactive surfaces for chronic neural interfaces. *Nat. Mater.* 11 (12): 1065–1073. https://doi.org/10.1038/nmat3468.

22 Luan, L., Wei, X., Zhao, Z. et al. (2017). Ultraflexible nanoelectronic probes form reliable, glial scar-free neural integration. *Sci. Adv.* 3 (2): e1601966. https://doi.org/10.1126/sciadv.1601966.

23 Royer, S., Zemelman, B.V., Barbic, M. et al. (2010). Multi-array silicon probes with integrated optical fibers: light-assisted perturbation and recording of local neural circuits in the behaving animal. *Eur. J. Neurosci.* 31 (12): 2279–2291. https://doi.org/10.1111/j.1460-9568.2010.07250.x.

24 Salatino, J.W., Ludwig, K.A., Kozai, T.D.Y., and Purcell, E.K. (2017). Glial responses to implanted electrodes in the brain. *Nat. Biomed. Eng.* 1 (11): 862–877. https://doi.org/10.1038/s41551-017-0154-1.

25 Canales, A., Jia, X.T., Froriep, U.P. et al. (2015). Multifunctional fibers for simultaneous optical, electrical and chemical interrogation of neural circuits in vivo. *Nat. Biotechnol.* 33 (3): 277. https://doi.org/10.1038/nbt.3093.

26 Park, S., Guo, Y.Y., Jia, X.T. et al. (2017). One-step optogenetics with multifunctional flexible polymer fibers. *Nat. Neurosci.* 20 (4): 612. https://doi.org/10.1038/nn.4510.

27 Chin, A.L., Jiang, S., Jang, E. et al. (2021). Implantable optical fibers for immunotherapeutics delivery and tumor impedance measurement. *Nat. Commun.* 12 (1): 5138. https://doi.org/10.1038/s41467-021-25391-z.

28 Guo, Y., Werner, C.F., Canales, A. et al. (2020). Polymer-fiber-coupled field-effect sensors for label-free deep brain recordings. *PLoS One* 15 (1): e0228076.

29 Jiang, S., Patel, D.C., Kim, J. et al. (2020). Spatially expandable fiber-based probes as a multifunctional deep brain interface. *Nat. Commun.* 11 (1): https://doi.org/10.1038/s41467-020-19946-9.

30 Guo, Y.Y., Jiang, S., Grena, B.J.B. et al. (2017). Polymer composite with carbon nanofibers aligned during thermal drawing as a microelectrode for chronic neural interfaces. *ACS Nano* 11 (7): 6574–6585. https://doi.org/10.1021/acsnano.6b07550.

31 Park, J., Jin, K., Sahasrabudhe, A. et al. (2020). In situ electrochemical generation of nitric oxide for neuronal modulation. *Nat. Nanotechnol.* 15 (8): 690–697. https://doi.org/10.1038/s41565-020-0701-x.

32 Montgomery, K.L., Iyer, S.M., Christensen, A.J. et al. (2016). Beyond the brain: optogenetic control in the spinal cord and peripheral nervous system. *Sci. Transl. Med.* 8 (337): 337rv5. https://doi.org/10.1126/scitranslmed.aad7577.

33 Lu, C., Park, S., Richner, T.J. et al. (2017). Flexible and stretchable nanowire-coated fibers for optoelectronic probing of spinal cord circuits. *Sci. Adv.* 3 (3): e1600955. https://doi.org/10.1126/sciadv.1600955.

34 Buchanan, K.L., Rupprecht, L.E., Kaelberer, M.M. et al. (2022). The preference for sugar over sweetener depends on a gut sensor cell. *Nat. Neurosci.* 25 (2): 191–200. https://doi.org/10.1038/s41593-021-00982-7.

35 Jiang, S., Song, J., Zhang, Y.J. et al. (2021). Nano-optoelectrodes integrated with flexible multifunctional fiber probes by high-throughput scalable fabrication. *ACS Appl. Mater. Interfaces* 13 (7): 9156–9165. https://doi.org/10.1021/acsami.0c19187.

36 Lee, Y., Canales, A., Loke, G. et al. (2020). Selectively micro-patternable fibers via in-fiber photolithography. *ACS Cent. Sci.* 6 (12): 2319–2325. https://doi.org/10.1021/acscentsci.0c01188.

37 Park, S., Yuk, H., Zhao, R. et al. (2021). Adaptive and multifunctional hydrogel hybrid probes for long-term sensing and modulation of neural activity. *Nat. Commun.* 12 (1): 3435. https://doi.org/10.1038/s41467-021-23802-9.

38 Koppes, R.A., Park, S., Hood, T. et al. (2016). Thermally drawn fibers as nerve guidance scaffolds. *Biomaterials* 81: 27–35. https://doi.org/10.1016/j.biomaterials.2015.11.063.

39 Shahriari, D., Loke, G., Tafel, I. et al. (2019). Scalable fabrication of porous microchannel nerve guidance scaffolds with complex geometries. *Adv. Mater.* 31 (30): 1902021. https://doi.org/10.1002/adma.201902021.

40 Jin, Y., Shahriari, D., Jeon, E.J. et al. (2021). Functional skeletal muscle regeneration with thermally drawn porous fibers and reprogrammed muscle progenitors for volumetric muscle injury. *Adv. Mater.* 33 (14): 2007946. https://doi.org/10.1002/adma.202007946.

41 Loh, Q.L. and Choong, C. (2013). Three-dimensional scaffolds for tissue engineering applications: role of porosity and pore size. *Tissue Eng. Part B* 19: 485–502.

42 Grena, B., Alayrac, J.B., Levy, E. et al. (2017). Thermally-drawn fibers with spatially-selective porous domains. *Nat. Commun.* 8: 364. https://doi.org/10.1038/s41467-017-00375-0.

43 Yu, L., Xuan, H., Guo, Y. et al. (2017). Porous polymer optical fiber fabrication and potential biomedical application. *Opt. Mater. Express* 7 (6): 1813–1819. https://doi.org/10.1364/OME.7.001813.

44 Al-Majed, A.A., Tam, S.L., and Gordon, T. (2004). Electrical stimulation accelerates and enhances expression of regeneration-associated genes in regenerating rat femoral motoneurons. *Cell. Mol. Neurobiol.* 24 (3): 379–402. https://doi.org/10.1023/B:CEMN.0000022770.66463.f7.

45 Ikeda, M., Uemura, T., Takamatsu, K. et al. (2014). Acceleration of peripheral nerve regeneration using nerve conduits in combination with induced pluripotent

stem cell technology and a basic fibroblast growth factor drug delivery system. *J. Biomed. Mater. Res. Part A* 102 (5): 1370–1378. https://doi.org/10.1002/jbm.a.34816.

46 Thompson, D.M., Koppes, A.N., Hardy, J.G., and Schmidt, C.E. (2014). Electrical stimuli in the central nervous system microenvironment. *Annu. Rev. Biomed. Eng.* 16: 397–430. https://doi.org/10.1146/annurev-bioeng-121813-120655.

47 Musick, K.M., Rigosa, J., Narasimhan, S. et al. (2015). Chronic multichannel neural recordings from soft regenerative microchannel electrodes during gait. *Sci. Rep.* 5: 14363. https://doi.org/10.1038/srep14363.

48 Kampasi, K., English, D.F., Seymour, J. et al. (2018). Dual color optogenetic control of neural populations using low-noise, multishank optoelectrodes. *Microsyst. Nanoeng.* 4 (1): 10. https://doi.org/10.1038/s41378-018-0009-2.

49 Lindenmann, N., Balthasar, G., Hillerkuss, D. et al. (2012). Photonic wire bonding: a novel concept for chip-scale interconnects. *Opt. Express* 20 (16): 17667. https://doi.org/10.1364/OE.20.017667.

9

Very-large-scale Integration for Fibers

Alexander Gumennik[1], *Jeffery Coulter*[1], *Louis A. van der Elst*[1], *Troy A. Leffel*[1], *Etgar C. Levi*[2], *Camila Faccini de Lima*[1], *Tyson Miller*[1], *and Mengxin Zheng*[1]

[1] *Indiana University Bloomington, Luddy School of Informatics, Computing, and Engineering, Department of Intelligent Systems Engineering, 700 N Woodlawn Ave, Bloomington, IN, USA*
[2] *Massachusetts Institute of Technology, Research Laboratory of Electronics, 77 Massachusetts Avenue, Cambridge, MA 02139, USA*

9.1 Introduction

The information medium of the twentieth century was revolutionized twice: first with the invention of the integrated circuit, boosting data processing, and second with the demonstration of low-loss fiber optics, bringing up-to-speed data communication. Those two components lay in the basis of the Internet in the form known to us: a giant network of data processing nodes interconnected by high-throughput communication links.

Similarly, in the twenty-first century, we find ourselves again on the verge of a transformative leap in the hardware used for data processing and distribution. With modern lithography reaching the ability to define features so minute that they are only a few tens of atoms in cross-section, Moore's law – the necessity of doubling the areal density of transistors on a microprocessor every two years – has factually reached its physical limit driving the emergence of a whole plethora of alternative computation schemes. From spintronic and photonic to neuromorphic and quantum, the novel computational hardware platforms operate on an increasingly demanding power budget, differ from each other in terms of protocols used for data processing, and even use physically disparate forms and degrees of freedom in which the data itself is encoded.

Yet, fiber optics – the workhorse of digital communication – still awaits the transformation that would enable it to interconnect those diverse platforms into one harmonized network. And the challenge is not only in the growing energy demands in computing across diverse data processing platforms requiring even more efficient long-haul communication links than the state-of-the-art fiber optics but also that

Optical and Electronic Fibers: Emerging Applications and Technological Innovations, First Edition.
Edited by Lei Wei.

the fiber of tomorrow will need to translate the data across those platforms. Imparting efficient data transduction and transformation capabilities into fiber optics is likely to require integrating photonic and optoelectronic devices, and systems into the fiber itself.

Standard fiber optics-manufacturing approaches, such as thermal draw, rely on pulling the fiber from a melt; but melt shaping of glass is prone to complex, hard-to-control fluid dynamics. The fabrication approach to the fiber-embedded photonics needs to defy the limitations of molten-phase multimaterial fiber processing in achieving the desired solid-state outcomes, aiming to embed active circuitry in fiber optics with submicron precision and tight material quality control. The ability to upgrade fiber optics manufacturing to deliver efficient interconnects for emerging computation platforms depends on this technological advance.

Is expectation for such an advance even realistic, one might ask? Well, at least one example of a comparable technological leap took place in integrated microelectronics. In microelectronics, the paradigm shift happened in the 1970s with the invention of the very-large-scale integration (VLSI) approach to microprocessors' design and fabrication. VLSI enabled a complete restructuring of the microelectronics industry from that taking place at small radio-electronic shops, where artisans would solder transistors onto the circuit board one by one and relocate it into giant clean-room fabs, such as Intel and TSMC, where an automated production line operates a microchip-on-a-silicon-wafer processing workflow in a closed loop, with the minimal human involvement. VLSI-Fi is a direct digital manufacturing (DDM) of microelectronic circuitry, allowing one to think about microprocessors in terms of circuit board design rather than the physics of semiconductors. Circuit design software used by electrical engineers today hides the physics "under the hood," reformulating it in terms of simple design rules, such as what should be the dimensions of the circuit components and the distances between them for optimal circuit performance. VLSI has resulted in the modern high-tech microprocessors industry, as we know it.

In order to materialize a microprocessor designed by VLSI, the photoresist is deposited on the silicon wafer surface, which is then exposed at selected locations using photolithography. Exposed areas are then functionalized by a series of chemical and thermal treatments. This defines a single-layer 2D cross-section of the microchip. The process is repeated layer by layer, resulting in a "pie" of connectorized layers stacked perpendicularly to the wafer surface in an additional 1D. Ion implantation into the semiconducting substrate and annealing of dopants functionalize each device electronically, providing a 0D structural and functional control on the level of an individual device. The result is the microprocessor's 3D **(or, rather, 2D+1D+0D)** structure.

Would it be possible to formulate a similar toolbox for architectural control of fiber-embedded structures? In other words, has the multimaterial fiber technology matured enough to deliver VLSI-Fi? Let us suggest here one of the possible pathways toward that end. In some ways, it's a "solution seeking a problem," because the multimaterial fiber technology is not yet mature enough to get a clear-cut solution to the problems of material quality and high-precision 3D architectural control of fiber-embedded structures. But it is close enough to maturity to discuss whether the remaining technical obstacles to bridging the gap to VLSI-Fi are addressable.

Let's, for instance, consider the collection of tools, demonstrated in Figure 9.1, designed to control the 3D architecture of in-fiber multimaterial structures:

- **(2D) In the fiber cross-sectional plane**: The structure will be defined by 3D printing the preform cladding, filling its voids with functional materials, such as metals and semiconductors, and scaling the cross-section geometry down by a thermal draw of the preform into a fiber.
- **(+1D) Along the fiber axis**: We will pattern the initially continuous functional material cores into periodic arrays of spheres using a material-selective, spatially coherent, capillary breakup.
- **(+0D) Internally to an individual active-material sphere**: We will control the crystallinity and electronic properties, forcing it to solidify from a melt along the externally imposed temperature profile. This will result in a segregation-driven dopant redistribution and crystallinity defined by the solidification-front stability.

In the following sections, we will consider each tool in the candidate toolbox for VLSI-Fi in Figure 9.1. We will consider in which aspects of architectural and

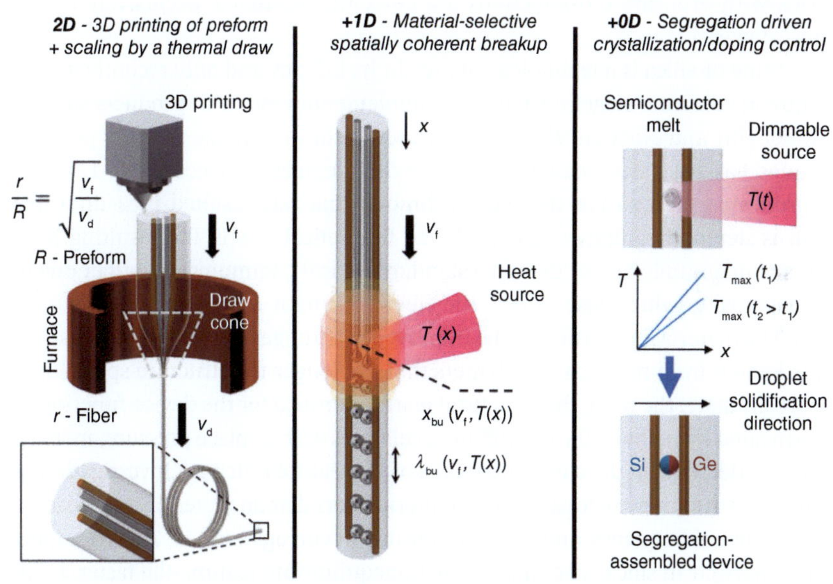

Figure 9.1 Set of candidate tools for VLSI-Fi. The manufacturing approach includes (from left to right) thermal draw of 3D-printed preforms, system assembly by a spatially coherent, material-selective capillary breakup (for instance, of an array of bispherical p–n Si diodes), controlled recrystallization, and dopants' redistribution (for example, of Ge mixed into Si) in an externally induced time-varying temperature gradient due to higher solubility of the dopant in a liquid than in solid phase of the solvent material. **Notations:** R-preform radius, r-fiber radius, R/r-draw scaling factor, v_f-feed speed, v_d-draw speed, x_{bu}-breakup location, λ_{bu}-breakup period, and $T(t, x)$-spatiotemporal temperature profile. *Source:* Adapted from: Faccini de Lima et al. [1], Figure 1.

morphological control of fiber-embedded structures they deliver and in which they fall short, what are the pathways to improvements on the latter, and what are the possible state-of-the-art alternatives to each of the suggested tools.

9.2 VLSI-Fi: State-of-the-art and Current Challenges

9.2.1 Architectural and Morphological Control of the Fiber Cross-section by a 3D Printing of the Preform and Its Thermal Draw

From the microelectronics realm, we know that the complexity of the computational tasks the microprocessor can perform is directly related to the complexity of the architectural arrangement of its elements. The more transistors one can cram into a microchip, the faster it will be able to calculate. Analogously, the fiber device functionality is directly related to the architectural complexity of the materials' arrangement in the preform [2]. Despite demonstrating an impressive library of complex sensing, transduction, and even computing capabilities in a fiber [3], the conventional fabrication techniques such as chemical vapour deposition [4, 5] and mechanical stacking [6] result in preforms of limited geometric complexity. Preforms with arbitrary cross-sections are currently attainable exclusively by 3D printing of glass [7].

3D printing of silica is a technological area in its infancy and only recently gets to the commercialization stage in some of its implementations, with products such as Glassomer ([8] and Nobula [9]. In the research settings, a range of glass printing techniques has been demonstrated with varying degrees of success [10–17]. The most promising and, so far, the only technique that has resulted in a drawable preform is stereolithography (SLA) [11]. Yet, the optical loss in the resulting fiber is orders of magnitude higher than in a standard optical communication fiber due to residual porosity, which also affects the fiber's strength and structural accuracy [18, 19]. The material properties can be improved by stronger in-draw self-annealing, but it is known from research in polymers that the longer the structure spends in a low-viscosity state, the more the structural features critical for the device functionality are smeared by reflow [20]. The reflow accelerates with Laplace pressure, increasing with the decrease in the curvature radius of the feature reflowed. Investigation of the reflow of structures corrugated on polymeric fibers demonstrates faster smearing for higher spatial frequency and larger amplitude of corrugation [21, 22]. We expect similar behaviour in silica: The smaller the nonequilibrium feature, the faster it will be smeared by the reflow.

Putting it in simple words, since fiber draw is a fluidic phenomenon, it exhibits two competing forces at play: while the viscosity of the drawn thread wants to preserve the geometry of its cross-section, the surface tension tries to defy it and reshape any non-equilibrium feature into a cylinder, as is schematically demonstrated in Figure 9.2a [21, 22], yet assists with annealing of pores [24–27]. The porosity inherent to the 3D-printed parts yields compromised mechanical and optical properties [11, 18], even after sintering. In a liquified state during the draw, this porosity is expected to get

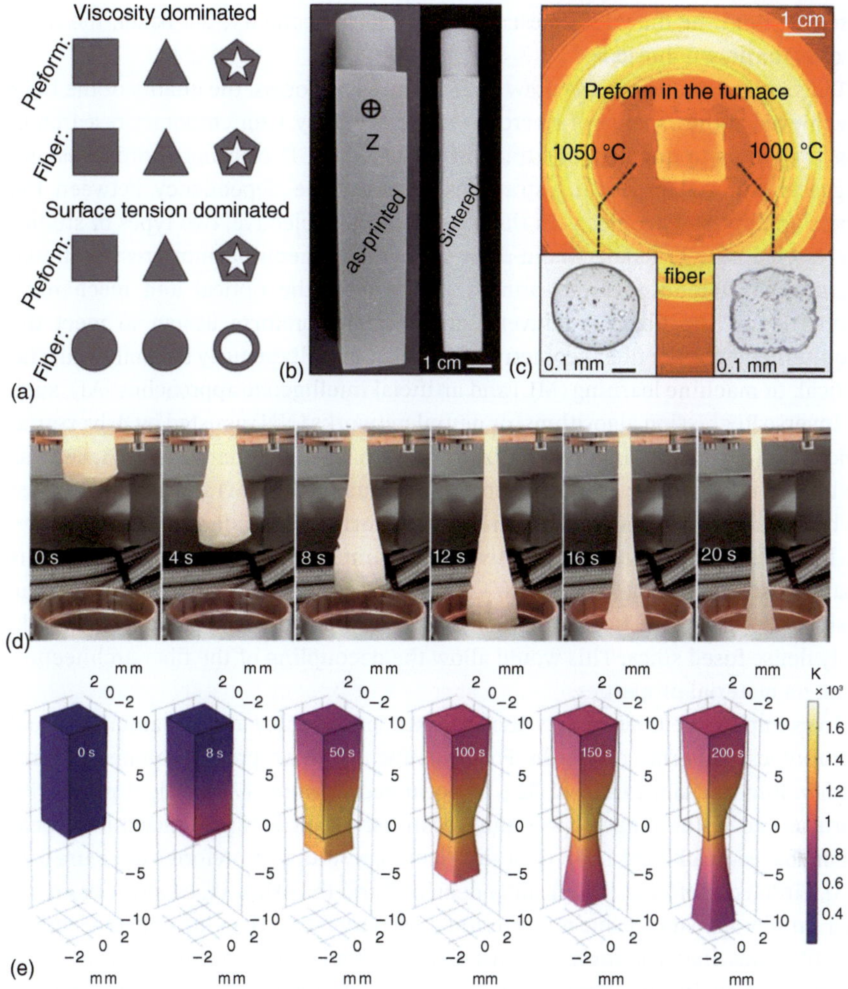

Figure 9.2 Retention vs. reshaping of the fiber cross-section and porosity self-annealing during the draw process. (a) Schematics of the typical shape-retention behavior of the nonequilibrium (noncylindrical) cross-sectional features in the draw process depending on the dominant force, signifying the square rod as a valid study system. (b) The as printed and sintered square rod preforms materialized by SLA using Form2 Printer in Formlabs Clear resin with volumetric silica infill of 32% of powdered Corning 7740. (c) Bottom view of the preform like that on (b) in the draw furnace during the drop-off heating. The insets demonstrate the optical micrograph on the fiber resulting from that preform drawn at 2 m/min with a scaling factor of 100 : 1, starting at 1150 °C and gradually reducing the temperature. (d) Experimental drop-off dynamics of a preform like that in (b) and (c) [23]. (e) Drop-off dynamics of a preform with viscosity–temperature dependence of Corning 7740 glass and dimensions qualitatively comparable to the preform in (b) simulated by finite element method (FEM). It is apparent from (d) and (e) that the drop-off dynamics of the cross-section geometry is dominated by the surface tension force. *Source:* Reproduced from Faccini de Lima et al. [1] / Society of Photo-Optical Instrumentation Engineers, Figure 9.2.

annealed; more effectively, the thinner the glass is liquified. Conversely, the fiber resulting from the draw [28] preserves the preform geometry; the better, the more viscous the drawn thread is.

Thus, unfortunately, in the draw of 3D-printed preforms, the quality of the fiber is achieved at the expense of its cross-section accuracy. Contemporary research in this area aims to quantify this tradeoff for existing 3D-printing approaches and suggests material processing strategies breaking the dependency between the fibers' quality and geometry. In the pursuit of this objective, two types of studies are carried out: 1) Relate the in-draw reflow of nonequilibrium cross-sectional features of stereolithography-printed preforms to the optical and mechanical properties of the fiber, and inverse-engineer the preform design to meet the requirements for the fiber geometry and quality simultaneously by analytical, statistical, or machine learning (ML) and artificial intelligence approaches (AI), such as Inverse Regression algorithms, or neural networks (NN) assisted by Adversarial Encoders [29] Examining materialization approaches alternative to SLA, such as by binder jetting (BJ) or freeform deposition modelling (FDM) and sintering either in hydrostatic conditions, such as in a muffle furnace or under uniaxial pressure, in hot presses, resistive, or inductive. An example of an inductive hot press is spark plasma sintering (SPS) press. The second type of study aims at sintering the residual porosity of 3D-printed glass to properties indistinguishable from a bulk, fully dense fused silica. This would allow the decoupling of the fiber architecture from its material properties.

To exemplify the tradeoff between cross-section accuracy and remaining residual porosity, using Form2 printer by Formlabs Inc., we have printed square preforms with volumetric silica infill of 32% of powdered Corning 7740 (polydisperse with particle size in the range of 2–60 μm, mean particle size 12 μm) mixed into Clear Resin, baked in a kiln at 1000 °C for three hours, resulting in preforms in Figure 9.2b [23]. Sintered preforms, significantly densified and thus shrunk, nevertheless retained a milky appearance due to the residual porosity rather than turning transparent. The fiber was drawn at 2 m/min with a scaling factor of 100:1, starting at 1150 °C and gradually reducing the temperature. The cross-sectional reflow and the residual porosity were monitored. We find that the fiber cross-section at 1050 °C is almost cylindrical, while at 1000 °C it becomes rectangular with slightly smeared corners, as in Figure 9.2c, indicating a switch from a surface-tension-dominated to a viscosity-dominated regime. The residual pores are larger and distributed sparser in 1050 °C fiber than in 1000 °C fiber, indicating a more efficient in-draw onset of pores annealing at higher temperatures [23], resulting in a less optically scattering fiber. In low-viscosity low-axial tension conditions (drop-off at the beginning of the fiber draw), the surface tension-dominated behavior is especially well pronounced, supported by experiments (Figure 9.2d) and simulations (Figure 9.2e) by Finite Element Method (FEM) in Comsol Multiphysics software [30].

Researchers exploring 3D printing of fiber preforms typically experiment with their cross-sections [31–34]. However, *axial* structuring of the preform broadens the range of functional modalities in a fiber. For instance, light shaping [35, 36] can be demonstrated in fibers where the cross-sectional refractive index (RI) distribution varies adiabatically along their axis. Such fibers allow the moulding of an input optical mode into an output mode with a different, user-engineered cross-sectional electromagnetic field.

This is one of the functional aspects that are likely to turn critical for preparing the fiber optics for the Internet of tomorrow. For instance, future quantum interconnect (QuIC) [37], cross-connecting nodes of quantum network implemented in dissimilar material platforms by flying qubits (photons) is required to couple them efficiently. Namely, optically encoded information outbound from one platform needs to undergo a transformation to match the optical mode of the inbound platform. Modes of different quantum platforms, from a laser frequency comb cavity, generating photonic qubits on demand [38], through photonic crystal fiber (PCF) used in cold atom traps [39, 40], to silicon-on-insulator (SOI) waveguide, standard to silicon photonics, are different from each other and from that of a single-mode fiber (SMF) interconnecting them long-haul, challenging such integration. Another emerging computational paradigm, photonic neuromorphic computing (PNC) [41–44], requires adiabatic light shaping components as well, such as tapers, couplers, multiplexers, and demultiplexers [41, 42, 44, 45], which have the potential for realization by adiabatically changing the cross-section of 3D-printed preform in an integrated fashion within the same fiber.

Light shaping fibers for QuIC/PNC need to be ultralow loss; thus, material quality is critical. We anticipate the improvement of losses in SLA-printed preforms concurrent with the maintenance of the geometry to be limited. When the magnitude of the reflow of a nonequilibrium feature becomes comparable to the dimensions of that feature, such as in fibers embedding photonic structures, inverse-engineered preform correction, exemplified in Figure 9.3a for the case of rectangular fiber, becomes impossible, since those features in the corrected preform start overlapping. Ultimately, we want to sinter the preform to transparency before the draw that, in the current case, needs to happen within viscosity-dominated regime preserving the cross-section.

A combination of BJ printing of powdered silica and sintering using an SPS press is likely to deliver the desired result. Baked-out BJ prints lack the layer-aligned fissures typical to SLA, thus resulting in preforms of higher density and better geometric accuracy. Towards that end, using ExOne Innovent+ BJ printer, we have printed preforms 2 cm in diameter and up to 10 cm long in a fused silica powder, polydisperse (3–70 μm, mean 23 μm). Those were debinded at 200 °C for two hours and baked out at 1200 °C for four hours in a muffle furnace. The resulting preforms with a variety of nonequilibrium hollow cores are shown in Figure 9.3b. The preforms are porous and white as the initial powder.

We then successfully sintered the same powder to translucency (Figure 9.3c) in Agus SPS-2250Rx at 1400 °C, under a pressure of 22 MPa, using a graphite dye with an inner diameter of 2 cm and graphite punches. In sintering, the SPS heats the material by transmitting electric pulses through the powder [46]. As a result, only the touchpoints at the boundaries between the powder grains, where the impedance is the highest, are heated up. SPS allows compacting the powder without melting it by fusing the grain boundaries only. This unique compaction mechanism is critical

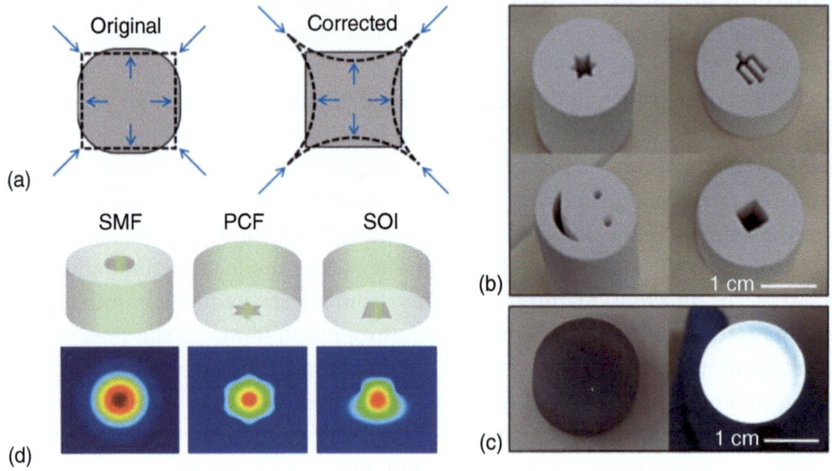

Figure 9.3 (a) Inverse engineering of the preform for in-draw reflow into the desired fiber cross-section (normalized to the draw scaling factor) exemplified for a rectangular fiber; "Original" – data point in the experimentally obtained training data set that can be used to engineer a corrected preform for the desired fiber profile through, for instance, inverse regression ML algorithms. Starting with square preform we expect to get a fiber in which the smallest curvature radii (corners) are smeared by reflow; "Corrected" – inverse-engineered output. The dashed line is the preform cross-section outline, and the grey area is the cross-section of the fiber. Blue arrows show the velocity field of the cross-section surface reflow. The first-order guess of the preform surface in the corrected profile would result from adjusting the zero-order rectangular surface of the preform in the original profile by rolling back the offsets that develop between the preform and the fiber geometries during the draw. (b) Variety of baked-out BJ printed preforms with nonequilibrium cross-sectional features; (c) Solid cylinder of fused silica powder used in BJ printer sintered to translucency in SPS press at 1400 °C, 22 MPa (top), and the same back-illuminated with a flashlight (bottom). (d) Design of preforms (top) for fibers shaping light into respective modes (bottom): "SMF" – circular core for standard single-mode fiber input and two types of outputs: "PCF" – a core with sixfold symmetry for coupling into a honeycomb photonic crystal fiber, and "SOI" – a trapezoidal core for coupling into a silicon on insulator waveguide. *Source:* Reproduced from Faccini de Lima et al. [1] / Society of Photo-Optical Instrumentation Engineers, Figure 9.3.

for the successful sintering of multimaterial preforms, where the thermal expansion mismatch between the core and cladding materials (such as silicon and silica) can cause stress and even cracking upon cooling the preform after it is sintered. We anticipate that preforms for light shaping fibers demonstrated in Figure 9.3d are relatively "low-hanging fruit" for the realization by the combination of BJ and SPS sintering. First, a hollow core preform is printed, then the core is filled with the powdered core material, which is the same as the cladding powder, doped to attain the desired refractive index. The result is a self-supported uniformly porous part, which can now be sintered to transparency using SPS press. Finally, for the structural support during the fiber draw, the 3D-printed section of the preform needs to be encapsulated in a glass tube by softening it in a hydrogen–oxygen flame and collapsing it around the 3D-printed part under a vacuum using a glass lathe.

For mode shaping, not only the cross-section of the fiber core needs to change adiabatically but also the core refractive index (RI), such that losses due to Fresnel reflection while coupling data to the outbound network node are prevented. Thus, the composition of the core material will also change adiabatically along the preform and, subsequently, the fiber axis. For instance, in the transition from SMF to SOI, the latter's mode has a higher RI. In 3D-printed preforms, the RI of the core can be axially varied by filling the core with a mixture of Ge-doped silica powder and silicon powder, the ratio of which will gradually change from 1 : 0 on the SMF side to 0 : 1 on the SOI side of the preform.

At this point, we would like to note that however complex the cross-section of the thermally drawn fiber is, it is still axially uniform. Namely, the standard way to think about the draw is that it is supposed to yield a fiber with the same cross-sectional geometry at any point along its length. Even with 3D printing, allowing for varying the preform cross-section axially, only adiabatic changes on the scale of meters can be achieved. Assuming a preform diameter $D = 1\,\text{cm}$ drawn into a fiber with diameter $d = 100\,\mu\text{m}$ thick fiber, due to the volume conservation, any axial feature is expected to elongate by a factor $(D/d)^2 = 10{,}000$. The typical smallest resolved feature of 3D printing is 100 μm, meaning that it will span a 1 m long section of the fiber.

The micrometric and submicrometric patterning of the fiber in the axial direction can be achieved so far exclusively by a postdraw processing step, such as an intricately amplified capillary breakup. Here the fiber is fed through the hot zone of a flame or a laser, while a short section of a core liquefies and breaks into droplets one by one in a process resembling water dripping from a faucet. We dedicate the next section to understand this process and suggest ways of controlling its outcomes deterministically in a predictable fashion.

9.2.2 Axial Fiber-structuring and Fiber-embedded Functional Systems Assembly by Material-selective Amplification of Spatially Coherent Capillary Instabilities

Beginning the first experiments a decade ago, it has been observed that the capillary breakup, normally highly complex, in a multimaterial fiber can become deterministic and predictable in some conditions, resulting in a strict monodisperse breakup period, as is depicted in Figure 9.1 [47]. In particular, this happens in conditions where the fiber is fed through a hot zone of a hydrogen–oxygen flame, locally liquifying the fiber, thus imposing conditions of axial viscosity gradient (AVG), as in the processes exemplified in Figure 9.4a [1, 50]. While fed through a localized liquefaction zone, initially continuous cores of a multimaterial fiber can get patterned into building blocks of photonic and optoelectronic circuitry in a material-selective spatially coherent manner (Figure 9.4b, c).

Moreover, such patterning can be performed on various dimensional scales, allowing for the axial patterning of fiber cores in micron and submicron regimens.

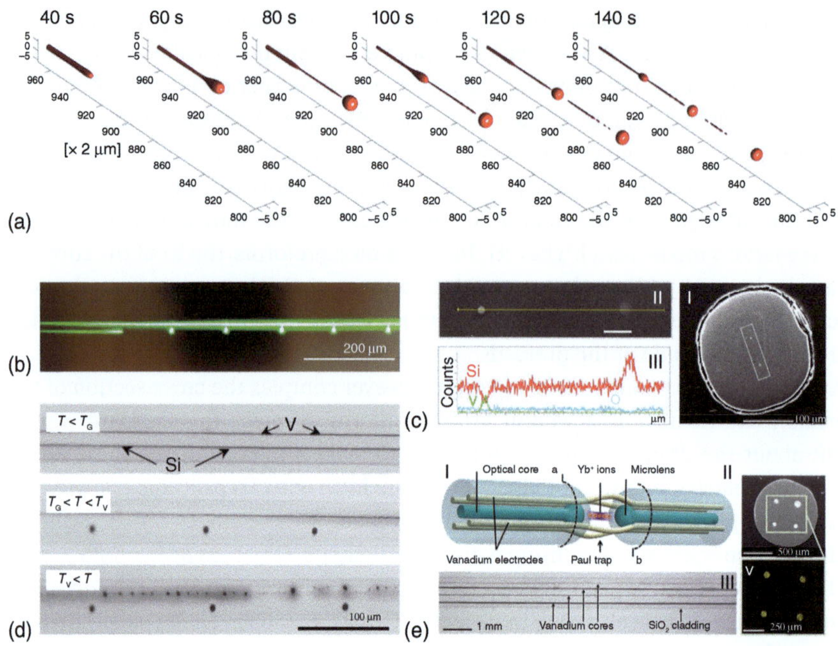

Figure 9.4 Capillary instability and its uses for the axial patterning of multimaterial fibers in the materialization of building blocks for in-fiber photonics (A) ab-initio simulation of AVG breakup in a 4 μm Si-core silica fiber, feed at speed v_f = 2 μm/s into a hot zone with the boundary temperature profile, mimicking a hydrogen–oxygen flame, $T = T_{max} - (T_{max} - T_{Si})e^{x/w,}$ with T_{Si} = 1414 °C being the melting point of Si, T_{max} = 2102 °C the maximal liquefaction zone temperature, and w = 1200 μm the liquefaction zone boundary width; the breakup period is ~78 μm. (b) Fabrication of arrayed frequency-resonant spherical whispering-gallery-mode switches in silicon with a flanking frequency-tuning vanadium electrode by a material-selective spatially coherent breakup of Si in a double-core silicon–vanadium fiber shown in (c). by feeding it through a hydrogen flame. (c) Double-core silicon–vanadium fiber: I – cross-section. The grey rectangle delineates the area that is demonstrated in panel II. II – zoom-in to the cores. The yellow line shows the path along which the electron dispersion spectroscopy (EDS) chemical profile analysis was collected for panel III. III – EDS of the cores' region. (d) Breakup resulting from feeding the fiber v_f = 2 μm/s through a hydrogen flame, emerging from an orifice of a torch 0.02" in diameter, in which the gas flow is gradually increased from 0.30 l/min to 0.6 l/min. The process switches from no breakup at the temperature T, smaller than $T_{G,}$ the softening point of the glass, through the selective breakup of Si when reaching the temperature above T_G yet below T_V the melting point of vanadium, to the breakup of both the cores when the temperature exceeds T_V. (e) A potential case use of the capillary instability in the cladding to axial patterning of fiber in the manufacturing of devices applicable to the quantum data processing: I – ion–photon quantum information converter (IP-QIC) for quantum-coherent translation of ionic-qubit information into photons (ion–photon qubit converter). II – the vanadium quadrupole fiber cross-section. Zoom-in: EDS map of the V quadrupole. III – the side view of the vanadium-quadruple fiber. The rather large dimensions of the fiber are necessary for the proper design of the ion trap. The thickness of the vanadium cores is derived from the impedances required for generating the trapping electric field, while the distances between them are derived from the required trapping potential-well dimensions. Panels A, B – *Source:* Camila Faccini de Lima (Chapter Contributor). Panels C, D, E - Reproduced from Faccini de Lima et al. [1], Society of Photo-Optical Instrumentation Engineers, Figures 9.5, 9.6, 9.7, *respectively.*

While the former is useful for the fabrication of electrically tunable whispering gallery mode spherical resonators, the latter can be applied, for example, to create 1D photonic gratings, serving as mirrors for resonant cavities, enhancing nonlinear optical interactions, critical for the amplification of quantum effects and useful for the creation of functional optoelectronic systems in fibers more generally.

A cylinder's surface is larger than a sphere's surface for the same volume of material. Thus, capillary instability is a process of surface energy minimization. Conventional mathematical descriptions of this process [51–53] going beyond the mere ab-initio solution of the Navier–Stokes equations [54], such as in Figure 9.4a, consider cylindrical liquid threads with axially uniform viscosity in which the instabilities develop at an accelerating rate from the initial white noise, simultaneously at all possible spatial frequencies. Such competition between various instability wavelengths yields a breakup characterized by statistical rather than deterministic outcomes [55–57].

Conversely, we hypothesize that in a fiber where the viscosity changes axially, following the temperature profile of the heating source, specific wavelength, defined by the temperature-gradient length-scale can be amplified selectively, yielding a fully deterministic breakup period. While axial viscosity gradient (AVG) breakup is already been extensively investigated for the assembly of photonic devices and systems, such as p–n diodes [47, 58], heterojunctions [48], arrayed photodetectors [49], photonic gratings [47], and optical resonators [59], necessitating spatially coherent, monochromatic outcomes [47], in our experience it is often characterized by incontrollable period or phase (Figure 9.4b).

The ongoing effort in the community is to acquire a physical intuition about AVG breakup, which will allow us to steer it towards deterministic outcomes. To this end, experimental, analytical, and numerical framework [60] ascertaining the hydrodynamic conditions at which the AVG capillary breakup is periodic, spatially coherent, and monodispersed is currently in development by a number of groups around the world. Additional efforts are invested into suggesting and experimentally evaluating conditions for miniaturization of the breakup to the nanometric regime useful for the fabrication of photonic gratings, specifically with tolerances making them applicable to QuIC/PNC. Additional control over breakup-assembled architecture has been achieved with guided Marangoni flow resulting from extreme spatiotemporal changes in viscosity gradient [58].

Capillary instability not just of the core–cladding interface can be harnessed for device assembly, but of the fiber outer surface, at the cladding–air interface as well. Figure 9.4e exemplifies an anticipated use case of such instability for the fabrication of a fiber-integrated ionic-to-photonic converter of quantum information [1, 50]. To materialize such a converter, a Paul trap can be created by etching a gap in the silica cladding of the vanadium quadrupole fiber using hydrofluoric acid, not affecting vanadium. Then the cladding ends caping the resulting gap (a and b on Figure 9.4e) will be heated to the temperature T such that $T_G < T < T_V$, where T_G is the softening point of silica, and the T_V is the melting point of vanadium. In these conditions, the capillary instability will be promoted in silica at the gap caps, and the surface tension will reshape the caps into spherical lenses. This can be done in an inert environment to prevent vanadium oxidation. The Yb^+ ion array will be then loaded into the Paul trap in ultra-high vacuum (UHV) conditions and trapped in it by an electric field generated using a quadrupole of electrodes. The ion array in this configuration will serve as a lightguide seamlessly integrated into the optical core of the fiber, concentric to it, maximizing the interaction between the ions and the propagating fiber mode. Currently, the efficiency of communication to ionic cubits in state-of-the-art ionic processors is limited by the free-space optics standardly used for these purposes. The numerical apertures (NA) of lenses used for such a communication don't exceed 0.6, limiting the efficiency of capturing the optical information resulting from ion–photon interactions. In the fiber, where the light is bound to the guided optical mode, this efficiency can, in principle, be maximized to unity.

Predictable, deterministic fiber patterning by design rather than exploration is imperative for turning the capillary breakup into an engineerable technology. Such predictability can be achieved by developing an intuitive analytical model that would provide a glimpse of a physical intuition into the process' fluid dynamics and help us understand the AVG breakup in simple terms, such as the balance of forces and conservation laws. Such a model will likely map the capillary number (Ca) into breakup position and period as one-to-one correspondence, defining the conditions for control over the periodicity of breakup through the engineering of the spatiotemporal temperature profile in the liquefaction zone.

9.2.3 Search for a Physically Intuitive Analytical Model Describing the AVG Breakup

Since the ab-initio solutions of the Navier–Stokes equations, as the one in Figure 9.4a, are computationally heavy and thus time-consuming, frequently, some simplifying assumptions are made. Those assumptions help accelerate the calculations on account of some loss in precision. One common simplification is a level-set assumption, which, in simple terms, advantages the attention given to the dynamics of the interfaces between the various liquids over the requirement of volume conservation. It is apparent from Figure 9.4a, depicting the simulation that uses such a level-set approximation that the spheres' volume decreases over time, while volume conservation is a very basic conservation law in experiments, valid for any realistic scenario. It is thus expected that ab-initio simulations executable in a reasonable time can be trusted only qualitatively. Even after the simplifications, the simulations are still quite heavy and time-consuming: one of the kinds depicted in Figure 9.4a typically takes two to three weeks to complete on a cluster with 32 CPUs. Running one simulation is bearable, but any optimization or parametric search turns into a "mission impossible."

Thus, quite a bit of attention is dedicated to searching for an intuitive analytical model describing AVG instability. Experimental observation of a periodic breakup in the AVG process suggests the following characteristic behavior; when a semi-infinite core of a fiber is fed into the flame at a speed v_f, it remains unchanged before reaching some axial position x_{bu}, at which the core material starts reshaping to form a droplet, growing to a critical size and pinching off the core at x_{bu}, as is demonstrated in Figure 9.1. The remainder of the pinching neck then retracts towards the detached droplet and the continuous core, and the process repeats.

The models being developed aim at mapping the capillary number (Ca) into breakup position and period as one-to-one correspondence. Ca categorizes the instability in terms of dimensional analysis: It is the ratio between the work for overcoming the viscous forces and the surface energy of the liquid thread. Ca is a qualitative measure of whether the system is in the jetting (viscosity dominated, $Ca \gg 1$) or in the dripping (surface tension dominated, $Ca \ll 1$) regime. We use the definition $Ca \equiv t_{pinch}/t_g$ [57], where t_{pinch} is the droplet pinch time and t_g is the droplet growth time to its critical pinching size. t_{pinch} is estimated, up to proportionality factor β, from the balance between viscosity and surface tension forces: $t_{pinch} = \beta a \eta_{clad}/\gamma$ [47], where a is a core radius, η_{clad} is the cladding viscosity, and γ is the interface tension between the core and the cladding. For any reasonable temperature, silica is orders of magnitude more viscous than any molten crystalline core; thus, we assume here that the viscose forces are attributed to η_{clad} exclusively. For the droplet to grow, the material is supplied into the droplet by feeding the core at a speed v_f, as in Figure 9.1, until the core section, a breakup period in length, is fed into the droplet, at which point the critical pinching size is reached. Thus, t_g is the ratio between the breakup period λ_{bu} and the feed speed v_f, i.e. $t_g = \lambda_{bu}/v_f$. The expression for capillary number becomes $Ca = \beta a\ \eta_{clad} v_f/\gamma\lambda_{bu}$.

Dimensional analysis for our configuration is equivalent to a claim of a steady-state process, in which every droplet pinches off with reaching the critical size at the same axial location x_{bu}. This can be formulated simply as $t_{pinch} = t_g$, or $Ca = 1$ at x_{bu}. One interesting conclusion is that AVG breakup cannot be thought of in common terms of either dripping or jetting but rather starts as a jetting process and switches to dripping as the core is fed deeper into the hot zone. Another conclusion is that $Ca = 1$ is simply not enough to define the pinch-off location, since for every axial viscosity $\eta_{clad}(x_{bu})$ there is a different breakup period λ_{bu} corresponding to $Ca = 1$., i.e. $Ca(\lambda_{bu}, x_{bu}) = 1$ is a single equation of two unknowns.

It is hypothesized that the condition that gives the second independent equation has to stem from the overlooked fact that the breakup location is unique [60]. Since we are interested in a breakup giving deterministic results, this breakup has one single period λ_{bu}, i.e. there are no pinches observed at other locations concurrently with λ_{bu}. Yet, to the date, no mathematical formulation of the last condition yielding quantitative agreement between the model, formulated with no free parameters, and the experimental behaviour of the breakup has been obtained. The complication stems from the fact that β in $Ca = \beta a\ \eta_{clad} v_f/\gamma\lambda_{bu}$ is a general function of the viscosity ratio between the core and the cladding. It is expected to converge to a constant asymptotically while this contrast increases [52], but that is yet to get corroborated for AVG configuration by ab-initio calculations, and experimental parametric search.

To conclude, AVG capillary breakup is a powerful tool for the assembly of arrays of devices and even entire integrated systems internal to the fiber, awaiting physically intuitive mathematical description. Yet there is another intriguing process taking place in the droplet pinched off the fiber core when cooling down and eventually solidifying upon leaving the localized liquefaction zone.

9.2.4 Engineering the Optoelectronic Properties, Crystallinity, Composition, and Internal Stress of the Breakup-assembled Devices by a Guided Solidification from Melt

Let's consider a droplet of molten material uniformly cooling down. After being cooled below its melting point, it will eventually solidify. Crystallization of a melt in spherically symmetric conditions is a spontaneous symmetry-breaking phase transition. In the absence of a "seed" crystal defining an energetically preferable site, the crystallization necessitates an undercooling below the material's melting point to get initiated [61, 62]. Once the crystallization starts, it protrudes the sphere omnidirectionally; the faster, the deeper the undercooling.

In AVG conditions, on the other hand, the cooling is not uniform. As we feed the fiber into the flame, the spheres pinch off the cores, eventually leaving the localized liquefaction zone. On the way out, the spheres not only cool down over time but also experience a lateral thermal gradient (LTG): each molten droplet is colder on the vanguard pole, distant from the heat source, than on the rearguard pole facing the heat source.

It has been experimentally observed that if the feed speed is slow enough such that the cooling can be considered quasistatic, the solidification front will be guided within the droplet along the direction of the externally imposed temperature gradient, as is demonstrated in Figure 9.5. In such conditions, the crystallization front will coincide with the isothermal surface defined by the droplet material melting temperature. Such a stable, directional crystallization will result in a monocrystalline morphology of the solidified material. Thus, quasistatic cooing in LTG conditions will yield a

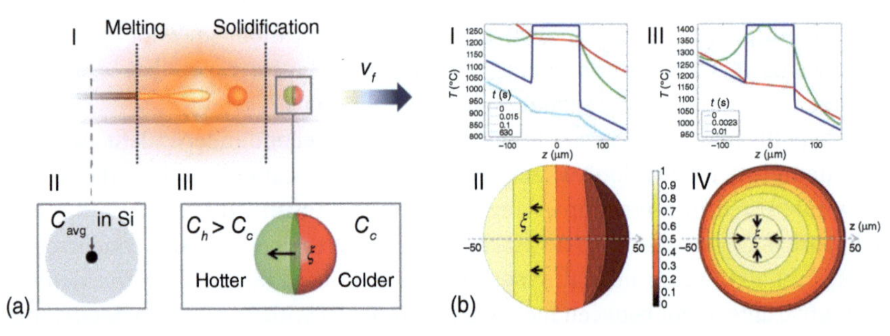

Figure 9.5 Solidification in AVG conditions (a) Postbreakup phenomena: I – AVG breakup and subsequent LTG solidification. II – a cross-section of prebreakup Si-core fiber with uniform doping C_{avg}. C_h, C_c – hot and cold pole doping, respectively. III – propagation of solidification front ξ creates lateral dopants distribution. (b) FEM simulation of temperature profile evolution (I, II) and solidification propagation front (III, IV) for I, III – $Si_{0.5}Ge_{0.5}$ and II, IV – Si spheres, 100 μm in diameter each, fed through LTG of 1 °C/μm at 10 μm/s. The simulation depicts mostly isothermal solidification after recalescence from an undercooling of 300 °C. Most of the sphere solidifies isothermally since only ~12% of an undoped silicon sphere releases enough latent heat in the isenthalpic phase to reheat the melt remainder back to the melting point. The color scale bar is the fraction of solid volume. Panel a.I – adapted from: [48] Figure 1a. Panels A.II, A.III – *Source:* Alexander Gumennik. Panel b – *source:* Etgar C. Levy.

monocrystalline sphere, with one of the crystal axes aligned along the temperature gradient, resulting in a uniform optical refraction described by the index ellipsoid of a known orientation [63]. Additionally, if the melt contains a dopant that is more soluble in liquidus than in solidus, the dopant will be increasingly segregated into the remaining liquid as the solidification progresses. Thus, solidification guided by the temperature gradient will provide spatial control of composition and, subsequently, electronic band structure, in addition to crystallinity.

Significant progress in the last few years is taking place in developing the experimental, analytical, and numerical framework for the identification of thermodynamic conditions in which the solidification is a gradual process, stably propagating along externally imposed thermal gradients [48, 59, 64, 65]. Identification of such conditions allows to engineer spatiotemporal temperature profiles, which, applied to the melt, would result in solidification-induced self-assembly of architectures typical to photonic devices, such as spherical whispering-gallery-mode optical resonators [66] and Si/Ge photonic gratings for the use in QuIC/PNC.

Quasistatic cooling is an idealized case, yet, generally, the realistic cooling conditions of a postbreak droplet are not quasistatic. Usually, the droplet reaches a significant undercooling below its melting point before the solidification is initiated. An undercooled droplet crystalizes in two distinct phases: isenthalpic phase, in which unstable growth of dendrites rapidly extracts the latent heat, causing the "recalescence" – reheating of the remaining liquid back to its melting point, followed by isothermal phase, in which the solidification front ξ propagates stably along the temperature gradient [61, 62, 67]. Resulting from recalescence, a transient spherically symmetric thermal gradient (SSTG) is created, in addition to LTG: The volume of the reheated droplet is hotter than its surface, fixed to an undercooling temperature by contact with the silica cladding. The larger the surface-to-volume ratio of the droplet, the faster it cools down to the temperature of the surrounding silica, following the recalescence. The magnitude of SSTG transient (compared to LTG) and its duration compared to the velocity of ξ will determine whether the stable portion of the crystallization will follow the internally imposed spherical symmetry or the externally imposed lateral symmetry. Additionally, the deeper the undercooling, the larger the fraction of the droplet that needs to solidify in an unstable, polycrystalline manner to release the latent heat needed to reheat the droplet back to the melting point. Thus, if we quench the droplet, the solidification will not follow any symmetry, resulting in polycrystalline morphology. To summarize, the solidification of a postbreakup droplet might follow three distinct symmetries – lateral, spherical, or random, dominated by LTG, SSTG, or quenching, respectively.

The presence of dopants in the melt further complicates the situation: in Si–Ge melt, for instance, Ge is pushed into the liquid as ξ propagates (Figure 9.5a). As a result, the melting point of the liquid remainder decreases, decelerating the solidification. Given this deceleration, transients, such as SSTG, are less influential. In other words, in the intrinsic-material droplet, the solidification is likelier to progress spherically and in a fast, transient fashion, while the doped sphere will solidify laterally in a slow, quasistatic fashion under the identical initial thermodynamic conditions, as is evident in Figure 9.5b.

The last provides a possible strategy for achieving a monocrystallinity in the breakup-assembled devices: a small content of dopant, such as Si, mixed into a droplet of the main constituent, such as Ge, under LTG will assist in initiating the crystallization, as the solidification begins, by the solidification-guided segregation, Si will be depleted into the liquid increasing the melting point of it and decelerating the process, eventually concentrating into a small Si-rich cap on the hot pole of the droplet; most of the sphere will solidify isothermally (i.e. the velocity of $\xi = v_f$ in Figure 9.5a) as a single-crystal Ge sphere. For instance, such a monocrystalline semiconducting sphere can be used as a whispering-gallery-mode optoelectronic resonator for switching applications in QuIC/PCN.

Both numerical and experimental studies support the viability of this strategy. FEM calculations for isothermal solidification in Si and $Si_{0.5}Ge_{0.5}$ spheres (Figure 9.5b) indicate composition-dependent symmetry. In an intrinsic Si sphere, the solidification is mostly spherical, completed in milliseconds, and only slightly distorted by LTG. In $Si_{0.5}Ge_{0.5}$, on the other hand, the solidification is mostly lateral, slightly affected by SSTG in the very beginning, and rapidly decelerating due to a segregation-driven decrease in the melting point. The solidification is resumed laterally, driven by feeding through LTG (velocity of $\xi = v_f$) similar to the Bridgman crystal growth process [68–70], which is quasistatic in its nature.

Similarly, in the AGV breakup experiments, the spheres resulting from feeding $Si_{0.5}Ge_{0.5}$ core fibers through a hydrogen–oxygen flame at 10 μm/s (Figure 9.6a.I) manifest a sequence of crystal morphologies overtaking each other as the solidification progresses. The solidification starts from a significantly undercooled state. Portion 1 in Figure 9.6a.II solidifies isenthalpically; it is polycrystalline with random composition. Following the recalescence, in Portion 2 of the sphere, SSTG kicks in since the boundary between 1 and 2 is curved, indicating partly spherically symmetric solidification. LTG overtakes the process in Portion III, where the composition distribution becomes axial. The Si–Ge phase diagram (Figure 9.6a.III) dictates a solid sphere's composition (Figure 9.6a.IV). It deviates from the calculated quasistatic scenario in Portion 1, in line with rapid isenthalpic solidification.

In an AVG breakup process scaled down to the nanometric regime submicron-sized Si spheres extracted from the silicon-in-silica photonic grating sample in Figure 9.7a, we observe a thin shell of nanocrystalline Si "fingers" extending from the dense diamond-cubic Si core (Figure 9.6b.I). This shell is possibly formed by a protrusion of Si from the solidifying sphere surface into nanocracks induced to the silica cladding by the Si–silica thermal expansion mismatch. If this is true, a spherical shell indicates SSTG-driven solidification. In addition, we have obtained fully polycrystalline $Si_{0.75}Ge_{0.25}$ spheres by melting the sphere with an 808 nm diode laser and quenching it, shutting off the laser abruptly (Figure 9.6b.II).

In summary, engineering of the spatiotemporal temperature profiles applied to a melt resulted in solidification-induced self-assembly of architectures typical of optoelectronic devices. The mentioned findings support the feasibility of this approach: the Janus particles in Figure 9.6a are structured as Si–Ge heterojunctions, forming an array of photodiodes for a fiber-integrated high-frequency readout of optical signals in QuIC/PIN [71]. By designing temperature profiles for the

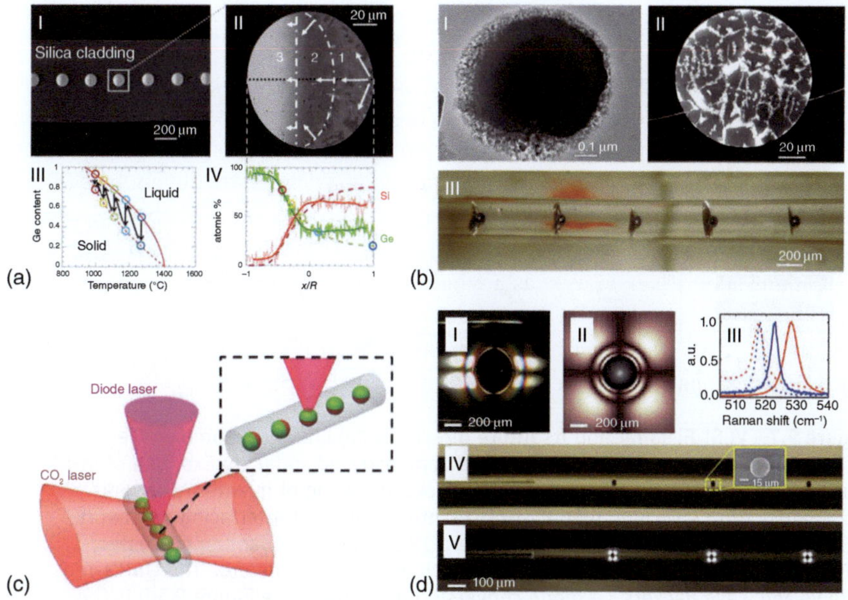

Figure 9.6 Solidification guided by the temperature gradient for the engineering of the internal architecture of a breakup-assembled device. (a) Postsolidification spheres resulting from the breakup of a $Si_{0.5}Ge_{0.5}$ core fiber: I – SEM in backscattering mode (heavier elements are brighter) of a fiber cross-section with an array of Janus particles. II – Morphology regions: 1 – cellular, 2 – uniform with curved boundary, 3 – uniform with the lateral change of composition. Arrows' length and direction indicate the magnitude and the direction of the propagation speed of the solidification front ξ (dash-dot line) III – Si–Ge phase diagram. Arrows show the progression of composition change at ξ. Circles of the same color show the segregation-driven difference in liquidus and solidus composition in the immediate vicinity of ξ. IV – EDS measurement of Si and Ge across the sphere median (black dashed line on panel II). Noisy solid lines are raw data, and smooth solid lines are locally averaged data. Dashed lines are the calculated contents of Si and Ge, assuming quasistatic solidification. (a) Morphology depends on solidification scenario: I – TEM of a 0.5 μm Si sphere solidified under SSTG conditions. The solid core cross-section in the diffraction measurements was found to be a diamond cubic silicon comprised of 3–5 large grains inclined towards the sphere centre. The shell is formed by Si nanocrystalline "fingers" with a typical granularity of ~10 nm. II – a cross-section of a quenched $Si_{0.75}Ge_{0.25}$ sphere. Brighter regions are Ge rich, and darker ones are Si rich. III – When the cladding is not thick enough to contain the pressure buildup due to the expansion of the $Si_{0.5}Ge_{0.5}$ spheres upon the solidification, the cladding cracks. The crack appears consistently on the hotter pole of the sphere, indicating a lateral solidification, resulting in a "focusing" of the shock wave. (c) Temperature-guided solidification. The cladding is softened by CO_2 laser, and then axially oriented Janus particles are reoriented by 808 nm laser. (d) Stress-altered properties of postbreakup spheres: I – stress pattern surrounding the 450-μm Si sphere as observed in cross-polarizing setup. II – the calculated stress pattern for the sphere in I resulting from a stress of 0.218 GPa. III – Raman shift in the continuous core sections of the samples in IV (dashed red) and I (dashed blue) compared with those of the spheres in IV (solid red) and I (solid blue). The core–sphere changes in Raman shifts for 35 and 450 μm spheres are $10.5 \pm 0.2\,cm^{-1}$ and $4.7 \pm 0.1\,cm^{-1}$, indicating hydrostatic stresses of 2.02 ± 0.07 GPa and 0.90 ± 0.06 GPa, respectively, assuming that the continuous cores aren't stressed. IV – A section of the fiber resulting from the process in Figure 9.5a, including a continuous core followed by broken-up spheres in an optical transmission microscope. The inset is an SEM image of an individual 35 μm diameter sphere after being released from the silica cladding by the dissolution of the fiber in hydrofluoric acid. V – Same as IV, while placed between crossed polarizers. Stress fields in the silica surrounding the spheres are evident. The absence of stresses in the continuous core region is evident as well. Panels A, B.II, B.III, C, D – Reproduced from Gumennik et al. [48] / National Academy of Sciences, Figures 1, 3c, S3, 2a, 5, respectively. Panel B.I – Reproduced from Gumennik et al. [47] / Springer Nature, Figure 9.4c.

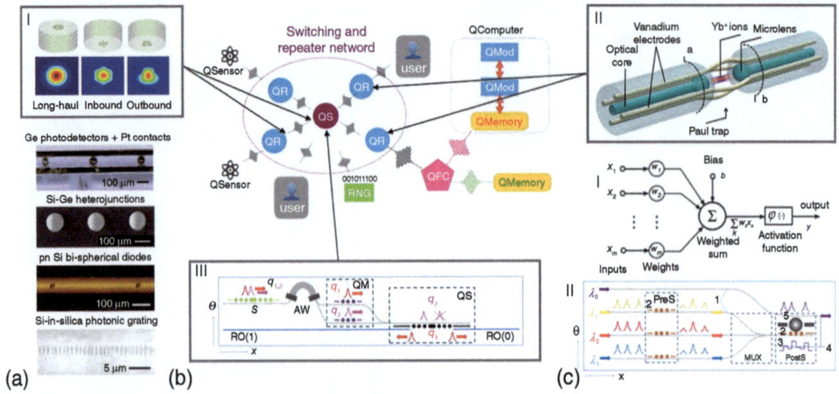

Figure 9.7 VLSI-Fi so far and its future directions. (a) Library of some of the optoelectronic and photonic architectures demonstrated so far. (b) Example of in-fiber photonics implementation to solve bottlenecks in scaling of quantum networks. Schematics from Awschalom et al. [37]. Quantum computer interfacing a network. QS-quantum switch; QR-quantum repeater; QMod-modular quantum processor; QFC-quantum frequency converter; RNG-random-number generator. The QuICs-red arrows and wave packets. I – mode conversion (for details, see Figure 9.3d) is the functional requirement for QuICs, II – qubit conversion (for details, see Figure 9.4e) is a functional requirement for QRs, III – Example of a fiber-embedded quantum circuit comprising a QS, arranged in the fiber circumferentially. x – longitudinal axis, θ – angular axis in cylindrical coordinates of the fiber. Initially continuous, the cores are selectively patterned into S-qubit source; QM-quantum memory; QS-quantum all-optical switch; AW-Arrayed waveguide demultiplexer; grey lines are optical cores; RO(0), RO(1) – single-photon readouts in superconducting nanowire core, such as V_3Si, which in-draw synthesis from vanadium core reacting with the silica cladding has been observed [1]. QS is a resonator, resonantly transmitting q_1. When switched by absorption of q$_2$, it gets out of resonance, and q_1 is reflected. (c) Example of PNC implementation in fiber photonics. I. Schematics of the artificial neuron (from Ref. [43])) II. VLSI-Fi implementation of a firing neural network node consisting of several presynaptic input neurons (PreS) and one postsynaptic output neuron (PostS). x, θ – fiber's cylindrical coordinates. The input spikes ($\lambda_1, \lambda_2, \lambda_3$) coupled to optical cores (1) are weighted using phase-changing material (PCM) cells (2) and summed up using a multiplexer (MUX). Suppose the summated power of the postsynaptic spikes (3) surpasses a certain threshold (4). In that case, the PCM cell on the spherical resonator will switch, changing coupling loss to the spherical resonator (5) and thus its resonant frequency. Initially, in resonance with λ_0, the resonator goes out of resonance, generating an output pulse (neuronal spike). PreS is implemented by drawing a core of germanium-based PCM (such as $Ge_2Sb_2Te_3$) in the proximity of the optical core ($Ge:SiO_2$), and its subsequent localized selective breakup into the array of spheres coupled to the optical core. PostS is implemented by drawing a thick core of SiGe in the vicinity of the optical core, mediated by a thin PCM core, that is locally broken up into an array of submicron-sized spheres coupled to the optical core. Then the SiGe core is broken up locally to form a large monocrystalline sphere of a few tens of microns in diameter. The diameter of this sphere should be tuned, on the one hand, to couple to the PCM array on the MUX output, on the other hand, to couple to the optical input of the PostS (λ_0). Panel A – Reproduced from Wei et al. [49] / JOHN WILEY & SONS, INC., Figure 1c, [47], Figures 9.5b, 9.3a, [48], Figure 1b. Panel B (center) – adapted from: [37]. Panels B.I, B.II – adapted from: Faccini de Lima et al. [1], Figures 9.3d, 9.5, respectively. Panel C.I – adapted from: [43], Figure 10. Panels B.III, C.II – source: Alexander Gumennik.

solidification to follow, one can induce the segregation in Si–Ge composition by design. For instance, exposure to the gradually dimming laser of $Si_{0.75}Ge_{0.25}$ spheres resulted in remelting and recrystallization of the spheres along the laser beam optical axis, as is schematically shown in Figure 9.6c [48]. The reorientation of the thermal gradient for the solidification to follow resulted in a subsequent reorientation of the doping gradient, initially aligned along the fiber axis following the AVG breakup.

Stresses developing around the sphere upon recrystallization, which we already mentioned in the context of micron spheres (Figure 9.6b.I), due to the tendency of cubic diamond semiconductors to expand upon solidification – a phenomenon known as "anomalous expansion" – change the band gap of the semiconductor by as much as multiple electronvolts [72]. In materials such as Si and Ge, this mechanism can be used to engineer the properties of the resulting optoelectronic devices. For instance, photodetectors in those materials, relying on the over-the-band gap excitation, become sensitive to the wavelength that nonstressed semiconductors do not necessarily pick. Also, if stressed nonuniformly, due to the symmetry breaking the semiconductors start manifesting a pronounced nonlinear optical behavior, nontypical to the stressed semiconductor [73, 74]. Additionally, germanium, normally an indirect band gap material, can be transformed into a direct band gap material under stress, which makes not only broader-band photoconduction photodetection but also lasing in it possible [75].

9.3 What's Next?

9.3.1 Conclusions and Future Directions

Photonic circuitry, in essence, is a combination of resonant structures for the photons to interact with, or, mediated by, interact with each other, and waveguides guiding the photons towards these structures, coupled appropriately for the right combinations of photons to interact.

Summarizing all said in this chapter, we can conclude that a critical mass of results has been generated in the last few years, proving that the methods presented here as possible candidates to comprise the future VLSI-Fi toolbox are capable of delivering the complete set of individual building blocks of the photonic circuitry. Specifically, AVG breakup provides versatile control over the creation of resonant structures in the fiber. Guided solidification of breakup-assembled structures ensures controllable optoelectronic properties for these structures. 3D printing of preforms enables the creation of user-designed combinations of adiabatically coupled waveguides interconnecting those photonic and optoelectronic structures.

Note that the ability to fabricate preforms with axially varying cross-sections by 3D printing not only yields mode shaping in a single core, as in Figure 9.3d, but also enables coupling multiple optical cores adiabatically, similarly to how it's done in building blocks of silicon photonics, such as 3 dB couplers, y-splitters, arrayed waveguide multiplexers, demultiplexers, etc. Combined with the ability to integrate multiple material platforms [76–82], VLSI-Fi forms a complete, non-CMOS approach to fabrication of photonic circuits, potentially surpassing in complexity those enabled by CMOS-compatible techniques [83], and addressing a number of standing challenges in quantum networks [37] and photonic neuromorphic computing [43].

The draw of fibers with cores, as thin as a single micron, separated by a few microns of cladding, has been demonstrated for a variety of functional core materials, including metals such as vanadium and platinum, and semiconductors such as Si and Ge. Thus, one could potentially package ~800 cores, each 1 μm – thick, into a typical fiber 500 μm in diameter, circumferentially close to its perimeter. With the preform thickness of 5 cm, which is the typical upper limit the furnaces of specialty draw towers can accept, the preform-to-fiber scaling factor of the cross-section diameter will be 100 : 1, which means that the cores will be defined by voids in the cladding 100 μm in diameter on the preform level. This is a reasonable resolution for commercial 3D printing. Conversely, an axial feature of 100 μm in the preform will elongate by a factor of 100^2 in the draw; thus, adiabatic changes in the fiber will take place on a typical scale of 1 m. We thus can conclude that using VLSI-Fi, in each 1 km of the fiber, one is already capable of creating a circuit with 800 devices per layer, 1000 layers deep, using the combination of material processing tools presented in this chapter.

Figure 9.7a shows the library of photonic and optoelectronic structures that have already been demonstrated using those tools, as well as examples of some standing challenges that can be solved in the near future using those tools. Those include possible implementations of yet-to-be-demonstrated elements of a quantum network in Figure 9.7b that would make them scalable. Data communication in single-photon bursts makes a loss of even a single photon significant. Thus, fiber photonics providing communication within the network will need to get upgraded for higher fidelity performance.

First, every quantum material platform that uses electromagnetic energy to communicate data has its unique mode. Thus, the QuIC needs to convert the modes of the outbound platform into a mode of long-haul communication and then into the mode of the inbound platform, to avoid mode coupling losses, as is shown in Figure 9.7b.I. The current state of the art for mode shaping involves complex and expensive approaches, such as two-photon stereolithographic 3D printing of mode converters on the cleaved end of fiber optic cable [84]. Adiabatic mode conversion, on the other hand, is naturally enabled by fibers with an axially varying cross-section as the one in Figure 9.2c.

Second, each material platform has its own form of a physical qubit, and the network portion, likely within the repeater, needs to be capable of converting qubits from the stationary qubits of the outbound platform into the "flying" qubits (the photons), which carry the information through the QuIC and reach until the destination node, where they have converted again into the stationary qubits of the inbound platform, as is depicted in Figure 9.7b.II. Qubit conversion has been demonstrated for multiple types of qubits: photonic-to-electronic [85], photonic-to-ionic [86], and photonic-to-superconducting [87]. Complexity of the setups and efficiency of the conversion are challenging in all those schemes, in addition to the challenge of integration of those converters into the rest of the quantum network. One promising and, likely, simpler alternative for the implementation of photon-to-ion qubit converters is the fiber-integrated Paul trap (Figure 9.4e).

Third, the quantum switch capabilities of the network should include routing. The quantum router (Figure 9.7b.III) becomes rather a full-scale integrated photonic circuit with qubit replication (including generation), delay (including memory) and the routing itself (i.e. channel switching) capabilities. And, if all that is implemented inside fibers, the whole network portion can become integrated and parallelized rather than composed of blocks coupled in series, each performing its individual functions.

It is important to note that the material-agnostic approach of VLSI-Fi is critical for quantum photonics implementation. The performance of elements of such circuitry is material-platform dependent: best-performance qubit sources are demonstrated in ferroelectrics [38] and nonlinear glasses [88], memory elements in garnets [89], and switches in high-dielectric constant materials, such as semiconductors [90]. In fibers, any material with the melting point below the draw temperature can be drawn as the core and coupled to any other core adiabatically in fibers resulting from 3D-printed preforms. This is usually not the case for CMOS-compatible fabrication bound to the substrate material and its derivatives.

Besides, it's known that in on-chip implementations of some of the quantum material platforms, such as trapped atomic and ionic arrays, the symmetry mismatch between the chip (planar) and the quantum material (for 1D ionic/atomic arrays – cylindrical) adversely affects decoherence times [91–93]. Fiber, on the other hand, is cylindrically symmetric and thus natural for integration with those platforms, as is evident from Figure 9.4e.

An additional realm of implementation for VLSI-Fi is in photonic neuromorphic computing elements on a fiber. In application to PNC, the potential of VLSI-Fi is not only in the ability to integrate multiple materials into structures typical to artificial neurons implementation [45, 94, 95], as the firing neural network in Figure 9.7c, but the ease of the scaling of the circuit's depth, inherent to the fiber. The last is critical for PNC, where greater depth results in better generalization for a wide variety of tasks [96]. At the same time, some technological steps are yet to be taken successfully to enable this scenario. For instance, the draw of optical phase changing materials (PCM) hasn't been demonstrated so far, although the chances are that some subset of Ge–Sb–Te family of PMCs, such as $Ge_2Sb_2Te_3$ can be drawn as a fiber core since Ge-core fibers have been drawn successfully.

VLSI-Fi, as a conceptual manufacturing engineering approach, is ready to be taken a level-up towards scaling and integration effort that will deliver the full-scale data processing photonic circuitry in fibers comprised of the multitude of building blocks that have already been demonstrated. One shouldn't expect that such scaling and integration will be trivial; it took more than two decades for the microelectronic industry, from the invention of the transistor in 1947 to a demonstration of the first VLSI circuit, to complete a comparable journey. But the first few hardest and boldest steps have already been taken towards this goal.

All in all, we are convinced that the future of VLSI-Fi is bright, and the potential impact that it can bring to integrating the emerging data processing platforms, such as quantum and neuromorphic, into the Internet of tomorrow will keep invoking enthusiasm for research and exploration of manufacturing engineering of fiber photonics for decades ahead.

References

1 de Faccini, Lima, C., Leffel, T., Zheng, M. et al. (2023). Creating fiber-embedded photonic circuitry by liquid-phase structuring of multi-material cores. In: *Advanced Fabrication Technologies for Micro/Nano Optics and Photonics XVI* (ed. G. von Freymann, E. Blasco, and D. Chanda). 12433 (15), 34–46. https://doi.org/10.1117/12.2644870.

2 de Faccini, Lima, C., van der Elst, L.A., Koraganji, V.N. et al. (2019). Towards digital manufacturing of smart multimaterial fibers. *Nanoscale Research Letters*. Springer New York LLC 209: 1–16. https://doi.org/10.1186/s11671-019-3031-x.

3 Yan, W., Dong, C., Xiang, Y. et al. (2020). Thermally drawn advanced functional fibers: new frontier of flexible electronics. *Materials Today*. Elsevier B.V. 168–194. https://doi.org/10.1016/j.mattod.2019.11.006.

4 Jablonowski, D.P. (1986). Fiber manufacture at AT&T with the MCVD process. *Journal of Lightwave Technology* 4 (8): 1016–1019. https://doi.org/10.1109/JLT.1986.1074870.

5 Lydtin, H. (1986). PCVD: a technique suitable for large-scale fabrication of optical fibers. *Journal of Lightwave Technology* 4 (8): 1034–1038. https://doi.org/10.1109/JLT.1986.1074872.

6 Abouraddy, A.F., Bayindir, M., Benoit, G. et al. (2007). Towards multimaterial multifunctional fibres that see, hear, sense and communicate. *Nature Materials* 6 (5): 336–347. https://doi.org/10.1038/nmat1889.

7 Luo, Y., Canning, J., Zhang, J., and Peng, G.-D. (2020). Toward optical fibre fabrication using 3D printing technology. *Optical Fiber Technology* 58: 102299. https://doi.org/10.1016/j.yofte.2020.102299.

8 Glassomer (n.d.). *Glass 3D Printing*. https://www.glassomer.com/index.php/products/glass-3d-printing (accessed 9 October 2022).

9 Nobula (n.d.). *Glass 3D printing made fast, affordable, and energy-efficient*. https://www.nobula3d.com/ (accessed 9 October 2022).

10 Camacho Rosales, A.L., Velázquez, M.M.A.N., Zhao, X., and Sahu, J.K. (2020). Optical fibers fabricated from 3D printed silica preforms. *SPIE-The International Society for Optical Engineering* 29. https://doi.org/10.1117/12.2543210.

11 Chu, Y., Fu, X., Luo, Y. et al. (2019). Silica optical fiber drawn from 3D printed preforms. *Optics Letters* 44 (21): 5358. https://doi.org/10.1364/ol.44.005358.

12 Grabe, T., Lammers, M., Wang, S. et al. (2021). Additive manufacturing of fused silica using coaxial laser glass deposition: experiment, simulation and discussion.

In: *SPIE-The International Society for Optical Engineering*, 32. https://doi.org/10.1117/12.2577205.

13 Klein, J., Stern, M., Franchin, G. et al. (2015). Additive manufacturing of optically transparent glass. *3D Printing and Additive Manufacturing* 2 (3): 92–105. https://doi.org/10.1089/3dp.2015.0021.

14 Kotz, F., Arnold, K., Bauer, W. et al. (2017). Three-dimensional printing of transparent fused silica glass. *Nature* 544 (7650): 337–339. https://doi.org/10.1038/nature22061.

15 Luo, J., Pan, H., and Kinzel, E.C. (2014). Additive manufacturing of glass. *Journal of Manufacturing Science and Engineering, Transactions of the ASME* 136 (6): https://doi.org/10.1115/1.4028531.

16 Nguyen, D.T., Meyers, C., Yee, T.D. et al. (2017). 3D-printed transparent glass. *Advanced Materials* 29 (26): 1701181. https://doi.org/10.1002/adma.201701181.

17 Sciancalepore, C. Federica Bondioli, Massimo Messori, Daniel Milanese (2021) 'Printing and characterization of three-dimensional high-loaded nanocomposites structures', *Material Design & Processing Communications*, p. e256. https://doi.org/10.1002/mdp2.256.

18 van der Elst, L., de Lima, C.F., Kurtoglu, M.G. et al. (2021). 3D printing in fiber-device technology. *Advanced Fiber Materials* 1: 3. https://doi.org/10.1007/s42765-020-00056-6.

19 Koraganji, V.N., de Lima, C.F., Zheng, M. et al. (2020). Effects of 3D printed preform annealing on structural and optical properties of fibers. In: *CLEO Pacific Rim Conference 2020, OSA Technical Digest (Optical Society of America, 2020)*, paper C6H.6. Optical Society of America.

20 Hamdorf, M. and Johannsmann, D. (2000). Surface-rheological measurements on glass forming polymers based on the surface tension driven decay of imprinted corrugation gratings. *Journal of Chemical Physics* 112 (9): 4262–4270. Available at: https://doi.org/10.1063/1.481002.

21 Nguyen Dang, T., Richard, I., Goy, E. et al. (2019). Insights into the fabrication of sub-100 nm textured thermally drawn fibers. *Journal of Applied Physics* 125 (17): 175301. Available at: https://doi.org/10.1063/1.5089022.

22 Nguyen-Dang, T., de Luca, A.C., Yan, W. et al. (2017). Controlled sub-micrometer hierarchical textures engineered in polymeric fibers and microchannels via thermal drawing. *Advanced Functional Materials* 27 (10): 1605935. https://doi.org/10.1002/adfm.201605935.

23 Zheng, M. (2021). Personal archives.

24 Iatsenko, A., Sych, O., and Tomila, T. (2015). Effect of sintering temperature on structure and properties of highly porous glass-ceramics. *Processing and Application of Ceramics* 9: 99–105. https://doi.org/10.2298/PAC1502099I.

25 Yong-Taeg, O., Fujino, S., and Morinaga, K. (2002). Fabrication of transparent silica glass by powder sintering. *Science and Technology of Advanced Materials* (4): 3, 297–301. https://doi.org/10.1016/S1468-6996(02)00030-X.

26 Yu, L., Xuan, H., Guo, Y. et al. (2017). Porous polymer optical fiber fabrication and potential biomedical application. *Optical Materials Express* 7 (6): 1813. https://doi.org/10.1364/OME.7.001813.

27 Zhang, J., Tu, R., and Goto, T. (2012). Fabrication of transparent SiO_2 glass by pressureless sintering and spark plasma sintering. *Ceramics International* 38 (4): 2673–2678. https://doi.org/10.1016/j.ceramint.2011.11.034.

28 Tao, G., Abouraddy, A.F., Stolyarov, A.M., and Fink, Y. (2015). Multimaterial fibers. In: *Lab-on-Fiber Technology*, 1–27. Springer https://doi.org/10.1007/978-3-319-06998-2.

29 Kudyshev, Z.A., Kildishev, A.V., Shalaev, V.M. et al. (2020). Machine-learning-assisted metasurface design for high-efficiency thermal emitter optimization. *Applied Physics Reviews* 7 (2): 021407. https://doi.org/10.1063/1.5134792.

30 Coulter, J., Leffel, T., Zheng, M. (2022). *Fiber-optim/draw-anim-1.gif at main · jefferycoulter/fiber-optim · GitHub*. https://github.com/jefferycoulter/fiber-optim/blob/main/animations/draw-anim-1.gif (accessed 17 October 2022).

31 Cook, K., Canning, J., Leon-Saval, S. et al. (2015). Air-structured optical fiber drawn from a 3D-printed preform. *Optics Letters* 40 (17): 3966. https://doi.org/10.1364/OL.40.003966.

32 Kaufman, J., Bow, C., Tan, F.A. et al. (2016). 3D printing preforms for fiber drawing and structured functional particle production. *Photonics and Fiber Technology 2016 (ACOFT, BGPP, NP)*. Washington, D.C.: OSA AW4C.1. https://doi.org/10.1364/ACOFT.2016.AW4C.1.

33 Talataisong, W., Ismaeel, R., Marques, T.H.R. et al. (2018). Mid-IR hollow-core microstructured fiber drawn from a 3D printed PETG preform. *Scientific Reports* 8 (1): 8113. https://doi.org/10.1038/s41598-018-26561-8.

34 Toal, P.M., Holmes, L.J., Rodriguez, R.X. et al. (2017). Microstructured monofilament via thermal drawing of additively manufactured preforms. *Additive Manufacturing* 16: 12–23. https://doi.org/10.1016/J.ADDMA.2017.03.009.

35 Brasselet, E., Gervinskas, G., Seniutinas, G. et al. (2013). Topological shaping of light by closed-path nanoslits. *Physical Review Letters* 111 (19): 193901. https://doi.org/10.1103/PhysRevLett.111.193901.

36 Forbes, A., de Oliveira, M., and Dennis, M.R. (2021). Structured light. *Nature Photonics*. Nature Research 253–262. https://doi.org/10.1038/s41566-021-00780-4.

37 Awschalom, D., Berggren, K.K., Bernien, H. et al. (2021). Development of quantum interconnects (QuICs) for next-generation information technologies. *PRX Quantum* 2 (1): 017002. https://doi.org/10.1103/PRXQuantum.2.017002.

38 Maleki, L., Savchenkov, A.A., Ilchenko, V.S. et al. (2003). Whispering gallery mode lithium niobate microresonators for photonics applications. In: *International Society for Optics and Photonics* (ed. A.R. Pirich, E.W. Taylor, and M.J. Hayduk), 1. https://doi.org/10.1117/12.488253.

39 Lindenfelser, F., Keitch, B., Kienzler, D. et al. (2018). An ion trap built with photonic crystal fibre technology. *The Review of Scientific Instruments* 86 (3): 033107. https://doi.org/10.1063/1.4914548.

40 Okaba, S., Takano, T., Benabid, F. et al. (2014). Lamb-Dicke spectroscopy of atoms in a hollow-core photonic crystal fibre. *Nature Communications* 5 (1): 1–9. https://doi.org/10.1038/ncomms5096.

41 Nahmias, M.A., Shastri, B.J., Tait, A.N. et al. (2018). Neuromorphic photonics. *Optics and Photonics News* 29 (1): 34. https://doi.org/10.1364/opn.29.1.000034.

42 Peng, H.T., Nahmias, M.A., de Lima, T.F. et al. (2018). Neuromorphic photonic integrated circuits. *IEEE Journal of Selected Topics in Quantum Electronics* 24 (6): https://doi.org/10.1109/JSTQE.2018.2840448.

43 Shastri, B.J., Tait, A.N., de Lima, T.F. et al. (2017). Principles of neuromorphic photonics. *Encyclopedia of Complexity and Systems Science* 1–37. https://doi.org/10.1007/978-3-642-27737-5_702-1.

44 Shastri, B.J., Tait, A.N., de Lima, T.F. et al. (2021). Photonics for artificial intelligence and neuromorphic computing. *Nature Photonics*. Nature Research 102–114. https://doi.org/10.1038/s41566-020-00754-y.

45 Feldmann, J., Youngblood, N., Wright, C.D. et al. (2019). All-optical spiking neurosynaptic networks with self-learning capabilities. *Nature* 569 (7755): 208–214. https://doi.org/10.1038/s41586-019-1157-8.

46 Munir, Z.A., Anselmi-Tamburini, U., and Ohyanagi, M. (2006). The effect of electric field and pressure on the synthesis and consolidation of materials: a review of the spark plasma sintering method. *Journal of Materials Science* 763–777. https://doi.org/10.1007/s10853-006-6555-2.

47 Gumennik, A., Wei, L., Lestoquoy, G. et al. (2013). Silicon-in-silica spheres via axial thermal gradient in-fibre capillary instabilities. *Nature Communications* 4: 2216. https://doi.org/10.1038/ncomms3216.

48 Gumennik, A., Levy, E.C., Grena, B. et al. (2017). Confined in-fiber solidification and structural control of silicon and silicon-germanium microparticles. *Proceedings of the National Academy of Sciences of the United States of America* 114 (28): 7240–7245. https://doi.org/10.1073/pnas.1707778114.

49 Wei, L., Hou, C., Levy, E. et al. (2017). Optoelectronic fibers via selective amplification of in-fiber capillary instabilities. *Advanced Materials* 29 (1): 1603033. https://doi.org/10.1002/adma.201603033.

50 Gumennik, A., Zheng, M., Koraganji, V.N. et al. (2021). Very large scale integration for fibers (VLSI-Fi) – US 2021/0333131 A1. Indiana University. https://patents.google.com/patent/US20210333131A1/en (accessed 8 November 2021).

51 Plateau, J. (1873). *Experimental and Theoretical Statics of Liquids Subject to Molecular Forces Only*. Paris: Gauthier-Villars.

52 Powers, T.R., Zhang, D., Goldstein, R.E. et al. (1998). Propagation of a topological transition: the Rayleigh instability. *Physics of Fluids* 10 (5): 1052–1057. https://doi.org/10.1063/1.869650.

53 Tomotika, S. (1935). On the instability of a cylindrical thread of a viscous liquid surrounded by another viscous fluid. *Proceedings of the Royal Society of London A: Mathematical, Physical and Engineering Sciences* 150 (870): 322–337. http://rspa.royalsocietypublishing.org/content/150/870/322 (accessed 11 July 2017).

54 Wang, F. (2019). *New Modeling of Compact, High-efficiency, and Widely-tunable Gas-phase Terahertz Lasers*. Massachusetts Institute of Technology https://dspace.mit.edu/handle/1721.1/123716 (accessed 22 June 2021).

55 Kaufman, J.J., Tao, G., Shabahang, S. et al. (2012). Structured spheres generated by an in-fibre fluid instability. *Nature* 487 (7408): 463–467. https://doi.org/10.1038/nature11215.

56 Rein, M., Levy, E., Gumennik, A. et al. (2016). Self-assembled fibre optoelectronics with discrete translational symmetry. *Nature Communications* 7: 12807. https://doi.org/10.1038/ncomms12807.

57 Utada, A.S., Lorenceau, E., Link, D.R. et al. (2005). Monodisperse double emulsions generated from a microcapillary device. *Science* 308 (5721): 537–541.

58 Zhang, J., Wang, Z., Wang, Z. et al. (2019). In-fibre particle manipulation and device assembly via laser induced thermocapillary convection. *Nature Communications* 10 (1): 1–10. https://doi.org/10.1038/s41467-019-13207-0.

59 Fokine, M., Theodosiou, A., Song, S. et al. (2017). Laser structuring, stress modification and Bragg grating inscription in silicon-core glass fibers. *Optical Materials Express* 7 (5): 1589–1597. https://doi.org/10.1364/OME.7.001589.

60 Faccini de Lima, C., Wang, F., Leffel, T.A. et al. (2023). Multimaterial fiber as a physical simulator of a capillary instability. *Nature Communications* 14: 5816. https://doi.org/10.1038/s41467-023-41216-7.

61 Jian, Z., Nagashio, K., and Kuribayashi, K. (2002). Direct observation of the crystal-growth transition in undercooled silicon. *Metallurgical and Materials Transactions A* 33 (9): 2947–2953. https://doi.org/10.1007/s11661-002-0279-y.

62 Nagashio, K., Kuribayashi, K., Okamoto, H. et al. (2005). Fragmentation of faceted dendrite in solidification of undercooled B-doped Si melts. *Metallurgical and Materials Transactions A* 36 (12): 3407–3413. https://doi.org/10.1007/s11661-005-0014-6.

63 Liang, H., Pan, Y., Zhao, S. et al. (1998). Two-dimensional state of stress in a silicon wafer. *Journal of Applied Physics* 71 (6): 2863. https://doi.org/10.1063/1.351017.

64 Coucheron, D.A., Fokine, M., Patil, N. et al. (2016). Laser recrystallization and inscription of compositional microstructures in crystalline SiGe-core fibres. *Nature Communications* 7: 13265. https://doi.org/10.1038/ncomms13265.

65 Healy, N., Fokine, M., Franz, Y. et al. (2016). CO_2 laser-induced directional recrystallization to produce single crystal silicon-core optical fibers with low loss. *Advanced Optical Materials* 4 (7): 1004–1008. https://doi.org/10.1002/adom.201500784.

66 Çirkinoğlu, H.O., Bayer, M.M., Gökay, U.S. et al. (2018). Silicon microsphere whispering gallery modes excited by femtosecond-laser-inscribed glass waveguides. *Applied Optics* 57 (14): 3687. https://doi.org/10.1364/ao.57.003687.

67 Levi, C.G. and Mehrabian, R. (1982). Heat flow during rapid solidification of undercooled metal droplets. *Metallurgical Transactions A* 13 (2): 221–234. https://doi.org/10.1007/BF02643312.

68 Dahlen, A., Fattah, A., Hanke, G. et al. (1994). Bridgman and Czochralski growth of Ge-Si alloy crystals. *Crystal Research and Technology* 29 (2): 187–198. https://doi.org/10.1002/crat.2170290206.

69 Kürten, M. and Schilz, J. (1994). Czochralski growth of SitGe1-x single crystals. *Journal of Crystal Growth* 139, 1–2: 1–5. https://doi.org/10.1016/0022-0248(94)90021-3.

70 Schilz, J. and Romanenko, V.N. (1995). Bulk growth of silicon-germanium solid solutions. *Journal of Materials Science: Materials in Electronics* 6 (5): 265–279. https://doi.org/10.1007/BF00125881.

71 Martinez, N.J.D., Gehl, M., Derose, C.T. et al. (2017). Single photon detection in a waveguide-coupled Ge-on-Si lateral avalanche photodiode. *Optics Express* 25 (14): 16130. https://doi.org/10.1364/oe.25.016130.

72 Healy, N., Mailis, S., Bulgakova, N.M. et al. (2014). Extreme electronic bandgap modification in laser-crystallized silicon optical fibres. *Nature Materials* 13 (12): 1122–1127. https://doi.org/10.1038/nmat4098.

73 Jacobsen, R.S., Andersen, K.N., Borel, P.I. et al. (2006). Strained silicon as a new electro-optic material. *Nature* 441 (7090): 199–202. https://doi.org/10.1038/nature04706.

74 Schriever, C., Bianco, F., Cazzanelli, M. et al. (2015). Second-order optical nonlinearity in silicon waveguides: inhomogeneous stress and interfaces. *Advanced Optical Materials* 3 (1): 129–136. https://doi.org/10.1002/ADOM.201400370.

75 Sánchez-Pérez, J.R., Boztug, C., Chen, F. et al. (2011). Direct-bandgap light-emitting germanium in tensilely strained nanomembranes. *Proceedings of the National Academy of Sciences of the United States of America* 108 (47): 18893–18898. https://doi.org/10.1073/PNAS.1107968108/SUPPL_FILE/PNAS.1107968108_SI.PDF.

76 Ballato, J., Hawkins, T., Foy, P. et al. (2010). Binary III-V semiconductor core optical fiber. *Optics Express* 18 (5): 4972. https://doi.org/10.1364/oe.18.004972.

77 Grodkiewicz, W.H. (1975) 'Fused silica fibers with metal cores', *Materials Research Bulletin*, 10(10), pp. 1085–1090. https://doi.org/10.1016/0025-5408(75)90219-6.

78 Homa, D., Suzuki, K., Enoki, T. et al. (2014). Electronic and magnetic fibers. *Materials Letters* 133: 135–138. https://doi.org/10.1016/j.matlet.2014.06.179.

79 Homa, D., Liang, Y., and Pickrell, G. (2013). Superconducting fiber. *Applied Physics Letters* 103 (8): 82601. https://doi.org/10.1063/1.4819074.

80 Scott, B.L. and Pickrell, G.R. (2013). Fabrication of GaSb optical fibers. In: *Processing and Properties of Advanced Ceramics and Composites V*, 65–70. Wiley https://doi.org/10.1002/9781118744109.ch7.

81 Ballato, J. and Snitzer, E. (1995). Fabrication of fibers with high rare-earth concentrations for Faraday isolator applications. *Applied Optics* 34 (6848-6854).

82 Zhang, T., Wang, Z., Srinivasan, B. et al. (2019). Ultraflexible glassy semiconductor fibers for thermal sensing and positioning. *ACS Applied Materials and Interfaces* 11 (2): 2441–2447. https://doi.org/10.1021/acsami.8b20307.

83 Shen, Y., Harris, N.C., Skirlo, S. et al. (2017). Deep learning with coherent nanophotonic circuits. *Nature Photonics* 11 (7): 441–446. https://doi.org/10.1038/nphoton.2017.93.

84 Vanmol, K., Baghdasaryan, T., Vermeulen, N. et al. (2020). 3D direct laser writing of microstructured optical fiber tapers on single-mode fibers for mode-field conversion. *Optics Express* 28 (24): 36147–36158. https://doi.org/10.1364/OE.409148.

85 Dahan, R., Gorlach, A., Haeusler, U. et al. (2021). Imprinting the quantum statistics of photons on free electrons. *Science* 373 (6561): https://doi.org/10.1126/SCIENCE.ABJ7128/SUPPL_FILE/SCIENCE.ABJ7128_SM.PDF.

86 Bock, M., Eich, P., Kucera, S. et al. (2018, 2018). High-fidelity entanglement between a trapped ion and a telecom photon via quantum frequency conversion. *Nature Communications* 9 (1): 1–7. https://doi.org/10.1038/s41467-018-04341-2.

87 Fan, L., Zou, C.L., Cheng, R. et al. (2018). Superconducting cavity electro-optics: a platform for coherent photon conversion between superconducting and photonic circuits. *Science Advances* 4 (8): https://doi.org/10.1126/SCIADV.AAR4994/SUPPL_FILE/AAR4994_SM.PDF.

88 Reimer, C., Kues, M., Roztocki, P. et al. (2016). Generation of multiphoton entangled quantum states by means of integrated frequency combs. *Science (New York, N.Y.)* 351 (6278): 1176–1180. https://doi.org/10.1126/science.aad8532.

89 Zhong, T., Kindem, J.M., Bartholomew, J.G. et al. (2017). Nanophotonic rare-earth quantum memory with optically controlled retrieval. *Science (New York, N.Y.)* eaan5959. https://doi.org/10.1126/science.aan5959.

90 Choi, H., Heuck, M., and Englund, D. (2017). Self-similar nanocavity design with ultrasmall mode volume for single-photon nonlinearities. *Physical Review Letters* 118 (22): 223605. https://doi.org/10.1103/PhysRevLett.118.223605.

91 Burgers, A.P., Peng, L.S., Muniz, J.A. et al. (2019). Clocked atom delivery to a photonic crystal waveguide. *Proceedings of the National Academy of Sciences of the United States of America* 116 (2): 456–465. https://doi.org/10.1073/pnas.1817249115.

92 Corzo, N.V., Raskop, J., Chandra, A. et al. (2019). Waveguide-coupled single collective excitation of atomic arrays. *Nature* 566 (7744): 359–362. https://doi.org/10.1038/s41586-019-0902-3.

93 Hood, J.D., Goban, A., Asenjo-Garcia, A. et al. (2016). Atom-atom interactions around the band edge of a photonic crystal waveguide. *Proceedings of the National Academy of Sciences of the United States of America* 113 (38): 10507–10512. https://doi.org/10.1073/pnas.1603788113.

94 Chakraborty, I., Saha, G., Sengupta, A., and Roy, K. (2018). *Toward Fast Neural Computing using All-Photonic Phase Change Spiking Neurons, Scientific Reports 2018 8:1*. Nature Publishing Group https://doi.org/10.1038/s41598-018-31365-x.

95 Cheng, Z., Ríos, C., Pernice, W.H.P., and Wright, C.D. (2017). On-chip photonic synapse. *Science Advances* 3 (9): e1700160. https://doi.org/10.1126/sciadv.1700160.

96 Deep Learning (n.d.). https://www.deeplearningbook.org/ (accessed 10 June 2021).

10

Inorganic Thermoelectric Fibers: Materials, Fabrication Methods, and Applications

Jiwu Xin[1], Yongke Wang[2], Yubo Luo[3], Qinghui Jiang[3], and Junyou Yang[3]

[1]*Nanyang Technological University, College of Engineering, School of Electrical and Electronic Engineering, 50 Nanyang Avenue, Singapore, 639798, Singapore*
[2]*University of Miami, ENGINEERING Department, Rosenstiel School of Marine and Atmospheric Science, 4808S Conway Rd, Orlando, FL, 32812, USA*
[3]*Huazhong University of Science and Technology, State Key Laboratory of Materials Processing and Die and Mould Technology, School of Materials Science and Engineering, 1037 Luoyu Rd, Wuhan, 430074, China*

10.1 Introduction

Globally, natural energy reservoirs, such as coal, oil, and gas, have greatly consumed in industrial power generation with the increasing electricity demand [1], thus developing ecofriendly and sustainable energy resources has become an urgent need. Meanwhile, portable electronic devices, such as mobile phones, watches, microcomputers, and health monitoring systems, have achieved rapid development and progress in miniaturization and integration, which promotes greatly the era of intelligent IOT (Internet of Things) [2]. In this scenario, thermoelectric (TE) materials have the potential to realize the conversion of waste heat between electricity and vice versa, which can be used to address the above-mentioned challenges [3–5]. The electricity generation (heat to electricity) and the thermoelectric cooling are originated from the Seebeck and Peltier effects. In the Seebeck effect, the energy conversion efficiency η and the maximum coefficient of cooling performance (COP) for TE devices can be expressed as Eqs. (10.1) and (10.2) [6]

$$\eta = \frac{T_h - T_c}{T_h} \times \frac{\sqrt{1+ZT} - 1}{\sqrt{1+ZT} + (T_c / T_h)} \tag{10.1}$$

$$\mathrm{COP}_{\max} = \frac{T_c}{T_h - T_c} \times \frac{\sqrt{1+Z_{\mathrm{device}}T_{\mathrm{ave}}} - (T_c / T_h)}{\sqrt{1+Z_{\mathrm{device}}T_{\mathrm{ave}}} + 1)} \tag{10.2}$$

Optical and Electronic Fibers: Emerging Applications and Technological Innovations, First Edition. Edited by Lei Wei.

where T_h is the hot side temperature, T_c is the cold side temperature, T_{ave} is the average temperature, and ZT and $Z_{device}T_{ave}$ are the dimensionless figure of merits for TE materials and devices, respectively. ZT is the dimensionless figure of merit of the materials employed, and $ZT = S^2\sigma T/(\kappa_e + \kappa_{lat})$, where σ, S, κ_e, κ_{lat} are corresponding to the electrical conductivity, Seebeck coefficient, electrical thermal conductivity, and lattice thermal conductivity, respectively [7]. So far, achieving a high *ZT* of bulk TE material requires high electrical performance and low thermal conductivity, which is often challenging to effectively decouple the electrical and thermal properties owing to the nature of carrier transports [8].

Flexible thermoelectrics (FTE) have recently gained much attention due to their portability and good compatibility with the IOT family originated by their intrinsic Seebeck as well as Peltier effect. FTE are usually composed of natural or synthetic thermoelectric fibers, which possess natural advantages such as good mechanical properties and flexibility. Many nanostructured thermoelectrics, including nanowires [9], nanotubes [10], superlattices [11, 12], and bulk materials, tend to achieve an extensive phonon scattering at boundaries and interfaces [13, 14], which results in inherent flexibility and low thermal conductivity. Currently, fiber-shaped form factors are widely used in engineering technology [15–17] to enable various equipment designing [18] or fiber-based thermoelectrics, specifically. Portable and wearable thermoelectric materials can be classified into organic fibers, inorganic fibers, and inorganic/organic hybrid fibers in terms of their compositions [19]. Among the accessible materials, inorganic fibers are superior compared to the organic fibers regarding the heat resistance, stability, and high thermoelectric performance. Herein, the development of inorganic thermoelectric devices achieved by $Bi_2(Te, Se)_3$, Ag_2Te, PbTe, SnSe, and $NaCo_2O_4$-based thermoelectric fiber are reviewed. In addition to the synthesis of electrospinning, two-step anodic oxidation process and dissolved phase deposition, etc., it also introduces the thermal drawing method to fabricate some excellent-performance fiber-based materials, which detailed by continuously pulling microscopic fibers from a macroscopic preform with a controllable manner. Moreover, the corresponding advanced synthesis, characterization, performance evaluation, and the potential application of wearable thermoelectric fabrics or devices in power generation are summarized in this chapter.

Inorganic fiber-based thermoelectrics exhibit more excellent thermoelectric performance and designable nanostructures compared with organic or inorganic/organic hybrid, which can be ascribed by their suitable electronic structures and low-lattice thermal conductivity. Thus, various types of low-dimensional thermoelectric materials such as nano- or micro-based fibers have been applied in miniatured thermoelectric devices or flexible thermoelectric generators.

10.2 $Bi_2(Te, Se)_3$-based Nanofibers

The Bi–Sb–Te alloy system has obtained extensive attention due to its excellent thermoelectric performance near room temperature. The Bi_2Te_3/Te multiple heterostructured nanowire arrays fabricated via a two-step anodization process can be understood by the precipitation reaction of the supersaturated $Bi_{0.26}Te_{0.74}$ alloy under a nanoconfined system [20]. Figure 10.1a depicts the FE-SEM and TEM images of Bi_2Te_3/Te nanowire arrays annealed at 300 °C, and the close-up of a coupled segments can be well indexed to Te (segment I) and Bi_2Te_3 (segment II), respectively [21]. By regulating the reaction conditions, we can obtain some fiber-based products with a unique morphology. By injecting Bi precursor into the colloidal suspension of PbTe/Te/PbTe nanowires (Figure 10.1b), the dumbbell-like heterostructured PbTe/Bi_xTe_{1-x}/PbTe nanowires were prepared, which have different characteristics of the Te/Bi core/shell nanowire by using single-crystal Te nanowires (see Figure 10.1c) [22]. Moreover, using a special drawing technique of glass-cladding fibers, the continuous Bi_2Te_3-core fibers with a bend radius of less than 5 cm was obtained as shown in Figure 10.1d. The fiber shows excellent continuity, and uniformity feature with the observed wetting angle is 45.3° (see Figure 10.1e), indicating that the wetting property of the two materials is superior at drawing temperature ~ 700 °C. The SEM image for the longitudinal section of the Bi_2Te_3 fibers has visualized, horizontally in the Figure 10.1f, which is parallel to the fiber symmetry axis, and reveals many alternately dark and bright parallel stripes with an average interval of 30–40 nm (blue circle region). Such characteristics have suggested a nanosheet structure of Bi_2Te_3 fibers [24]. Another fiber synthesis process was idealized by the focused ion beam (FIB) technique, where a single nanowire was trimmed down from 750 to 285 nm by FIB shown in Figure 10.1g such as Bi–Sb–Te nanowires. Such *in-situ* techniques have various advantages to control the reduction of nanowires size, measurement of intrinsic thermoelectric properties to avoid the easier oxidation of vast surfaces in air [25]. In general, these nanowire or nanofiber structured materials ultimately need to be designed with an appropriate size to poses a certain strength and flexibility. Prototype of miniatured modules consisting of up to four uni-couples composed by p-type $Bi_{0.5}Sb_{1.5}Te_3$ and n-type $Bi_2Te_{2.7}Se_{0.3}$ sintered fibers is displayed in Figure 10.1h. As a result, the as-extruded fibers were very flexible and could be formed into various shapes, which can be used in various applications of portable and wearable devices (inset of Figure 10.1h) [26]. Moreover, a single bendable thermoelectric fiber was formed from the nanoscale to microscale (Figure 10.1j) through thermal drawing process, which was specified by continuously pulling microscopic fibers from a macroscopic preform in a controllable manner; this technique usually requires that the T_g (glass transition temperatures) of the cladding materials need to be higher than the T_m (melting points) of the functional core materials. Through regulating the preform feeding and the fiber drawing speed, the fiber core diameter can be controlled within a wide range (from tens nanometers to several millimeters), and the built-in stress between the cladding and core materials induced by the mismatch of thermal expansion coefficients can be optimized. The SEM image of cross-section indicates that the fiber possesses excellent flexibility, and a clean interface between the thermoelectric core and the glass cladding (Figure 10.1i). Furthermore, those p-type $Bi_{0.5}Sb_{1.5}Te_3$ and n-type Bi_2Se_3 TE fibers were constructed in a wearable fabric for flexible electronics on large areas (Figure 10.1k) [27].

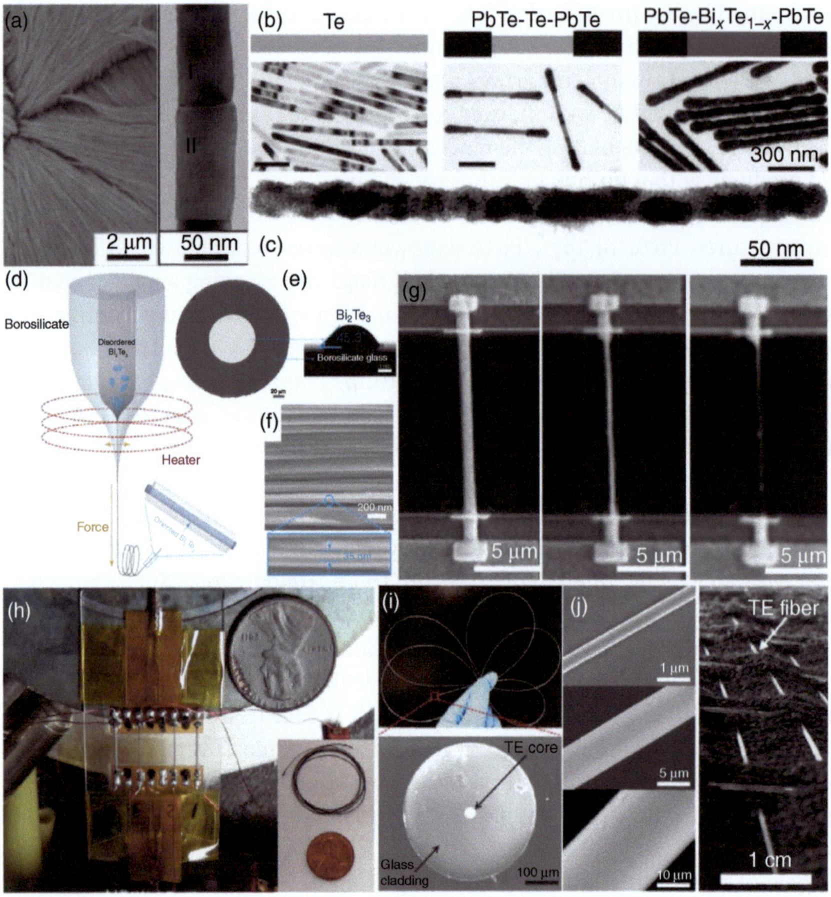

Figure 10.1 (a) Scanning electron microscopy (SEM) images of the Bi_2Te_3/Te multisegmented nanowires based on the top-view and close-up observation. Source: Reproduced with permission [21]. Copyright 2007, American Chemical Society. (b) Schematic sketch and microstructure characterization for the formation of PbTe/Bi_xTe_{1-x}/PbTe dumbbell-like, heterostructured nanowires. Source: Reproduced with permission [22]. Copyright 2010, Wiley. (c) TEM images of the interfaces of an individual Te/Bi core/shell nanowire. Source: Reproduced with permission [23]. Copyright 2010, Wiley. (d) Sketch of the molten-core drawing process. (e) Electron micrograph image of the cross-section and (f) longitudinal section of the Bi_2Te_3 core fiber sample. Source: Sun et al. [23] / AIP Publishing / CC BY 4.0. (g) The SEM images of pristine and trimmed Bi–Sb–Te nanowire with sizes of 750, 490, and 285 nm, respectively. Source: Reproduced from [25] © 2016, Springer Nature / CC BY 4.0. (h) A miniature uni-couple fabricated from the sintered p-type/$Bi_{i0.4}Sb_{1.6}Te_3$ and n-type/$Bi_2Te_{2.7}S_{e0.3}$ fibers. Source: Reproduced with permission [26]. Copyright 2016, Springer Nature. (i) Single TE fiber and (j) its cross-sectional SEM image. (k) A large-area wearable TE device constructed by TE fibers. Source: Reproduced with permission [27]. Copyright 2017, Elsevier.

10.3 PbTe-based Fibers

Thermoelectric materials coated by the nanocrystals can be combined with the glass fibers, which can be mass-produced in an unlimited length and induce a lower thermal conductivity. For example, flexible glass fibers were coated with PbTe nanocrystals via a scalable solution-phase deposition method [28]. From the low-resolution TEM image of Figure 10.2a, it can be easily seen that the PbTe uniform nanocrystals were dispersed in an average size around 10 nm. The solution-phase deposition process (Figure 10.2b) consists of the following procedures by dipping bare fluffy glass fibers into PbTe nanocrystal solution, removing the capping ligands of its surface, and discarding hydrazine with nitrogen [29, 31, 32]. The coated glass fibers have a uniform diameter of 10 μm with 300 nm of PbTe nanocrystal coating (Figure 10.2c), while the cubic nanocrystals are distributed with an edge length of 12.9 ± 1.1 nm presented in Figure 10.2d. Moreover, to obtain the thermoelectric parameters of glass fibers coated with PbTe nanocrystals, a single fiber with self-heated 3ω technique can be understood from Figure 10.2i [30, 33]. On the other hand, the samples with various nanocrystal (having various coating thicknesses produced by varying coating cycles) were displayed in the cross-section of Figure 10.2e–h. Based on these synthesized glass fibers and measurement platforms, the experimental results suggest that the glass fibers with a small volume fraction of PbTe nanocrystals lead to

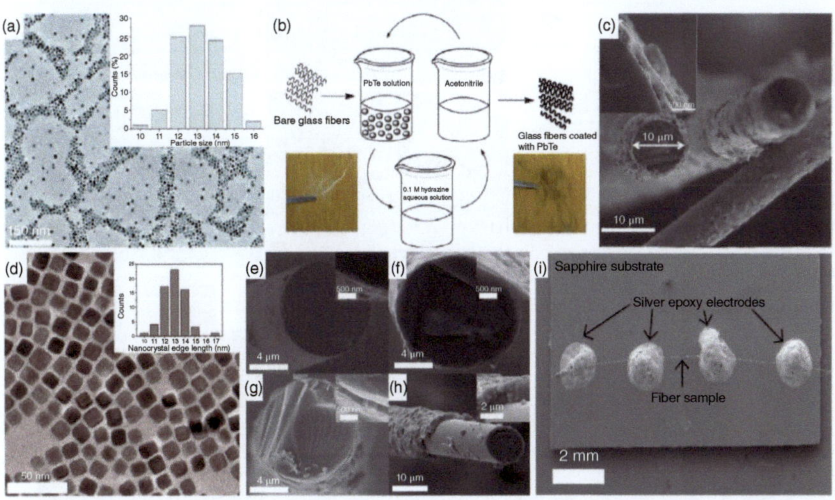

Figure 10.2 (a) TEM image of PbTe nanocrystals and upper inset is the size distribution of PbTe nanocrystals. (b) Scheme of the coating procedure to fabricate the bare glass fibers and PbTe nanocrystal-coated glass fibers. (c) SEM image of glass fibers coated by PbTe nanocrystal with a thickness ~300 nm. Source: Reproduced with permission [29]. Copyright 2012, American Chemical Society. (d) Low-resolution TEM image of PbTe nanocrystals with the inset of size distribution. (e–h) Cross-section SEM images of cut ends of glass fibers based on PbTe samples. (i) SEM image of the PbTe fiber sample used for 3ω measurements. Source: Reproduced with permission [30]. Copyright 2013, American Chemical Society.

an increased thermal conductivity and reduced with a higher volume fraction. These results are useful for optimizing the thermal measurement for some fiber- or wire-shaped samples with low thermal conductivity [30], which thereby further promote the applications of the 3ω measurement technique in fiber- or wire-shaped thermoelectrics.

10.4 Ag_2Te-based Fibers

It is known that the flexibility of inorganic semiconductors is too difficult to transfer into bendable or foldable films by themselves. Some flexible substrates such as polyvinylidene fluoride (PVDF) and glass fiber are necessary to support the inorganic films, which have been emerged as a novel flexible substrate of thermoelectric films owing to its feature of light, foldable, and breathable [28, 34]. Miao et al. reported a novel glass-fiber-aided cold-press method for achieving flexible n-type Ag_2Te films on a copy paper substrate [35]. As shown in Figure 10.3a, the dispersion of Te nanowires (NWs) was firstly obtained through a hydrothermal reaction [36] and

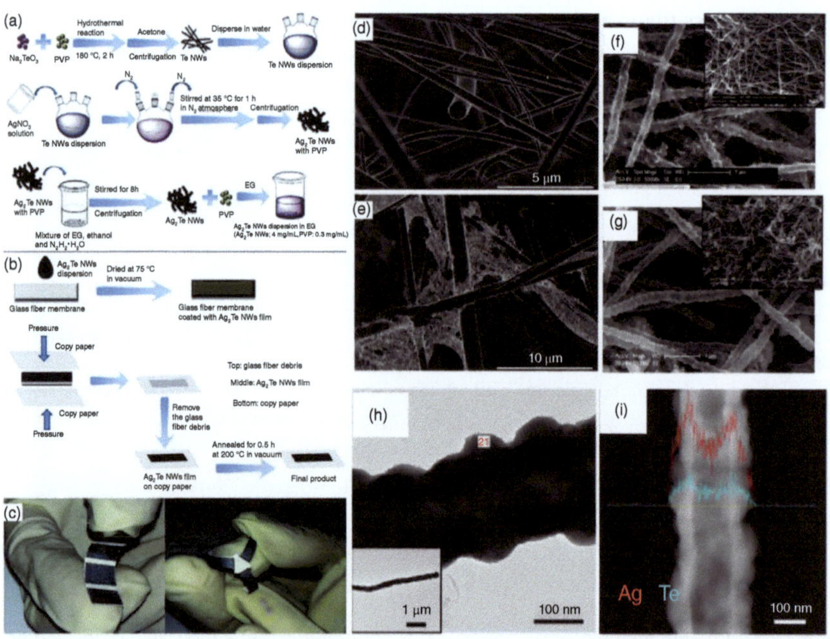

Figure 10.3 (a) Schematic diagram of demonstrating Ag_2Te nanowires and (b) the detailed process of fabricating paper coated with Ag_2Te films. (c) One piece of paper constructed by four sheets of parallel (9 mm × 5 mm) of Ag_2Te nanowires films and a clover-like shape with three sheets (20 mm × 3 mm). (d) Representative SEM images of Ag_2Te glass fiber sheet and (e) coated with Ag_2Te nanowires dispersions. Source: Reproduced permissions from Royal Society of Chemistry [35]. (f, g) High-definition SEM images of synthesized Ag_xTe_y hollow nanofibers and inserted are low-magnification pictures. (h) The TEM pictures and (i) line-scan EDS analysis of Ag_xTe_y nanofibers. Source: Reproduced with permission [34]. Copyright 2015, American Chemical Society.

then by pouring $AgNO_3$ solution into Te NWs to form Ag_2Te NWs. After removing the surface ligands under vigorous stirring, the pure Ag_2Te NWs were finally collected by centrifugation and washing process. Based on the synthesis of Ag_2Te NWs, the corresponding dispersion droplets were dried on a glass fiber sheet and then sandwiched between copy papers. After pressing at a certain pressure, the Ag_2Te NWs film was tightly adhered to the surface of the copy papers (Figure 10.3b). For better understanding, four sheets of parallel Ag_2Te NWs films on a paper substrate and three sheets of films with a clover-like shape were selected as shown in Figure 10.3c. The SEM images (Figure 10.3d,e) of the formed film reveal a porous structure of interlaced glass fiber sheet, and the substrate is hardly covered by the Ag_2Te NWs. Different from the above-mentioned method, another technique of synthesizing ultralong Ag_xTe_y nanofibers was carried out via combining the electrospinning with galvanic displacement reaction (GDR), by putting electrospinning Ni nanofibers [37] into TeO_2, $AgNO_3$, and HNO_3 solution at room temperature. Thus, the branched and continuous Ag_xTe_y nanofibers facilitate the regulation of concentrated Ag^+ for electrolytes. As shown in Figure 10.3f, a low Ag^+ concentration of 0.01 mM was beneficial to obtain the nanofibers with similar morphology to the pristine Te nanofibers [38]. Conversely, the growth of branched structures (see Figure 10.3g) was observed with higher Ag^+ concentrations (i.e. 0.2 mM), leading to the preferential deposition in the (001) direction, which may be attributed to the limited mass transfer of $HTeO_2{}^+$ ions as clarified by Zhang et al. [34]. The bright-field TEM analysis of an as-formed nanofiber (Figure 10.4h) shows a nodular surface and hollow structure, and the line-scan EDS (Figure 10.4i) further reveals that the hollow structure was composited by a nominal Ag_xTe_y.

10.5 SnSe-based Fibers

The rigid single-crystal SnSe is the best ever with a high TE performance of 2.8; however, it is difficult to apply in flexible and wearable devices currently. With the help of the thermally drawing process, an ultralong single-crystal SnSe wire with a diameter from micro- to the nanoscale has been achieved by Zhang et al. The single-crystal SnSe fiber is polycrystalline, highly flexible, ultralong, and mechanically stable, as shown in Figure 10.4a,b. Flexible SnSe fiber with variable diameters can be achieved by the CO_2 laser taper process without changing its single-crystal features (Figure 10.4c). Such fabrics woven by functional SnSe fibers enable the breathability of covered areas when put on the human body (Figure 10.4d). For example, a shirt composed of flexible thermoelectric SnSe fibers can induce an output voltage of 30 mV accompanied by a temperature difference between skin and environment due to the Seebeck effect. Commonly, miniatured thermoelectric devices require nanoscale materials. As for SnSe nanowires, the SEM image (Figure 10.4e) of the deposited nanowires fabricated by a catalyst-assisted thermal VLS (vapor–liquid–solid) process reveals that SnSe nanowires were formed uniformly on the surface of the substrates with different diameters and lengths. Such NWs are also consistent with the X-ray spectroscopy (EDS) quantified results of

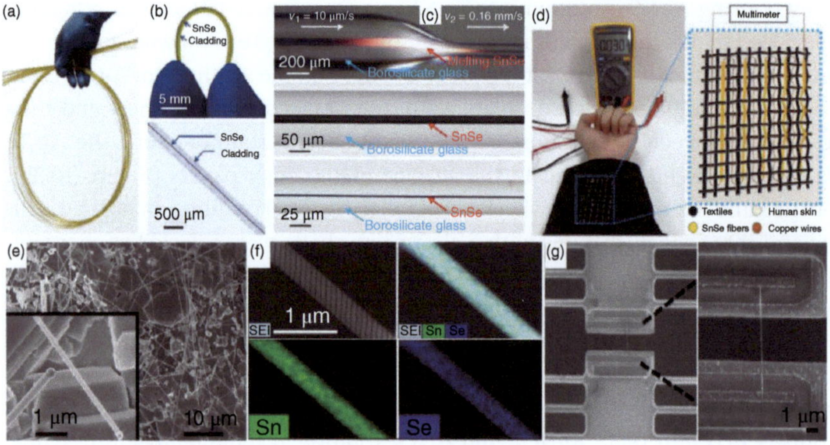

Figure 10.4 (a) Photograph of SnSe fibers. (b) Single SnSe fiber and its optical microscope image without polymer coating. (c) Fabrication of single-crystal SnSe-core fiber based on CO_2 laser taper process. (d) Demonstrations of multidimensional SnSe fabric for converting heat to electricity. Source: Reproduced with permission [39]. Copyright 2020, JOHN WILEY & SONS, INC. (e) Scanning electron microscopy (SEM) image of the as-grown SnSe nanowires. (f) Energy-dispersive X-ray spectroscopy mapping for a measured SnSe nanowire (480 nm diameter). (g) The low- and high-magnification SEM images of SnSe nanowire from left to right transferred onto the microdevice. Source: Reproduced from [40] © 2018, Springer Nature / CC BY 4.0.

Sn:Se with an atomic ratio ~1 : 1 (Figure 10.4f). For such nanowires or nanofibers, it is usually to detach from substrates and transport to microdevices (FIB system) for thermoelectric properties measurement as shown in Figure 10.4g.

10.6 $NaCo_2O_4$-based Fibers

$NaCo_2O_4$ is one of the potential oxide thermoelectric materials that carries a variety of attractive features of nontoxic and good chemical stability in the air or high temperature, which possesses a layered crystalline structure similar to $Ca_3Co_4O_9$ and $Bi_2Sr_2Co_2O_x$. Recent reports reveal that the ZT value of 1.2 and 0.8 for single-crystal $NaCo_2O_4$ and polycrystalline $NaCo_2O_4$. Also, literature reveals that polymer or ceramic fibers have been prepared by the convenient electrospinning. The lengths and diameters of internal fibers prepared by this approach can be achieved with a few microns to tens of nanometers through controlling the types of polymers and synthesis conditions [41, 42]. Drawing support from a sol–gel-based electrospinning technique [43], the thermoelectric nanofibers with grain size as small as 10 nm have been achieved in $NaCo_2O_4$ system, which can extend the applications of flexible thermoelectrics. As shown in Figure 10.5a–d, different solvents of methanol and water treated with annealing lead to distinctive morphology of $NaCo_2O_4$ nanofibers. The nanofibers in Figure 10.5a,c show the continuous fibrous morphology with smooth surface before annealing, and the surfaces of nanofibers become rough and

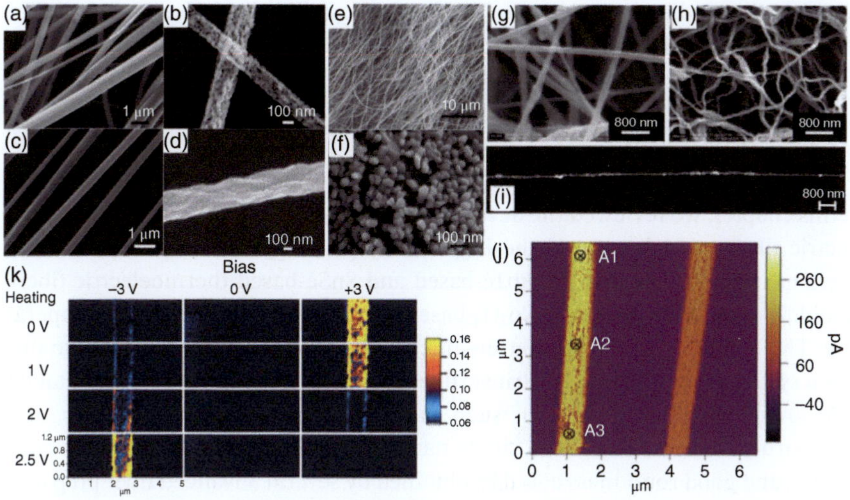

Figure 10.5 (a–d) SEM images of synthesized $NaCo_2O_4$ nanofibers using two different solvents before and after annealing. (d) Sketch the molten-core drawing process. (e) SEM image of as-spun nanofiber mat before annealing. (f) SEM images of nanofibrous mat annealed at 750 °C for 10 minutes. Source: Reproduced with permission [44]. Copyright 2010, American Chemical Society. (g) SEM micrographs of sodium acetate/cobalt acetate/PAN composite fibers and (h) $NaCo_2O_4$ fibers. (i) A single-aligned $NaCo_2O_4$ fiber coated on Si substrate. Source: Reproduced with permission [43]. Copyright 2006, Elsevier. (j) The measurement of *I–V* curve for nanofiber. (k) Electrostatic force microscopy (EFM) contrast variation in $NaCo_2O_4$ nanofiber vs. bias voltage with different heating voltages. Source: Reproduced with permission [44]. Copyright 2010, American Chemical Society.

porous after calcination due to the volatile PVP (polyvinylpyrrolidone) polymers (Figure 10.5b,d). To further verify this influence of temperature, the SEM images of as-spun mat without annealing and another one with 750 °C heating for 10 min are displayed in Figure 10.5e,f, respectively. Such results demonstrate that it is important to control the calcination time for three-dimensional growth mode of grains only. Interestingly, another vivid case of synthetic $NaCo_2O_4$ nanofibers is built on the calcination of electrospinning $CH_3COONa/Co(CH_3COO)_2$/PAN precursor mixture, indicating an opposite effect of temperature on nanofiber growth. As shown in Figure 10.5g,h, compared with the original composite nanofibers ($CH_3COONa/Co(CH_3COO)_2$/PAN) in Figure 10.5g, the diameters of $NaCo_2O_4$ nanofibers became even smaller, about 30–80 nm after annealing at 400 °C (Figure 10.5h). Furthermore, by increasing the calcination time to five hours, a single aligned $NaCo_2O_4$ nanofiber collected over a gap formed between two strips of silicon substrate [45] exhibits the long-range orderly filamentous characteristic (Figure 10.5i). The corresponding electrical and thermal transport properties of the $NaCo_2O_4$-based fiber system can be measured by the local *I–V* curve on three spots of the nanofiber. However, the difference in color contrast reflects the disparity of overall electrical conductivity for $NaCo_2O_4$ fiber (see Figure 10.5j). Meanwhile, the electrostatic force microscopy (EFM) mapping equipped with a thermal probe could display the changes at surface potential (electrical conductivity) of the fiber under different heating voltages (temperature gradient), as shown in Figure 10.5k; such a phenomenon helps us to better observe the thermoelectric performance of a single fiber.

10.7 Conclusion

In this chapter, we reviewed the development of several classic inorganic thermoelectric fibers, which covered the $Bi_2(Te, Se)_3$- and Ag_2Te-based thermoelectric fibers near room temperature, PbTe-based and SnSe-based thermoelectric fibers at middle-temperature, and $NaCo_2O_4$-based thermoelectric fiber at high temperature. This article emphasizes that the micro- or nanofibers corresponding to different systems have their own intrinsic fiber morphologies observed in the detailed SEM and TEM characterizations such as layered Bi_2Te_3, SnSe nanosheets, and cubic structure of PbTe. Also, a single nanofiber or coated nanofibers with continuous and good roundness could be obtained by several advanced fiber prepared techniques. Specifically, the convenient electrospinning method was implemented to prepare the target fiber product with some auxiliary solvents in Ag_xTe_y and $NaCo_2O_4$ fibers. Both the two-step anodization process and solution-phase deposition method are promising to obtain the segmented fibers and coated glass fibers due to their scalable and multifunctional feature. Unlike the FIB or self-heated 3ω technique, to achieve a single nanowire and to measure their thermoelectric properties simultaneously, the thermal drawing based on glass-cladding fibers can produce large-scale and high-quality submicro fibers. Such kinds of fibers exhibit flexible, bendable, and wearable charm for fabrics and portable electronics. Conclusively, understanding this rapid development and advanced synthesis of fiber-based inorganic thermoelectrics is beneficial to finding the current challenges. In addition, we believe these thermoelectric fibers could play a greater role in the future of wearable and smart electronics.

Generally, inorganic thermoelectric fibers are fascinating because of their intrinsic flexibility, better performance, and varied regulations compared with organic fibers. However, their disadvantages such as high cost and poor flexibility limit their practical applications in wearable or portable thermoelectrics. Therefore, organic/inorganic hybrid fiber-based thermoelectrics may be an advantageous hotspot in the future research of thermoelectric fibers. So far, it is challenging for the researcher to effectively balance the flexibility and performance of fiber-based materials. The above-mentioned challenges pointing to develop a new fabrication and processing technology that can optimize the thermoelectric properties such as nanostructuring, low-dimensional band structure, as well as introducing good flexibility and high continuity of the fibers. As for the applications of thermoelectric fibers, the heat conversions in wearable devices or fabrics need to be focused on in the future research. For these materials, it is urgently required to realize their low cost, stability, flexibility, and durability. Besides, multifunctional materials integrated thermoelectric fibers with such as biosensing fibers and optical fibers are favorable to develop a comprehensive energy conversion system. Referring to the advanced energy systems, there will be many obstacles and challenges in the commercialization of fiber-based thermoelectrics; but it is also fascinated by their promising applications in wearable and smart sensing areas.

References

1 Shi, X.-L., Zou, J., and Chen, Z.-G. (2020). *Chem. Rev.* 120: 7399.
2 Dargusch, M., Liu, W.D., and Chen, Z.G. (2020). *Adv. Sci.* 7: 2001362.
3 Xin, J., Basit, A., Li, S. et al. (2021). *Sensors* 21: 3437.
4 Xiao, Y. and Zhao, L.-D. (2020). *Science* 367: 1196.
5 Xin, J., Li, S., Yang, J. et al. (2020). *Nano Energy* 67: 104292.
6 Yang, J., Li, G., Zhu, H. et al. (2022). *Joule* 6: 193.
7 Li, S., Xin, J., Basit, A. et al. (2020). *Adv. Sci.* 7: 1903493.
8 Xin, J., Yang, J., Jiang, Q. et al. (2019). *Nano Energy* 57: 703.
9 Zhang, G., Fang, H., Yang, H. et al. (2012). *Nano Lett.* 12: 3627.
10 Ohnishi, M., Shiga, T., and Shiomi, J. (2017). *Phys. Rev. B* 95: 155405.
11 Heremans, J.P., Jovovic, V., Toberer, E.S. et al. (2008). *Science* 321: 554.
12 Wan, C., Tian, R., Kondou, M. et al. (2017). *Nat. Commun.* 8: 1.
13 Xin, J., Li, W., Li, S. et al. (2022). *Sci. China Mater.* 65: 1353.
14 Zhou, M., Li, J.-F., and Kita, T. (2008). *J. Am. Chem. Soc.* 130: 4527.
15 Canales, A., Jia, X., Froriep, U.P. et al. (2015). *Nat. Biotechnol.* 33: 277.
16 Chen, M., Wang, Z., Zhang, Q. et al. (2021). *Nat. Commun.* 12: 1.
17 Yan, W., Page, A., Nguyen-Dang, T. et al. (2019). *Adv. Mater.* 31: 1802348.
18 Lu, Y., Qiu, Y., Cai, K. et al. (2020). *Energy Environ. Sci.* 13: 1240.
19 Shi, X.L., Chen, W.Y., Zhang, T. et al. (2021). *Energy Environ. Sci.* 14: 729–764.
20 Masuda, H. and Fukuda, K. (1995). *Science* 268: 1466.
21 Wang, W., Lu, X., Zhang, T. et al. (2007). *J. Am. Chem. Soc.* 129: 6702.
22 Zhang, G., Yu, Q., Wang, W., and Li, X. (2010). *Adv. Mater.* 22: 1959.
23 Zhang, G., Wang, W., and Li, X. (2008). *Adv. Mater.* 20: 3654.
24 Sun, M., Qian, Q., Tang, G. et al. (2018). *APL Mater.* 6: 036103.
25 Chien, C.-H., Lee, P.-C., Tsai, W.-H. et al. (2016). *Sci. Rep.* 6: 1.
26 Ren, F., Menchhofer, P., Kiggans, J., and Wang, H. (2016). *J. Electron. Mater.* 45: 1412.
27 Zhang, T., Li, K., Zhang, J. et al. (2017). *Nano Energy* 41: 35.
28 Shamsa, M., Liu, W., Balandin, A. et al. (2006). *Appl. Phys. Lett.* 89: 161921.
29 Liang, D., Yang, H., Finefrock, S.W., and Wu, Y. (2012). *Nano Lett.* 12: 2140.
30 Ma, W., Luther, J.M., Zheng, H. et al. (2009). *Nano Lett.* 9: 1699.
31 Luther, J.M., Beard, M.C., Song, Q. et al. (2007). *Nano Lett.* 7: 1779.
32 Luther, J.M., Law, M., Beard, M.C. et al. (2008). *Nano Lett.* 8: 3488.
33 Finefrock, S.W., Wang, Y., Ferguson, J.B. et al. (2013). *Nano Lett.* 13: 5006.
34 Zhang, M., Park, H., Kim, J. et al. (2015). *Chem. Mater.* 27: 5189.
35 Gao, J., Miao, L., Liu, C. et al. (2017). *J. Mater. Chem. A* 5: 24740.
36 Gao, J., Liu, C., Miao, L. et al. (2016). *RSC Adv.* 6: 31580.

37 Zhang, M., Kim, J., Kim, S. et al. (2014). *Chem. Mater.* 26: 2557.

38 Park, H., Jung, H., Zhang, M. et al. (2013). *Nanoscale* 5: 3058.

39 Zhang, J., Zhang, T., Zhang, H. et al. (2020). *Adv. Mater.* 32: 2002702.

40 Hernandez, J.A., Ruiz, A., Fonseca, L.F. et al. (2018). *Sci. Rep.* 8: 1.

41 Li, D. and Xia, Y. (2004). *Adv. Mater.* 16: 1151.

42 Yan, X., Yu, M., Ramakrishna, S. et al. (2019). *Nanoscale* 11: 19166.

43 Maensiri, S. and Nuansing, W. (2006). *Mater. Chem. Phys.* 99: 104.

44 Ma, F., Ou, Y., Yang, Y. et al. (2010). *J. Phys. Chem. C* 114: 22038.

45 Li, D., Wang, Y., and Xia, Y. (2003). *Nano Lett.* 3: 1167.

Index

Optical and Electronic Fibers: Emerging Applications and Technological Innovations, First Edition. Edited by Lei Wei.